One Health Perspectives on Climate Change and Emerging Occupational Health Risks

Ioannis Adamopoulos
Hellenic Open University, Greece & University of West Attica, Greece

Aida Vafae Eslahi
Qazvin University of Medical Sciences, Iran

IGI Global
Scientific Publishing
Publishing Tomorrow's Research Today

Vice President of Editorial	Melissa Wagner
Director of Acquisitions	Mikaela Felty
Director of Book Development	Jocelynn Hessler
Production Manager	Mike Brehm
Cover Design	Jose Rosado

Published in the United States of America by
IGI Global Scientific Publishing
701 East Chocolate Avenue
Hershey, PA, 17033, USA
Tel: 717-533-8845 | Fax: 717-533-7115
Website: https://www.igi-global.com E-mail: cust@igi-global.com

Library of Congress Cataloging-in-Publication Data

LCCN: 2026011391 (CIP Data Pending)
ISBN13: 979-8-3373-8470-2
Isbn13Softcover: 979-8-3373-8471-9
EISBN13: 979-8-3373-8472-6
British Cataloguing in Publication Data
A Cataloguing in Publication record for this book is available from the British Library.

Table of Contents

Detailed Table of Contents

a low-carbon economy does not come at the cost of worker health. Furthermore, the changing climate functions as a risk multiplier that exacerbates occupational health inequities, disproportionately burdening socially, economically, or politically marginalised populations, including migrant and Indigenous workers. Ultimately, comprehensive policy frameworks and "Just Transition" principles must be continually developed to guarantee procedural, distributive, and intergenerational justice across all global labour sectors.

Muhammad Zia Ul Haq, Department of Agronomy, University of Agriculture, Faisalabad, Pakistan
Rana Nadeem Abbas, Department of Agronomy, University of Agriculture, Faisalabad, Pakistan
Rizwan Maqbool, Department of Agronomy, University of Agriculture, Faisalabad, Pakistan
Safdar Ali, Department of Agronomy, University of Agriculture, Faisalabad, Pakistan
Umair Gull, Department of Agronomy, University of Agriculture, Faisalabad, Pakistan
Rukhsar Saleem, Department of Agronomy, University of Agriculture, Faisalabad, Pakistan

The present chapter examines the intersection of climate change, agrochemical exposure, and occupational health through a One Health framework. Agricultural workers face escalating risks as rising temperatures and elevated carbon dioxide levels drive herbicide resistance and increased pesticide application. These climatic shifts simultaneously enhance human physiological vulnerability, where heat stress accelerates dermal absorption and impairs the renal clearance of toxins. Furthermore, socio-economic factors—including substandard housing and piece-rate pay—limit recovery and exacerbate health disparities among migrant populations. The authors advocate for integrated mitigation strategies, such as climate-smart personal protective equipment (PPE), updated occupational exposure limits (OELs), and an agroecological transition to protect the interconnected health of humans, animals, and the environment.

Climate change is fundamentally reshaping occupational health risks, yet standardized monitoring frameworks incorporating climate-induced risks remain underdeveloped. This chapter addresses this gap through a bibliometric analysis of 1,314 Scopus articles using VOSviewer, and a proposed Performance Monitoring Framework (PMF) for occupational health under climate change. Co-occurrence analysis of 72 keywords confirmed a critical absence of indicator-based monitoring research. Co-authorship analysis revealed a fragmented global network, with the Global South significantly underrepresented despite disproportionately high climate risks. The PMF is organized around three pillars: an accounting foundation aligned with GRI 403, ESG, and IFRS S1/S2; climate risk metrics covering heat stress, air pollution, zoonoses, UV radiation, and mental health; and OHS performance indicators for early warning and outcomes. Aligned with One Health principles, the framework supports organizations and policymakers in building climate-resilient occupational health systems.

Extreme heat is intensifying across the U.S. Southwest and is increasingly consequential for maternal and child health. This narrative review synthesizes evidence on heat exposure and outcomes across pregnancy, postpartum and lactation, and early childhood through an environmental and reproductive justice lens. We searched biomedical and interdisciplinary databases from 2000 to the present and supplemented them with gray literature on heat guidance, workplace standards, and housing and energy policies. Evidence links higher temperatures with increased risks of preterm birth and stillbirth and identifies postpartum, infant feeding, and pediatric morbidity during heat events and outages. Inequities reflect uneven neighborhood heat, energy insecurity, occupational exposure, and barriers to cooling and responsive care. We propose an interprofessional agenda that integrates clinical heat risk screening and tailored plans, heat-event care coordination with referral pathways, and longer-term structural investments in housing, energy reliability, and enforceable workplace protections

Global climate change is having a variety of effects on students' health and wellbeing. Considering the serious issues it causes; climate change is at the top of the global agenda. This study aimed to examine knowledge of climate change and its psychosocial effects among students. Quantitative design approach was used. Sample size of 291 students aged 18-24 years were involved in the study out of the 320 targeted population. Questionnaires and focus group discussion tools were used during the study. Data was analysed using the mean, median, and mode for the Likert scale questions. cross-tabulations were also done to compare two different variables. Results indicated students affected psychologically were at the average of 37.5%, while climate change effect on mental health has an overall impact mean (SD) of 4.38 (0.851). The findings of this study suggest a more comprehensive approach to education that can produce resilient, well-rounded students.

Climate change amplifies existing social and health inequities. Marginalized and low-income populations bear the greatest burden of heat-related illness and cancer risk. Climate stress is mediated by biological mechanisms (epigenetics), social structures (housing, occupation), and environmental exposures (heat, pollution).

Preface

Climate change has emerged as one of the most significant challenges of the twenty-first century, affecting not only ecosystems and environmental sustainability but also the health, safety, and well-being of workers across all sectors of society. Rising global temperatures, extreme weather events, changing patterns of infectious diseases, environmental degradation, and the transition toward low-carbon economies are creating unprecedented occupational risks that demand new approaches to research, policy, and practice. These challenges are increasingly complex because they occur at the intersection of human health, animal health, and environmental systems, reinforcing the importance of the One Health approach as a framework for understanding and addressing emerging occupational health threats.

One Health Perspectives on Climate Change and Emerging Occupational Health Risks brings together multidisciplinary scholarship that explores the evolving relationship between climate change and occupational health from diverse geographical, disciplinary, and methodological perspectives. The chapters collectively demonstrate how climate-related hazards extend beyond traditional workplace risks to encompass biological, chemical, psychosocial, reproductive, and transgenerational health consequences. They also highlight the growing need for integrated surveillance systems, resilient public health infrastructures, equitable policies, and evidence-based interventions capable of protecting workers in a rapidly changing climate.

A central theme throughout this volume is the recognition that occupational health can no longer be examined in isolation. Climate change acts as a risk multiplier, exacerbating existing vulnerabilities and disproportionately affecting workers in agriculture, forestry, construction, public health inspection, disaster response, healthcare, and other climate-sensitive sectors. The One Health perspective adopted throughout the book emphasizes that safeguarding worker health is inseparable from protecting environmental integrity and preventing risks that emerge at the human–animal–environment interface.

The contributions included in this volume span conceptual analyses, empirical investigations, narrative reviews, bibliometric studies, and policy-oriented discussions. Together, they provide a comprehensive overview of current knowledge while identifying critical gaps that require further research and action. The volume also addresses emerging topics such as climate-related mental health challenges, maternal and child health equity, reproductive toxicology, zoonotic and vector-borne diseases, occupational exposure to agrochemicals, and the role of data visualization in strengthening climate literacy and informed decision-making.

CHAPTER OVERVIEW

Chapter 1 examines how climate change is transforming occupational hazards for public health inspection services. It explores the convergence of physical, biological, chemical, and psychosocial risks arising from extreme heat, flooding, wildfires, emerging infectious diseases, and climate-related disasters while emphasizing the importance of a One Health approach to workforce protection.

Chapter 2 presents a systems-thinking perspective on the interactions between climate change and occupational health and safety. It discusses how climate adaptation and mitigation measures may create unintended occupational risks and highlights the importance of justice-oriented policies that support equitable and sustainable transitions for workers.

Chapter 3 focuses on the intersection of climate change, agrochemical exposure, and occupational health in agriculture. It examines how rising temperatures and changing environmental conditions influence pesticide use, worker vulnerability, and health outcomes while proposing integrated strategies for risk reduction.

Chapter 4 explores the development of accounting-based occupational health indicators for climate-related risks. Through a bibliometric review and the introduction of a performance monitoring framework, the chapter addresses the need for standardized metrics that support climate-resilient occupational health systems.

Chapter 5 reviews the impacts of extreme heat on maternal and child health, particularly within vulnerable populations. The chapter synthesizes evidence linking heat exposure to adverse pregnancy and childhood outcomes and advocates for integrated clinical, occupational, and policy interventions.

Chapter 6 examines the psychological impacts of climate change education among university students. It investigates the relationship between climate awareness and mental well-being, highlighting the need for educational approaches that foster resilience while addressing climate-related psychological distress.

Chapter 7 analyzes reproductive toxicities and transgenerational cancer risks associated with climate-related occupational hazards. Drawing on advances in exposome science and epigenetics, the chapter explores how environmental exposures influence long-term and intergenerational health outcomes.

Chapter 8 addresses workplace mental health challenges associated with climate change, including stress, burnout, anxiety, and reduced occupational well-being. It discusses organizational strategies for strengthening workforce resilience and supporting psychological health in climate-sensitive work environments.

Chapter 9 explores the occupational risks associated with vector-borne and zoonotic diseases in the context of climate change. Through a One Health lens, it examines changing disease ecologies, vulnerable occupations, and integrated approaches to surveillance and prevention.

Chapter 10 highlights the role of visualization technologies in enhancing understanding of climate change and its environmental impacts. The chapter demonstrates how interactive data visualization can improve climate literacy, public engagement, and evidence-informed policymaking.

Collectively, these chapters underscore the urgent need for interdisciplinary collaboration to address the occupational dimensions of climate change. By integrating perspectives from public health, environmental science, occupational safety, epidemiology, psychology, agriculture, accounting, and data science, this volume contributes to the growing body of knowledge needed to build healthier, safer, and more resilient workplaces in an era of environmental transformation.

We hope that this book serves as a valuable resource for researchers, policymakers, occupational health professionals, public health practitioners, educators, and students seeking to understand and address the emerging occupational health risks associated with climate change. Most importantly, we hope it encourages continued dialogue and action toward protecting workers while advancing the broader goals of environmental sustainability, health equity, and One Health resilience.

Chapter 1

The Global Climate Crisis and Emerging Occupational Hazards Among Public Health Inspection Services: A One Health Perspective

Ioannis Adamopoulos
http://orcid.org/0000-0002-4942-7123
Hellenic Open University, Greece & University of West Attica, Greece

Aida Vafae Eslahi
http://orcid.org/0000-0001-5399-0552
Qazvin University of Medical Sciences, Iran

ABSTRACT

Climate change is fundamentally transforming occupational hazards for public health inspection services. This chapter examines how rising temperatures, extreme weather events, wildfires, and changing disease ecology create converging physical, biological, chemical, and psychosocial risks for health inspectors working at the human-animal-environment interface. Physical hazards include extreme heat stress, flooding-related injuries, and wildfire smoke exposure. Biological hazards encompass expanding vector-borne diseases, zoonotic spillovers, and waterborne pathogens (leptospirosis, cholera). Psychosocial hazards include disaster-induced trauma, eco-anxiety, moral distress, and occupational burnout. Current preparedness remains inadequate due to limited climate literacy in training, fragmented surveil-

DOI: 10.4018/979-8-3373-8470-2.ch001

lance systems, outdated regulations, and insufficient PPE guidance for extreme conditions. Using a One Health framework, this chapter argues that inspector safety is inseparable from system effectiveness.

1. INTRODUCTION

There is an increasing awareness that climate change represents one of the biggest health challenges of the twenty-first century, impacting on the health and well-being of people, animals and the environment (Watts et al. 2021; Romanello et al. 2022). A stern warning was issued by the Intergovernmental Panel on Climate Change that the warming of the climate system is unequivocal, that human activities are the primary driver and that we have already experienced changes in disease vector distribution, extreme weather event frequency and intensity and reductions in the environmental resources needed to sustain human health (IPCC 2022). Despite substantial research into the immediate and downstream effects of climate change upon the health of populations (Patz et al. 2014; Rocque et al. 2021), little research has examined the occupational health impacts of the environmental change experienced by the workforce that governs and enforces health and safety standards.

Public health and occupational health and safety inspection services perform a vital and unique role at the interface of environmental monitoring, health protection and disease prevention, and worker safety enforcement (Schulte and Chun 2009) with health inspectors performing field investigations, enforcement of standards, and supervision of high risk working environments and acting as responders to outbreaks or environmental disasters (Levy et al. 2017). The working environment of these inspection services, however, are changing dramatically due to rising global temperatures, increased and extreme heatwaves, flood events, wildfires and changing patterns in the epidemiology of climate-sensitive infectious diseases (Kjellstrom et al. 2016; Adam-Poupart et al. 2015).

Historically, inspection services operated with relatively stable seasonal and work site predictable exposure patterns (Schulte et al. 2016) that no longer exists today. Health inspectors are facing an unpredictable working environment and hazard scenario that overwhelms traditional exposure assessment and risks have exceeded assumptions used in current standards (Applebaum et al. 2016) while climate change remains a backdrop issue in existing training programs and protocols for inspection services (Bennett and McMichael 2010). The One Health paradigm explicitly recognizes that human, animal and environmental health are inextricably linked and provides a useful framework for the study of climate change and occupational risks faced by public health and safety inspectors (Zinsstag et al. 2011; Mackenzie and Jeggo 2019) since inspectors can not be isolated from their environment.

The health of inspectors depends on the health of people, animals and the environments that inspection services regulate (Coker et al. 2011). This chapter will address how climate change is transforming the occupational risks of public health and safety inspection services in terms of their changing physical, biological, chemical and psychosocial work hazards from a One Health perspective. We discuss shifting work practices, a lack of institutional and professional preparation for climate change and the importance of inspectors as essential partners in One Health while addressing adaptive measures for inspector work protection and professional training. This discussion will not be restricted by income status, considering applicability globally and especially in countries already severely impacted by climate change (Sorensen et al. 2018).

2. METHODS

2.1 Search strategy and selection criteria

The development of this chapter was derived from the results of a systematic narrative literature review of peer-reviewed scientific articles published from 2000 to 2026. The literature was retrieved through searches of three major scientific databases: PubMed/MEDLINE, Web of Science and Scopus. Keyword searches utilized combinations of 'climate change' or 'global warming' with 'occupational health' or 'workplace exposure' and the terms of specific diseases. Complementary searches of the grey literature were executed by interrogating Google Scholar as well as the websites of the World Health Organization (WHO), the International Labour Organisation (ILO) and of the Intergovernmental Panel on Climate Change (IPCC). The retrieval of both scientific and grey literature sources was considered important in order to combine the most recent scientific findings with those of international organizations on authoritative technical guidelines.

2.2 Inclusion and exclusion criteria

Papers were eligible for inclusion if they met the predetermined criteria: addressed at least one climate sensitive infectious disease, considered workers/occupationally exposed and if the study was in the English language, of the original research, systematic review, meta-analysis or case series with at least adequate epidemiological/ exposure data to allow sensible interpretation. Papers excluded were editorials/ commentaries which contain no primary or synthesised data, conference abstracts with incomplete methodology and outcome data, and animal studies where relevance to human health was unable to be established. There was no exclusion based on geographical region: low, middle and high income country research was eligible as climate change, and occupation health problems associated with climate change, are a global issue.

2.3 Data extraction and synthesis

Information was extracted for each included source, which included climate driver investigated, disease/pathogen, occupation/occupational sector investigated, pathway of exposure and outcomes reported (seroprevalence figures, ORs, attack rates in outbreak scenarios, Quantitative Microbial Risk Assessment data, etc.). The evidence was synthesised narratively and summarized by themes in tables (Tables 1-3), by disease category and by occupational sector, this approach of narrative synthesis was used to integrate varied evidence types whilst retaining context important for understanding the climate-occupational-disease relationship.

3. RESULTS

3.1 One Health and occupational risk in inspection services

3.1.1 The One Health framework as an analytical lens

The One Health concept is not merely a rhetorical device but an operational framework for understanding and managing health risks at the human-animal-environment interface (Zinsstag et al., 2011; Destoumieux-Garzón et al., 2018). One Health is defined by the WHO as "an approach to designing and implementing programs, policies, legislation and research in which multiple sectors work together in communication and collaboration to achieve better health outcomes for all people" (WHO 2017). Climate change is essentially a threat multiplier, aggravating known risks and defining new risk pathways across all these spheres (Carlson et al. 2022; Altizer et al. 2013).

The One Health framework is most relevant for public health inspection services since inspectors are often found exactly at these human/animal/environmental junctures (Schulte et al. 2016). In food inspection at a processing plant, for example, the same inspection requires evaluating structure (environment), animal products (animal health) and consumer health (human health) (Levy et al. 2017). Climate change provides an additional layer across all three spheres.

3.1.2 Climate change as a threat amplifier across linked pathways

Accumulating research indicates that climate change increases occupational risk in three interlinked pathways directly applicable to inspection work (Kjellstrom et al., 2016; Bennett & McMichael, 2010).

Ecological pathways are a direct result of disruption of ecosystem integrity, contamination of water sources, and degradation of air quality (Patz et al., 2014). For example, an inspector conducting field work following a flood will face higher potential exposure to waterborne pathogens, chemical pollutants from industrial spills, and indoor air quality hazards exacerbated by moisture and mold growth (Levy et al., 2016; Paterson, et al., 2014).

Biological pathways relate to changes in pathogen viability, expansion of vector ranges and increased frequency of zoonotic transmission (Ogden et al., 2014; Carlson et al., 2022). Climate-sensitive pathogens, including *Leptospira* spp., *Vibrio cholerae* and several enteric viruses multiply in hotter and wetter environments that result in direct increase in exposure risk when performing sanitation and water quality inspections (Lau, Smythe, & Weinstein, 2010; Pascual et al., 2002). Meanwhile the expansion of tick-borne pathogens, including *Borrelia burgdorferi* (Lyme disease) and tick-borne encephalitis virus, to higher latitudes and altitudes increases the risk to inspectors working in new areas (Medlock & Leach, 2015).

Finally, occupational pathways are described as changes to duration, intensity, and feasibility of protection (Applebaum et al., 2016; Schulte & Chun, 2009). For instance, climate change may increase duration of peak vector activity, levels of exposure and working conditions which prevent the steady use of personal protective equipment and safe work practices (Morris et al., 2022). For an outdoor worker experiencing extreme heat events, this creates a dilemma where thermal comfort is challenged by the necessity of maintaining a layer of Impermeable protection (Kjellstrom et al., 2016).

Figure 1 illustrates these pathways, showing how climate drivers cascade through ecosystem degradation, pathogen amplification, and surveillance gaps to produce disease burden, with One Health interventions as countermeasures.

Figure 1. One Health conceptual pathway from climate change to occupational disease burden in public health inspection services.

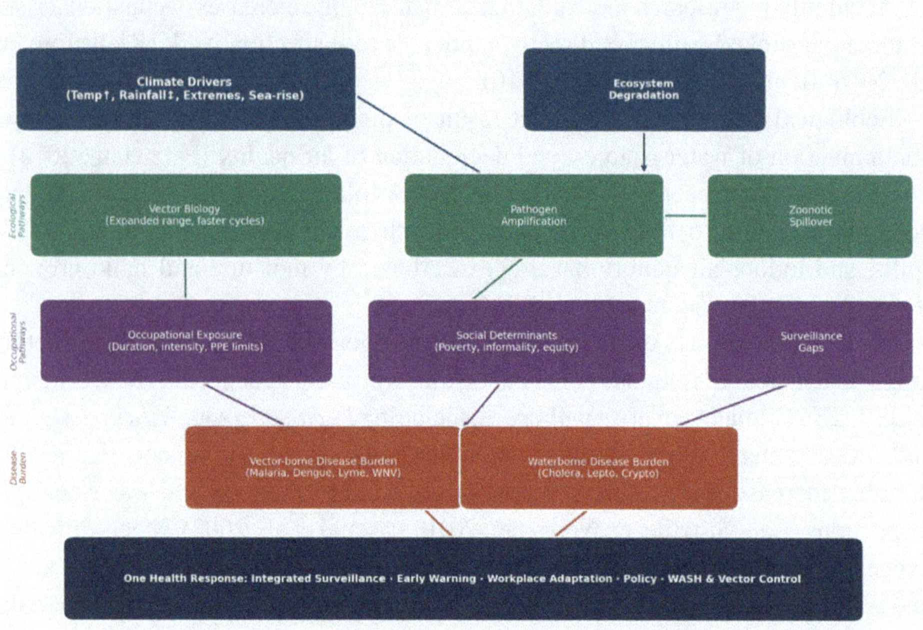

3.1.3 The interdependence of inspector safety and system effectiveness

This chapter is premised on the idea that the safety of individual inspectors, their health, and the overall quality and reliability of the inspection service are one and the same (Schulte et al., 2016). Unhealthy inspectors (due to heat, communicable disease or psychological trauma) directly reduce the health services the inspection system can provide (Levy et al., 2017). Investing in inspector safety is, from a One Health perspective, not an added service to inspection itself, but an enabler to that service under conditions of climate change (Bennett & McMichael, 2010).

3.2 Climate-driven physical and environmental hazards for inspectors

3.2.1 Extreme heat and heat-related illness

The warming of global average temperatures and the increase in the incidence of heatwaves may pose perhaps the most evident and inescapable physical hazard for inspectors carrying out field work (Kjellstrom et al., 2016; Adamopoulos et al., 2025a). Inspectors who carry out duties in agricultural, construction, waste management and disaster recovery work usually work outdoors or in unconditioned indoor spaces where workplace temperatures can meet or exceed safe limits (Spector et al., 2019).

The range of physiological responses in workers to occupational heat stress includes, but is not limited to, heat rash and cramps as well as heat exhaustion and the fatal heatstroke (Flouris et al., 2018). In addition, thermal stress reduces cognitive performance, decision-making ability and alertness which are all vital for effective and accurate inspection to determine health and safety violations (Hancock et al., 2007). Flouris et al. (2018) found that occupational heat stress caused significant increases in workplace injuries across a number of sectors, with effect sizes comparable to sleep deprivation and fatigue in a systematic review.

The hazards posed by heat to inspectors also include those experienced in the field not just those workers exposed in hot work environments. Those inspections in confined spaces, laundry facilities, kitchens or waste water treatment plants can present health and safety risks to inspectors from prolonged heat exposure which will be exacerbated by global warming trends (Sorensen et al., 2018). It is worth noting that personal protective equipment can be a major risk factor in relation to heat illness as it has a tendency to reduce evaporative cooling, for example some impermeable suits and respirators/chemical resistance gloves (Morris et al., 2022).

Table 1 summarizes the four principal groups of climate related work related hazards for public health inspectors-physical, biological, chemical and psychosocial. Included is the specific climate change element causing the hazard and observed and predicted outcomes for public health inspectors.

Table 1. Summary of climate-driven occupational hazards for public health inspectors.

Hazard category	Specific hazards	Climate change driver	Health effects on inspectors
Physical	Extreme heat exposure	Rising global temperatures, increased heatwaves	Heat rash, heat exhaustion, heatstroke; reduced cognitive performance; increased workplace injury risk
Physical	Flood-related hazards	Increased extreme precipitation events	Drowning, slips/falls, electrocution, structural collapse injuries
Physical	Wildfire smoke exposure	Increased wildfire frequency/ severity	Respiratory illness, cardiovascular effects, reduced lung capacity
Biological	Vector-borne diseases	Geographic expansion of vectors (ticks, mosquitoes, sandflies)	Lyme disease, dengue, malaria, leishmaniasis, tick-borne encephalitis

continued on following page

Table 1. Continued

Hazard category	Specific hazards	Climate change driver	Health effects on inspectors
Biological	Zoonotic spillover	Ecological disruption, habitat change	Leptospirosis, avian influenza, emerging zoonotic pathogens
Biological	Waterborne pathogens	Flooding, extreme rainfall, warming waters	Cholera, cryptosporidiosis, giardiasis, enteric viral infections
Chemical	Industrial and agricultural toxins	Flood mobilization of contaminants	Variable depending on agent; acute and chronic toxicity
Chemical	Wastewater contaminants	Sewage system failure during floods	Endotoxin exposure, chemical byproducts, mixed chemical exposures
Psychosocial	Disaster-induced trauma	Response to climate-amplified disasters	Post-traumatic stress symptoms, anxiety, depression
Psychosocial	Eco-anxiety	Daily witnessing of environmental degradation	Persistent dread, cumulative psychological burden
Psychosocial	Moral distress	Recognizing hazards but lacking authority to act	Powerlessness, frustration, ethical conflict
Psychosocial	Occupational burnout	Cumulative physical and psychological stressors	Emotional exhaustion, depersonalization, reduced personal accomplishment

3.2.2 Flooding, water quality, and chemical hazards

Increasing in both occurrence and magnitude, extreme precipitation events, and therefore, subsequent flooding events are anticipated under a changing climate (IPCC, 2022). In the context of post-flood assessment, there are myriad of physical and chemical exposures to the public health inspector. Standing water and saturated environments present a risk for drownings, slips and falls, electrocution by energized submerged structures and building structural failure (Paterson et al., 2014). Furthermore, industrial chemicals, petroleum compounds, agricultural toxins and other waste materials that are located in areas where the flood waters recede may lead to varied and uncertain chemical exposures as the inspector conducts assessments of damage and sanitation (Levy et al., 2016). Contamination from failing waste water treatment plants and combined sewer overflows bring enteric pathogens, endotoxins, and chemical byproducts to homes and businesses, providing potential exposures to the public health inspector (Patz et al., 2014).

3.2.3 Wildfires and air quality hazards

The frequency, severity, and length of wildfire seasons has increased across many areas due to climate change (Xu et al., 2020; Adamopoulos & Syrou, 2024). Public health inspectors may be assigned to areas impacted by wildfire smoke in order to evaluate structural safety of buildings, determine air quality inside temporary shelters and evacuate housing centers, or inspect food services and environmental regulations at shelters and during restoration processes (Reid et al., 2016).

Wildfire smoke consists of an intricate combination of particles (PM2.5 and PM10), carbon monoxide, various volatile organic compounds, and other dangerous contaminants (Black et al., 2017). Chronic and acute exposure to wildfire smoke has been found to exacerbate respiratory illnesses and trigger cardiovascular effects as well as lead to decreased lung capacity (Reid et al., 2016; Cascio, 2018). Inspectors assigned to these areas have long-term exposure to high atmospheric particulate concentrations, without any protective equipment or gear for respiratory health (Morris et al., 2022).

3.3 Biological and zoonotic risks in inspection activities

3.3.1 Zoonotic spillover during animal-related inspections

Climate change is leading to an increase in ecological disruption of both pathogens and animal hosts, thereby creating an environment with higher risk for zoonotic spillover (Carlson et al., 2022). Occupational health inspectors working in the various food production facility inspections, agricultural facility inspections, live animal market inspections, processing plant inspections have increased occupational exposure risks to zoonotic pathogens (Coker et al., 2011).

There are numerous mechanisms by which climate change has been hypothesized to promote zoonotic risk. Rising temperatures and changing precipitation regimes can alter ranges and dynamics of animal hosts (Altizer et al., 2013). Land use change and habitat fragmentation in combination with climatic stress promote interactions between wild animals and humans/domesticated animals creating opportunities for spillover (Patz et al., 2014). Inspectors are likely some of the earliest occupational groups exposed when a zoonotic agent emerges (Mackenzie & Jeggo, 2019). Zoonotic agents that are of concern to inspectors include: *Leptospira* spp. shed from animals through urine and able to survive in the environment for long periods of time in hot, humid conditions (Lau et al., 2010); enteric viruses such as norovirus and hepatitis E that have been observed to amplify in warmer conditions (Levy et al., 2016); and highly pathogenic avian influenza viruses whose environmental viability and transmission depends on temperature and humidity (Gilbert et al., 2014).

3.3.2 Expansion of vector-borne disease risk

Vector-borne diseases (VBDs) are a rapidly emerging occupational risk to public health inspectors working in field settings in peri-urban, agricultural, and natural settings (Semenza, Rocklv, & Ebi, 2022). Climate change causes geographic expansion of disease vectors, extending ranges to higher altitudes and higher latitudes and introducing inspectors to organisms that were previously absent or extremely rare in their geographic locations (Campbell et al., 2015).

Mosquito-borne diseases (MBDs) such as malaria, dengue, chikungunya, and West Nile virus, are sensitive to climate variables (Paz & Semenza, 2013). The transmission potential of dengue in Europe, for instance, has increased by approximately 29.7% since the 1981–2010 baseline period, thereby enabling local transmission in regions where it was previously restricted almost exclusively to returned travelers (Semenza et al., 2022; Gossner, Ducheyne, & Schaffner, 2021). Agricultural, construction, and military/disaster response outdoor workers are at particular occupational risk because they spend long hours outdoors in environments with a high potential of vector activity (Schulte & Chun, 2009).

Tick-borne diseases (TBDs) pose similar occupational risks to public health inspectors and other outdoor workers. Lyme borreliosis and tick-borne encephalitis are now considered occupational risks to foresters, animal husbandry workers, agricultural workers, and others whose outdoor jobs involve conducting inspections in vegetated areas (Van Hecke, Cameron, & Cruickshank, 2018; Rizzoli et al., 2011). Forestry workers have consistently high seroprevalance of *B. burgdorferi* in comparison to the general population, with the higher levels exceeding 20% in some European studies (De Keukeleire, Vanwambeke, Somers, & Luyasu, 2017). Farmers and veterinarians are also at substantially higher risk of occupational exposure than that of low-risk controls with multiple studies documenting seroprevalance multiple times that of controls (Piacentino & Schwartz, 2002).

Leishmaniasis is a risk to outdoor construction workers, military personnel and others working in semi-arid or arid regions that may be subjected to increased sandfly activity with future climate change and expansion of sandfly habitats (Ready, 2010; Gonzlez et al., 2010).

Table 2 described VBDs of occupational or potential occupational significance to public health inspectors. Information for each disease included vector species, climate-sensitive transmission components, geographically documented and projected geographic expansion, and particular inspection situations associated with a heightened exposure risk. Diseases are classified by vector (tick-borne, mosquito-borne, and sandfly-borne) respectively. Geographic expansion information is based on trends observed from long-term surveillance and ecological niche modeling data from cited literature.

Table 2. Vector-borne diseases of occupational relevance to public health inspectors

Disease	Vector	Climate-sensitive factors	Geographic expansion	Occupational risk context for inspectors
Lyme borreliosis	*Ixodes* ticks (deer ticks)	Warmer temperatures; longer growing seasons; expanded tick habitat to higher latitudes/altitudes	Northern Europe, Canada, northern US states	Inspectors conducting field inspections in forested, grassy, or vegetated areas; forestry and agricultural site inspections
Tick-borne encephalitis	*Ixodes* ticks	Extended tick activity season; geographic range expansion	Central/ Eastern Europe, Scandinavia, Siberia	Inspectors in rural, forested, and agricultural settings; animal facility inspections
Dengue	*Aedes* mosquitoes	Rising temperatures; increased transmission potential; vector range expansion	Southern Europe, Americas, Southeast Asia	Inspectors in peri-urban and agricultural areas during outbreak investigations; standing water inspections
Malaria	*Anopheles* mosquitoes	Temperature-dependent parasite development; changed rainfall patterns	Highland regions of Africa, Asia, Latin America	Agricultural and rural area inspectors; disaster response in endemic regions
Chikungunya	*Aedes* mosquitoes	Climate-driven vector expansion	Europe, Americas, Asia	Similar to dengue; inspectors in outbreak response settings
West Nile Virus	*Culex* mosquitoes	Temperature and precipitation effects on vector populations	North America, Europe, Mediterranean	Outdoor inspectors in agricultural and peri-urban settings
Leishmaniasis	Sandflies	Expanded sandfly habitats; deforestation; land use change	Southern Europe, North Africa, Middle East, Latin America	Construction site inspectors; military/ disaster response in arid/ semi-arid regions; dusk/ dawn field work

3.3.3 Outbreak investigations and sanitation inspections

Public health inspectors conduct the "first response" in outbreak investigations and sanitary assessments during public health crises (Levy et al. 2017) and there are climate change impacts altering the epidemiology of infectious diseases, thereby exacerbating the frequency and complexity of public health crisis responses (Patz et al. 2014). Inspectors may be called upon to respond to foodborne disease outbreaks attributed to warming agricultural seasons, waterborne disease outbreaks resulting from increased extreme rainfall events or respiratory illness outbreaks at emergency shelters (Levy et al. 2016).

Public health inspectors engaged in outbreak responses face greater risks of exposure due to time pressure during investigative measures, risk of contact with infected premises and a compromised ability to implement optimal infection control measures at field sites (Paterson et al. 2014). Inspectors engaged in sanitary assessments of flood affected areas face potential failure of sewage systems, contamination of water sources and elevated concentrations of fecal indicator organisms in their water sources (Levy et al. 2016). All of the aforementioned scenarios place public health inspectors squarely at the nexus of humans, animals and the environment where One Health-based protections are most direly needed (Coker et al. 2011).

Table 3 is a list of relevant WBDs for a Public Health Inspector, and more specifically for use during a post flood inspection, sanitation inspection, and when observing a wastewater facility.

Table 3. Waterborne diseases of occupational relevance to public health inspectors.

Disease	Pathogen	Climate-Sensitive Factors	Primary Exposure Route for Inspectors	Occupational Context
Leptospirosis	*Leptospira* spp. bacteria	Flooding; heavy rainfall; warm temperatures; survival in wet environments	Contact with floodwater, mud, sewage, or animal urine-contaminated water	Post-flood assessments; sanitation inspections; wastewater facility inspections; agricultural site visits
Cholera	*Vibrio cholerae* bacteria	Sea-level rise; coastal flooding; warm ocean temperatures	Exposure to contaminated coastal or flood waters	Coastal disaster response; port and fisheries inspections; drinking water quality assessments
Cryptosporidiosis	*Cryptosporidium* spp. parasite	Heavy precipitation; runoff events; flooding	Ingestion of oocysts from contaminated water or soil	Wastewater treatment plant inspections; agricultural site inspections; drinking water quality assessments
Giardiasis	*Giardia* spp. parasite	Flooding; contaminated water sources	Exposure to fecally contaminated water	Sanitation inspections; rural water system assessments
Enteric viral infections	Norovirus, enterovirus, hepatitis E	Warmer temperatures; flooding; sewage overflow	Contact with contaminated water or surfaces during inspections	Outbreak investigations; post-flood sanitation assessments; food facility inspections

3.4 Psychosocial hazards and mental health impacts

3.4.1 Disaster-induced stress and trauma

The psychosocial aspects of climate-related occupational hazards are frequently overlooked in the curriculum of inspection service training programs (Morris et al., 2022). When engaged in flood, wildfire, hurricane, and other climate-amplified disaster response activities, public health inspectors can become significantly psychologically distressed (Paterson et al., 2014). Encounters with human suffering, destruction of property, and collapse of community structures inherent in emergency response may be related to elevated rates of post-traumatic stress symptoms, anxiety, and depression among disaster responders (Benedek et al., 2007).

When inspectors themselves reside within affected communities, personal and professional roles can collapse on each other, leading to additional psychological stress (Applebaum et al., 2016). Public health inspectors, unlike designated emergency responders, are often not trained in psychological first aid or trauma-informed approaches and may have little access to mental health resources (Schulte et al., 2016).

3.4.2 Eco-anxiety and moral distress

The feeling of a persistent dread of an environmental catastrophe-'eco-anxiety'-that is rooted in the observation of a changing environment, has come to be widely accepted as a psychological state impacting health and public health professionals who work in environmental protection and monitoring (Clayton et al., 2017; Pihkala, 2020). The nature of the work for public health inspectors may result in a cumulative psychological burden on health and public health professionals, because they witness climate-related hazards, non-compliance, and the inadequacy of legal and regulatory mechanisms on a daily basis.

The term 'moral distress', "a state in which one knows or believes that a certain moral choice is the right one but feels powerless to act on it," has previously been identified in the field of medicine (Jameton, 1984). An analogous form of moral distress among inspectors can occur when one recognizes health and safety hazards associated with a changing climate but lacks the authority, resources or institutional backing to compel action (Schulte & Chun, 2009). Furthermore, this burden is potentially intensified as inspectors perceive that those who work in these affected settings-workers, and indeed the communities in which the inspectors are also based-bear the greatest brunt of climate-related health impacts (Sorensen et al., 2018).

3.4.3 Occupational burnout and workforce retention

The cumulative impact of occupational physical, biological, and psychological stressors related to climate change-driven inspection work could be one explanation for higher prevalence of occupational burnout among public health inspectors (Maslach et al., 1996; Adamopoulos et al., 2022a, 2022b). Job burnout (which is defined by emotional exhaustion, depersonalization, and lack of personal accomplishment) has been correlated with impaired work performance, high absenteeism rates, and worker turnover (Leiter & Maslach, 2017).

Inspection programs are already experiencing recruitment and retention difficulties in many areas, especially in remote and rural locations which may be more significantly affected by climate change (Schulte et al., 2016). Climate change, by increasing workload while working conditions become worse, could be expected to further aggravate worker shortages and threatening work systems viability over the long term (Applebaum et al., 2016; Adamopoulos, 2025; Adamopoulos et al., 2025b, 2025c).

Figure 2 compares current versus projected (2°C warming) psychosocial hazard scores, highlighting worsening risks across PTSD, burnout, eco-anxiety, and work-life conflict.

Figure 2. Psychosocial hazard profile of public health inspectors: Current vs. projected climate scenarios.

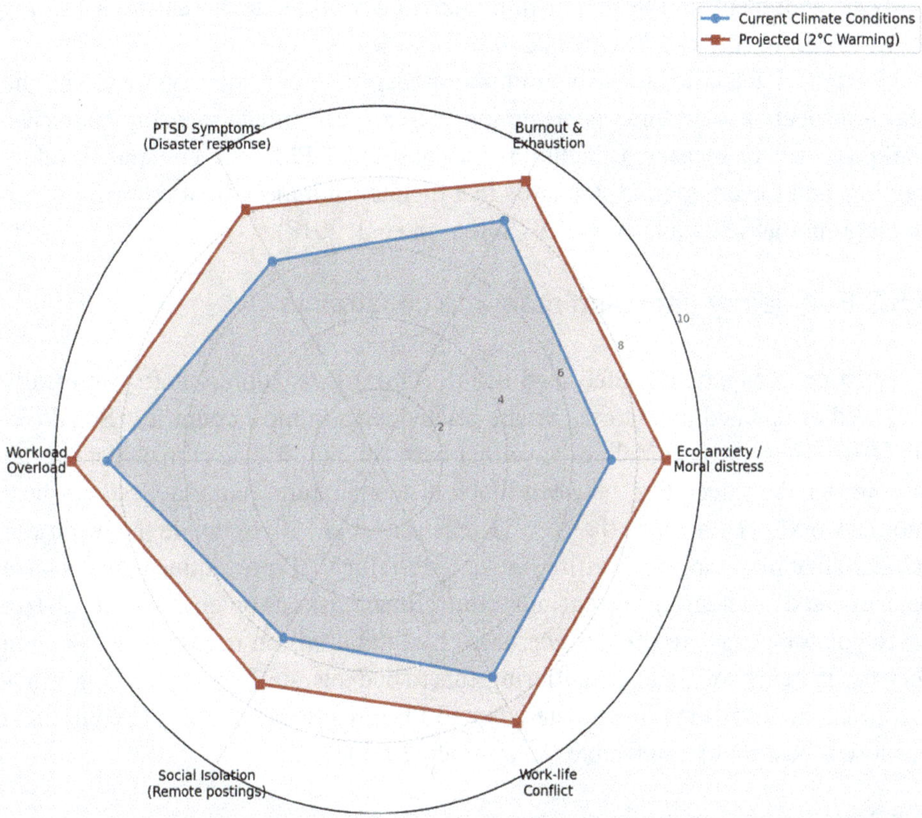

The presented figure is generated using Python version 3.14 (https://www.python.org).

3.5 Social and institutional pathways of vulnerability

3.5.1 Unequal exposure among inspector subpopulations

Climate-related biological and physical changes only lead to actual disease burden among inspectors to the extent they are buffered by social pathways, including differences in access to resources, training, and personal protective equipment (PPE) (Sorensen et al. 2018; Adamopoulos et al., 2023). Some subgroups within the inspection workforce are at elevated risk of the effects of climate change. Inspectors early in their careers, for instance, may be placed in the most dangerous field operations without appropriate training or adequate protective equipment (Levy et al. 2017). Inspectors in low- and middle-income countries face a multitude of additional barriers to worker protection and capacity including lack of resources, outdated equipment, and weaker institutional frameworks than those in high-income countries (Bennett & McMichael 2010).

Gender-related risks are also an important aspect to consider. While the public health inspection workforce is becoming increasingly female in many countries, workplace safety measures, including the design of PPE, the content of safety training, and the ergonomic standards that dictate equipment fit, are largely based on anthropometric data from males (Sorensen et al. 2018).

3.5.2 Regulatory gaps and resource constraints

Existing occupational safety and health (OSH) legislation is not systemically reviewed or updated to address climate-related risks in most countries (Schulte et al., 2016). Heat stress guidelines are either based on out-of-date climate data or fail to consider the interaction between ambient temperature, humidity, radiant heat and the workload placed on the body (Kjellstrom et al., 2016); while the respirator standard that may apply to wildfire smoke or PPE for VBD prevention may not have been updated considering current and future climate risks (Applebaum et al., 2016).

Regulatory gaps are often exacerbated by limitations on resources. Inspection services in many countries are suffering cuts to budgets, staff, and increasing workload, making investment in climate-adapted PPE, real-time monitoring equipment or more robust training unviable (Levy et al., 2017).

3.5.3 Weak integration of surveillance systems

The challenge in climate risk management is to have an integrated surveillance system connecting environmental, animal health and occupational health data (Coker et al., 2011; Destoumieux-Garzn et al., 2018). Yet these systems remain in siloses, with little information sharing or interconnectivity (Mackenzie & Jeggo, 2019). In many cases, the role of public health inspection services within climate adaptation efforts is being overlooked; their role is that of on-the-ground first responders in detecting new hazards (Schulte et al., 2016).

The integration of surveillance does not lead to an effective alert system when climate sensitive hazards emerge. Front-line inspectors may be the first to notice an emerging hazard, a number of febrile illnesses occurring in workers in a region; an animal die-off, or a pattern of flooding-related breakdowns in sanitary systems; without an easy or efficient way to notify relevant systems or receive early warnings about emerging health hazards (Levy et al., 2017).

3.6 Training gaps, surveillance, and preparedness challenges

3.6.1 Limited climate literacy in inspection training

While the landscape of risk continues to change, most of the training frameworks provided for inspection services are relatively stagnant (Schulte et al. 2016). As reported by Adam-Poupart et al. In their review of the field, training materials for occupational safety and health training courses did not contain much of the climate change material and when they did contain them, they were not substantial and were generally not action-oriented for field workers.

At minimum climate literacy for an inspector would entail understanding key meteorological risk indices (heat index, air quality index, forecast floods); recognizing regional climate sensitive disease trends; and grasping climate adaptation concepts (Kjellstrom et al. 2016); Most existing inspection training, according to the existing baseline studies, failed to capture these desired inspector competencies (Bennett and McMichael 2010; Adamopoulos et al., 2022c).

3.6.2 Inadequate guidance on PPE performance under extreme conditions

Personal protective equipment is manufactured and tested under normal environmental conditions, not with the extreme temperatures, humidity and particle concentrations likely in a climate-changed world (Morris et al., 2022). Inspectors need practical recommendations regarding how personal protective equipment performance decreases in extreme heat, work-rest cycles that can mitigate heat stress while wearing the equipment, and personal protective equipment choice in terms of thermal comfort versus hazards of work (Kjellstrom et al., 2016). Regarding vector-borne diseases prevention, guidance is needed to inspectors concerning appropriate use of repellents, permethrin-treated clothing, and tick checks in differing conditions (Piacentino & Schwartz, 2002). For wildfire smoke, guidance on choice between, use of, and fit-testing N95 respirators versus elastomeric half-mask respirators versus powered air-purifying respirators, as well as proper donning and doffing procedures in field conditions is needed (Reid et al., 2016).

3.6.3 Psychosocial support mechanisms

The increased inspection work caused by climate conditions has major psychological hazards. However, few inspection services possess robust psychosocial support measures (Applebaum et al., 2016). Mental health support such as critical incident stress debriefing, mental health care access, or peer support programs (key elements of disaster responder support) is largely inaccessible to public health inspectors (Paterson et al., 2014).

Proactive prevention of mental health problems and resilience enhancement strategies are also less prominent in the inspection arena. Staff training in coping skills, stress management, and awareness of burnout could equip inspectors with necessary resources to manage ongoing climate-related stress (Schulte et al., 2016).

3.6.4 Weak integration of environmental and occupational surveillance

In this section, it explains how the failures in surveillance can affect preparedness. Inability to foresee where and when climate sensitive health hazards will arise, as a result of non-integration of systems, means that inspection services can not prepare accordingly (Coker et al., 2011). With a system which integrates predictions on heat waves with health surveillance in the workplace for example, a response system can be developed to send inspectors to vulnerable areas ahead of predicted hot conditions, providing information regarding rest and work periods, along with provision of cooling vests and other protective gear. Likewise, with integration of vector surveillance data and inspection scheduling the authorities can time visits to high risk areas outside peak times of exposure, or make specific provisions regarding protective clothing for those working in potentially high risk areas (Semenza et al., 2022). The absence of integration puts the inspection service at risk to preventable harms.

3.7 The strategic role of inspection services in One Health governance

3.7.1 Inspectors as early detectors of climate-sensitive risks

Within a One Health framework, the role of public health inspection services extends beyond that of simple enforcement; they are integral sentinel systems which identify early warning signs of climate-sensitive health hazards (Zinsstag et al., 2011). Inspectors in food processing plants may observe a trend in pest invasion or food spoilage which identifies a shift in climate conditions, inspectors in animal health services may detect emerging zoonoses prior to an outbreak in the human population (Coker et al., 2011), and inspectors of sanitation infrastructure may identify risks for contamination during flood events prior to WBDs outbreak in the wider community (Paterson et al., 2014). The full potential of this sentinel role is not yet being realized, with limited mechanisms for documenting inspector observations, aggregating them, and acting on them (Schulte et al., 2016). Investing in improved training, reporting, and inter-agency coordination can significantly enhance the sentinel capacity of public health inspection services as a cost-effective approach to anticipating the impacts of climate on health (Bennett & McMichael, 2010).

3.7.2 Coordination across occupational, public, and environmental health sectors

To manage climate-sensitive risks effectively, communication across sectors that have traditionally operated separately is essential (Destoumieux-Garzn et al. 2018). Public health inspection services are perfectly placed to enable such cross-sectoral cooperation due to routine engagement with occupational health, environmental protection, agriculture and veterinary services and emergency services (Levy et al. 2017). One Health governance structures highlight the necessity of cross-sectoral working platforms such as co-operative risk assessment and surveillance networks, joint response protocols etc (Mackenzie & Jeggo 2019) which can be delivered by inspectorates at the operational level (Gibbs 2014).

3.7.3 Informing adaptive regulation and preventive strategies

Static regulatory approaches are likely becoming increasingly outmoded due to the changing environment brought about by climate change. In order to effectively achieve adaptation to climate change, regulations will need to be dynamic, anticipatory, and responsive to changing risk patterns (Schulte et al., 2016). Inspectors of public health facilities and their workers are at a prime position to contribute to regulations which can effectively adapt to climate change due to their direct experiences with impacts on the facilities and worker populations they inspect (Applebaum et al., 2016). System for taking inspector feedback and feeding it into regulatory updates such as regular after-action reviews, hazard reporting systems and inspector membership in standard-setting groups will be instrumental to make climate adapted regulations more relevant and effective (Levy et al., 2017). Reconceptualizing inspector insights as a source of knowledge, and not as simply work dedicated to enforcement, will mark a crucial cultural change (Schulte & Chun, 2009).

3.8 Adaptation strategies and policy implications

3.8.1 Climate-adapted training and capacity development

Reecent work by Adamopoulos et al. (2026) provides a systematic framework for classifying and mitigating climate-related occupational risks specific to public health inspectors, complementing the adaptation strategies outlined below.

A very quick and low cost way to adapt is the increased capacity of the inspection workforce. Training programmes need to be revised to cover 1) climate science basics and local projected climate changes; 2) identification and management of heat illness; 3) emergency management for floods, wild-fires, and extreme weather; 4) new bio-threats to the workplace that result from changing climates; 5) selection and appropriate use of protective clothing in a range of situations, including the hottest extremes; 6) basic psychological first aid and stress management (Schulte et al., 2016; Morris et al., 2022). Training should be competency based, revised routinely, and offered in a variety of formats that includes workshops and on-line learning, as well as simulated scenarios (Bennett and McMichael, 2010). Training must be on-going, and embedded into professional continuing education.

3.8.2 Digital tools and real-time environmental monitoring

Technology can improve inspector safety on climate-exacerbated field work (Applebaum et al., 2016). Mobile applications capable of receiving and presenting current data on heat index, air quality, flood warnings, and vector activity can improve the flexibility of dynamic risk assessment during inspections (Kjellstrom et al., 2016). Integrated into scheduling systems, early warning systems can facilitate the proactive management of risks, such as canceling unnecessary outdoor inspections during heat alerts or assigning higher-level PPE during periods of elevated vector populations (Semenza et al., 2022).

Wearable sensors monitoring the inspector's heat strain through physiological cues (heart rate, core body temperature, sweat rate) can alert the inspector in real-time when safe exposure levels are reached (Flouris et al., 2018). Geographic information systems (GIS) can map climate hazards onto inspection routes, facilitating route optimization to limit exposure (Campbell et al., 2015).

Simulation technologies, such as VR and MR, have shown potential for training inspectors on safely operating in extreme heat settings, flood prone workplaces, and hazardous emergency situations without introducing an actual risk into training scenarios (Morris et al., 2022). Furthermore, scenario-based simulations can help inspectors train on decision-making under climate stressed conditions, strengthening cognitive resilience and enhancing hazard detection (Schulte et al., 2016).

3.8.3 Formal recognition of climate-related occupational hazards

Health and safety legislation (Schulte & Chun, 2009) needs to formally recognize climate change as a workplace hazard, with implications in terms of authorized use of resources to ameliorate hazards, the grounding for regulations' modernization, and workers' compensation rights and ethical duties of the employer to prevent workers' exposure to climate-related occupational hazards (Applebaum et al., 2016). Operationalization of formal recognition needs to include: 1) integration of climate-related hazards into occupational health and safety risk assessment framework; 2) incorporation of climate adaptation standards into inspection service guidelines and policies; 3) availability of climate-safe personal protective equipment (PPE) and work-rest environments; 4) medical surveillance programs for climate sensitive diseases; and 5) development of occupational injury and illness compensation schemes that account for climate-related issues (Kjellstrom et al., 2016; Sorensen et al., 2018).

3.8.4 Cross-sector collaboration with a One Health approach

This complex set of climate-change risks cannot be dealt with by one agency in isolation, and inter-sectoral cooperation, using a One Health perspective, is needed (Mackenzie & Jeggo, 2019). Inspection services should formalize cooperation with public health agencies on disease surveillance and outbreak response, environmental agencies on monitoring of air quality, water quality and vector abundance, veterinary services on surveillance of animal populations for zoonotic diseases, meteorological services for the provision of forecasts and warnings, emergency services to co-ordinate disaster response and preparedness efforts, trade unions and worker organisations for participation on protective measures and academic bodies to provide research expertise and training (Coker et al., 2011; Destoumieux-Garzn et al., 2018). Guidelines on inter-sectoral co-operation are already available, for instance the WHO/ILO Joint Policy Brief on climate change and occupational health (WHO & ILO, 2021). In countries where national or regional One Health platforms have already been set up inspection services should participate, as existing co-ordinating structures will already exist.

3.8.5 Investment in climate-resilient inspection systems

Lastly, it is essential that we invest long term in guarding inspection services from the effects of climate change (Bennett & McMichael, 2010). The elements of an inspection system that can endure the impacts of climate change include an appropriate number of employees so they are not over worked in case of emergency from a climate event; new equipment suitable for extreme heat and new, climate compatible PPE; dependable communication so workers are kept informed of emerging climate hazards and response coordination can be implemented; vehicles and workplaces that can be maintained at moderate temperatures where workers can rest and recover from extensive field work during prolonged heat events; access to occupational health services that include medical monitoring and support services for mental health issues and reliable systems for providing workers' compensation and disability for injuries and illnesses due to climate factors (Schulte et al., 2016; Applebaum et al., 2016).

4. DISCUSSION

4.1 Summary of principal findings

This chapter provided a systematic overview of shifting occupational risks for public health inspection services under climate change using a One Health framework. Four main conclusions can be drawn:

First, a range of hazard domains are being reshaped by climate change. Physical hazards such as extreme heat, floods, wildfire and poor air quality will become more frequent and intense (Reid et al. 2016; Paterson et al. 2014). Biological hazards such as zoonoses and VBDs are spreading and re-emerging in novel ways at human-animal-environment interface points (Carlson et al. 2022; Semenza et al. 2022). Chemical hazards are also being unpredictable mobilised by flood events and fires (Levy et al. 2016), while psychosocial risks including trauma, eco-anxiety, moral distress, and burnout can pose an equally serious challenge (Clayton et al. 2017; Pihkala 2020).

Second, inspector safety and system efficacy are interdependent. An unwell inspector can not undertake surveillance, enforcement and response activities (Schulte et al. 2016). Within a One Health framework, worker health is a prerequisite for system effectiveness under climate change (Bennett & McMichael 2010).

Third, current preparation and response systems are inadequate. Static inspector training often fails to address the implications of climate change (Schulte & Chun 2009). Surveillance systems lack interdisciplinary integration and data sharing (Coker et al. 2011). Regulations are failing to keep pace with changing hazard profiles and PPE frequently proves insufficient in extreme events (Morris et al. 2022).

Fourth, inspection services are underestimated potential sentinel surveillance and climate adaptation assets. Their position at human-animal-environment interface points enables early identification of novel threats (Zinsstag et al. 2011) and they are able to utilize regulatory leverage and interagency relationships to advocate for climate adaptive practices and inter-agency cooperation (Gibbs 2014; Destoumieux-Garzn et al. 2018).

4.2 Interpretation within the One Health framework

The One Health approach was crucial for understanding climate-induced hazards for public health inspectors on three interrelated grounds.

First, One Health recognizes that environmental disruption creates interactive risk pathways, not isolated hazards (Zinsstag et al., 2011). For example, a flood simultaneously mobilizes chemicals, creates mosquito breeding sites, destroys infrastructure, and inundates wastewater facilities forcing inspectors to face merged, compounding risks.

Second, One Health highlights that climate risks are unequally distributed across inspector subgroups (Sorensen et al., 2018). Early-career inspectors, those in low- and middle-income countries, and female inspectors (whose PPE is often designed for male body types) face disproportionately higher exposure risks and fewer resources, with climate change exacerbating existing structural inequalities.

Third, One Health's systems approach shows that protecting inspector health cannot be separated from protecting the human, animal, and environmental systems they regulate (Coker et al., 2011). A depleted inspection service due to injury, illness, or burnout cannot enforce regulations, detect zoonotic risks, or ensure food and water security. Conversely, interventions that support inspector well-being (climate-resilient PPE, real-time monitoring, mental health services) yield population-level public health co-benefits.

4.3 Comparison with existing literature

This chapter builds upon and complements work across multiple domains. Within occupational health, this discussion supports and extends Schulte et al.'s (2016) proposed framework for categorizing worker climate-related hazards (they proposed eight types), as it operationalized each category of hazard specifically to the occupational circumstances of inspectors. Kjellstrom et al.'s (2016) work on occupational heat strain, which focused mainly on agricultural and construction workers, is readily transferable to inspectors working outdoors; additionally, we draw attention to a significant source of heat strain from PPE that is less of a concern in some other working populations.

Within infectious disease epidemiology, our discussion further solidifies and expands the findings of Carlson et al. (2022) on climate-driven zoonotic disease spillover risk; however, this chapter identifies more narrowly defined occupational exposure pathways that receive little mention in the wider climate-and-infection literature. Within VBD research, we supplement Semenza et al.'s (2022) findings on expanding ranges across Europe by addressing potential occupational exposure for inspectors performing field investigations in peri-urban and rural areas.

Within the field of disaster mental health, the psychological hazards discussed in this chapter-eco-anxiety, moral distress, and disaster-induced trauma-have been documented amongst designated first responders (Benedek et al., 2007), but the psychosocial consequences for public health inspectors are an under-researched subject and are a significant gap in the literature on responder mental health; inspectors are not typically included in either disaster responder surveillance for mental health or support networks after disasters.

4.4 Implications for policy and practice

Leaders in inspection services and public health agencies must prioritize recognizing climate change as a workplace hazard under their safety management plans, worker compensation systems, and resource allocation priorities. Without this formal recognition, no action is possible on the use of adapted PPE, reform of training programs, or the use of mental health support services. Formal recognition should be translated into practical use through updated risk assessment protocols and revised standard operating procedures for conditions of extreme heat, cold and precipitation; and climate resilience needs to be monitored and used as a performance indicator.

Professional schools and certification agencies need to revise training programs and curricula. Climate knowledge should be integrated, and it must be applied through identification of hazards and appropriate responses, selection and use of extreme condition PPE; and basic psychological first aid and stress management techniques.

Regulators and standard setting bodies need to move beyond static models for occupational exposure limits, PPE standards, and work-rest guidance. Occupational exposure limits need to integrate extreme heat-humidity combinations, above historical maxima; PPE needs to be rated under high heat-particulate conditions, and static guidelines need to be supplemented by dynamic risk assessments for unpredictable conditions.

Occupational health services and worker compensation systems must recognize climate related occupational diseases and injuries. This requires revising clinical definitions and reporting for climate-sensitive conditions-leptospirosis, Lyme disease, heatstroke, post-disaster psychiatric symptoms-as compensable as long as the causal linkage can be made. Surveillance protocols, and particularly occupational medical surveillance for inspector populations, must identify potential climate-sensitive conditions, particularly emergent risk from VBDs.

One Health governing bodies and cross-sectoral coordination platforms must incorporate the sentinel public health inspectorate's potential for detecting climate sensitive threats into their warning systems. Mechanisms should be established to feed observations by inspectors-unusual die-offs, cluster of illnesses, failure of infrastructure for disease prevention- into a process for integration with animal and environmental health intelligence.

4.5 Research gaps and future directions

There are large evidence gaps in each category discussed that warrant further research. First, and most important, there is an alarming dearth of empirical, epidemiological studies that are specific to the health impacts of climate change on inspection work populations. There are very few estimates on the prevalence of heat-related illness, VBDs seropositivity, post-disaster mental illness, and work-related burnout among inspectors. Prospective cohort studies that connect actual measurement of inspector exposures to health outcomes are required. Second, data on the effectiveness of adaptations specifically tailored to inspector work is needed. How well do adapted PPE training, real-time environmental monitoring, and real-time environmental health alerts modify inspector behavior? How are effective mental health interventions adapted and delivered to inspectors suffering post-disaster trauma? Intervention and implementation science are severely lacking.

Third, an economic evidence base regarding costs of adaptation is lacking: What are the productivity losses, healthcare costs and labor turnover expenses that are tied to inspector health issues related to climate change? How does this compare to investment costs in adapted PPE, reforms to training curriculum, and adoption of new surveillance systems?

Fourth, qualitative research on inspectors' lived experiences of climate hazards, moral injury, and structural barriers is crucial for understanding intervention development. How do inspectors themselves prioritize climate risks and what informal adaptations have they already made?

5. CONCLUSION

The fundamental operation of public health and occupational safety inspection services is being re-imagined by climate change. From more frequent and intense extreme weather to changes in infectious disease patterns to widespread environmental degradation, these factors directly influence inspectors' exposures to hazards during field operations. Physical hazards like extreme heat, flooding, and wildfires are becoming more frequent and severe, while biological hazards such as zoonotic spillovers and expansion of VBD range create new exposure risks for inspectors working across the human-animal-environment interface. Emerging psychosocial hazards such as climate-induced stress, eco-anxiety, and moral distress are also presenting significant but overlooked risks for this workforce.

It is here that the One Health lens proves especially useful. Inspector safety cannot be considered in isolation from the broader environment in which they work. Inspectors are necessarily tethered to the health of the populations, animal species, and ecosystems for which they are responsible. As such, they are not only exposed to risks stemming from disjunctions at the human-animal-environment interface but they also represent a lever point; investment in inspector safety yields co-benefits in the form of population and environmental health.

Current levels of preparedness are severely lacking. Training programs for inspectors generally reflect static environments, with poor climate literacy and limited guidance for field work. Surveillance systems for early detection of climate-sensitive hazards remain fragmented and disconnected from inspection operations. Regulatory frameworks have not yet been updated to reflect climate risk, while the capacity for PPE, equipment, and staffing is frequently insufficient. Psychosocial support for inspection work is rarely built into formal systems.

However, despite these challenges, public health inspection services can and should be envisioned as crucial adaptation assets. As sentinels working on the front lines of climate-sensitive health risks, inspectors are ideally positioned to detect emergent threats. They occupy a unique position to foster collaboration across disparate agencies of government responsible for occupational health and safety, public health, the environment, and animal health, while enforcement authority provides leverage to catalyze adoption of climate-adaptive practices across thousands of.

To realize this potential, action must be taken on multiple fronts. Training programs must be redesigned to include climate literacy, hazard recognition skills, and capacity for adaptive decision making. Technology, including real-time environmental monitoring, early warning systems, and simulation-based learning, can improve inspector safety and efficiency.

REFERENCES

Adam-Poupart, A., Labrèche, F., Smargiassi, A., Duguay, P., Busque, M. A., Gaucher, C., & Fournier, M. (2015). Climate change and occupational health and safety in a temperate climate: Potential impacts on vector-borne diseases. *International Journal of Environmental Research and Public Health, 12*(10), 13232–13250.

Adamopoulos, I. (2025). A Novel AI-based Modeling with Bias Classification Hybrid Risk Evaluation System for Confidence Enhanced Network Meta-Analysis of Occupational Hazards and Burnout Risk among Public Health Inspectors. Mesopotamian Journal of Artificial Intelligence in Healthcare. 219–233. DOI: 10.58496/MJAIH/2025/021

Adamopoulos, I., Lamnisos, D., Syrou, N., & Boustras, G. (2022a). Public health and work safety pilot study: Inspection of job risks, burn out syndrome and job satisfaction of public health inspectors in Greece. *Safety Science, 2022*(147), 105592. DOI: 10.1016/j.ssci.2021.105592

Adamopoulos, I., Lamnisos, D., Syrou, N., & Boustras, G. (2022b). Inspection of job risks, burn out syndrome and job satisfaction of Greek public health inspectors. *Safety and Health at Work, 2022*(13), S294. DOI: 10.1016/j.shaw.2021.12.1670

Adamopoulos, I., Lamnisos, D., Syrou, N., Boustras, G. (2022c). Training needs and quality of public health inspectors in Greece during the COVID-19 pandemic. European Journal of Public Health 2022, 32, ckac131.373. DOI: 10.1093/eurpub/ckac131.373

Adamopoulos, I., Syrou, N. (2024). Climate Change, Air Pollution, African Dust Impacts on Public Health and Sustainability in Europe, European Journal of Public Health, Volume 34, ckae144.1374. DOI: 10.1093/eurpub/ckae144.1374

Adamopoulos, I., Syrou, N., Lamnisos, D., Boustras, G. (2023). Cross-sectional nationwide study in occupational safety & health: Inspection of job risks context, burn out syndrome and job satisfaction of public health inspectors in the period of the COVID-19 pandemic in Greece. Safety Science, Elsevier, 2023, 158, 105960. . 105960DOI: 10.1016/j.ssci.2022

Adamopoulos, I., Syrou, N., Lamnisos, D., Valamontes, A., Karantonis, J. T., Tsirkas, P., & Dounias, G. (2026). *Classifying and mitigating occupational risks for public health inspectors in the context of the global climate crisis.* European Journal of Sustainable Development Research., DOI: 10.29333/ejosdr/17273

Adamopoulos, I., Syrou, N., Mpourazanis, G., Constantinidis, T. C., & Dounias, G. (2025a). The Association of the Global Climate Crisis with Environmental Risks and the Impact of Heat Stress on Occupational Safety, Health, and Hygiene. Med. Sci. Forum, 33, 2. DOI: 10.3390/msf2025033002

Adamopoulos, I., Valamontes, A., Tsirkas, P., & Dounias, G. (2025b). Predicting Workplace Hazard, Stress and Burnout Among Public Health Inspectors: An AI-Driven Analysis in the Context of Climate Change. *European Journal of Investigation in Health, Psychology and Education, 15*(5), 65. DOI: 10.3390/ejihpe15050065 PMID: 40422294

Adamopoulos, I., Valamontes, A., Tsirkas, P., & Dounias, G. (2025d). Predicting Workplace Hazard, Stress and Burnout Among Public Health Inspectors: An AI-Driven Analysis in the Context of Climate Change. *European Journal of Investigation in Health, Psychology and Education, 15*(5), 65. DOI: 10.3390/ejihpe15050065 PMID: 40422294

Altizer, S., Ostfeld, R. S., Johnson, P. T., Kutz, S., & Harvell, C. D. (2013). Climate change and infectious diseases: From evidence to a predictive framework. *Science, 341*(6145), 514–519. DOI: 10.1126/science.1239401 PMID: 23908230

Applebaum, K. M., Graham, J., Gray, G. M., LaPuma, P., McCormick, S. A., Northcross, A., & Perry, M. J. (2016). An overview of occupational risks from climate change. *Current Environmental Health Reports, 3*(1), 13–22. DOI: 10.1007/s40572-016-0081-4 PMID: 26842343

Benedek, D. M., Fullerton, C., & Ursano, R. J. (2007). First responders: Mental health consequences of natural and human-made disasters for public health and public safety workers. *Annual Review of Public Health, 28*(1), 55–68. DOI: 10.1146/annurev.publhealth.28.021406.144037 PMID: 17367284

Bennett, C. M., & McMichael, A. J. (2010). Non-heat related impacts of climate change on working populations. *Global Health Action, 3*(1), 5640. DOI: 10.3402/gha.v3i0.5640 PMID: 21191440

Black, C., Tesfaigzi, Y., Bassein, J. A., & Miller, L. A. (2017). Wildfire smoke exposure and human health: Significant gaps in research for a growing public health issue. *Environmental Toxicology and Pharmacology, 55*, 186–195. DOI: 10.1016/j.etap.2017.08.022 PMID: 28892756

Campbell, L. P., Luther, C., Moo-Llanes, D., Ramsey, J. M., Danis-Lozano, R., & Peterson, A. T. (2015). Climate change influences on global distributions of dengue and chikungunya virus vectors. *Philosophical Transactions of the Royal Society B*, 370(1665), 20140135.

Carlson, C. J., Albery, G. F., Merow, C., Trisos, C. H., Zipfel, C. M., Eskew, E. A., Olival, K. J., Ross, N., & Bansal, S. (2022). Climate change increases cross-species viral transmission risk. *Nature*, *607*(7918), 555–562. DOI: 10.1038/s41586-022-04788-w PMID: 35483403

Cascio, W. E. (2018). Wildland fire smoke and human health. *The Science of the Total Environment*, *624*, 586–595. DOI: 10.1016/j.scitotenv.2017.12.086 PMID: 29272827

Clayton, S., Manning, C. M., Krygsman, K., & Speiser, M. (2017). *Mental health and our changing climate: Impacts, implications, and guidance*. American Psychological Association.

Coker, R., Rushton, J., Mounier-Jack, S., Karimuribo, E., Lutumba, P., Kambarage, D., Pfeiffer, D. U., Stärk, K., & Rweyemamu, M. (2011). Towards a conceptual framework to support one-health research for policy on emerging zoonoses. *The Lancet. Infectious Diseases*, *11*(4), 326–331. DOI: 10.1016/S1473-3099(10)70312-1 PMID: 21376670

De Keukeleire, M., Vanwambeke, S. O., Somers, B., & Luyasu, V. (2017). Lyme borreliosis in forestry workers: A systematic review. *European Journal of Public Health*, *27*(3), 514–519.

Destoumieux-Garzón, D., Mavingui, P., Boetsch, G., Boissier, J., Darriet, F., Duboz, P., Fritsch, C., Giraudoux, P., Le Roux, F., Morand, S., Paillard, C., Pontier, D., Sueur, C., & Voituron, Y. (2018). The one health concept: 10 years old and a long road ahead. *Frontiers in Veterinary Science*, *5*, 14. DOI: 10.3389/fvets.2018.00014 PMID: 29484301

Flouris, A. D., Dinas, P. C., Ioannou, L. G., Nybo, L., Havenith, G., Kenny, G. P., & Kjellstrom, T. (2018). Workers' health and productivity under climate change: The emerging role of heat stress. *Temperature*, *5*(3), 210–221.

Gibbs, E. P. J. (2014). The evolution of One Health: A decade of progress and challenges for the future. *The Veterinary Record*, *174*(4), 85–91. DOI: 10.1136/vr.g143 PMID: 24464377

Gilbert, M., Golding, N., Zhou, H., Wint, G. R., Robinson, T. P., Tatem, A. J., Lai, S., Zhou, S., Jiang, H., Guo, D., Huang, Z., Messina, J. P., Xiao, X., Linard, C., Van Boeckel, T. P., Martin, V., Bhatt, S., Gething, P. W., Farrar, J. J., & Yu, H. (2014). Predicting the risk of avian influenza A H7N9 infection in live-poultry markets across Asia. *Nature Communications*, *5*(1), 4116. DOI: 10.1038/ncomms5116 PMID: 24937647

Gossner, C. M., Ducheyne, E., & Schaffner, F. (2021). Increased risk of autochthonous dengue transmission in Europe. *Eurosurveillance*, *26*(21), 2100431.

Hancock, P. A., Ross, J. M., & Szalma, J. L. (2007). A meta-analysis of performance response under thermal stressors. *Human Factors*, *49*(5), 851–877. DOI: 10.1518/001872007X230226 PMID: 17915603

IPCC. (2022). *Climate Change 2022: Impacts, Adaptation and Vulnerability*. Contribution of Working Group II to the Sixth Assessment Report of the Intergovernmental Panel on Climate Change. Cambridge University Press.

Jameton, A. (1984). *Nursing practice: The ethical issues*. Prentice-Hall.

Kjellstrom, T., Briggs, D., Freyberg, C., Lemke, B., Otto, M., & Hyatt, O. (2016). Heat, human performance, and occupational health: A key issue for the assessment of global climate change impacts. *Annual Review of Public Health*, *37*(1), 97–112. DOI: 10.1146/annurev-publhealth-032315-021740 PMID: 26989826

Lau, C. L., Smythe, L. D., & Weinstein, P. (2010). Leptospirosis: An emerging disease in travelers? *Travel Medicine and Infectious Disease*, *8*(1), 33–39. DOI: 10.1016/j.tmaid.2009.12.002 PMID: 20188303

Leiter, M. P., & Maslach, C. (2017). Burnout and engagement: Contributions to a new vision. *Burnout Research*, *5*, 55–57. DOI: 10.1016/j.burn.2017.04.003

Levy, B. S., Wegman, D. H., Baron, S. L., & Sokas, R. K. (2017). *Occupational and environmental health: Recognizing and preventing disease and injury* (7th ed.). Oxford University Press. DOI: 10.1093/oso/9780190662677.001.0001

Levy, K., Woster, A. P., Goldstein, R. S., & Carlton, E. J. (2016). Untangling the impacts of climate change on waterborne diseases: A systematic review of relationships between diarrheal diseases and temperature, rainfall, flooding, and drought. *Environmental Science & Technology*, *50*(10), 4905–4922. DOI: 10.1021/acs.est.5b06186 PMID: 27058059

Mackenzie, J. S., & Jeggo, M. (2019). The One Health approach—Why is it so important? *Tropical Medicine and Infectious Disease*, *4*(2), 88. DOI: 10.3390/tropicalmed4020088 PMID: 31159338

Maslach, C., Jackson, S. E., & Leiter, M. P. (1996). *Maslach Burnout Inventory manual* (3rd ed.). Consulting Psychologists Press.

Medlock, J. M., & Leach, S. A. (2015). Effect of climate change on vector-borne disease risk in the UK. *The Lancet. Infectious Diseases*, *15*(6), 721–730. DOI: 10.1016/S1473-3099(15)70091-5 PMID: 25808458

Morris, K. I., Chan, A., Morris, K. J. K., & Walker, T. R. (2022). Climate change and workers' health: A scoping review. *International Journal of Environmental Research and Public Health, 19*(15), 9445. PMID: 35954801

Ogden, N. H., Radojević, M., Wu, X., Duvvuri, V. R., Leighton, P. A., & Wu, J. (2014). Estimated effects of projected climate change on the basic reproductive number of the Lyme disease vector Ixodes scapularis. *Environmental Health Perspectives, 122*(6), 631–638. DOI: 10.1289/ehp.1307799 PMID: 24627295

Pascual, M., Bouma, M. J., & Dobson, A. P. (2002). Cholera and climate: Revisiting the quantitative evidence. *Microbes and Infection, 4*(2), 237–245. DOI: 10.1016/S1286-4579(01)01533-7 PMID: 11880057

Paterson, J., Berry, P., Ebi, K., & Varangu, L. (2014). Health and safety aspects of flood cleanup. *Canadian Medical Association Journal, 186*(3), 199.

Patz, J. A., Frumkin, H., Holloway, T., Vimont, D. J., & Haines, A. (2014). Climate change: Challenges and opportunities for global health. *Journal of the American Medical Association, 312*(15), 1565–1580. DOI: 10.1001/jama.2014.13186 PMID: 25244362

Paz, S., & Semenza, J. C. (2013). El Niño and climate change—Contributing factors in the dispersal of West Nile virus in the United States. *Global Health Action, 6*(1), 21830.

Piacentino, J. D., & Schwartz, B. S. (2002). Occupational risk for Lyme disease in the United States. *Journal of Occupational and Environmental Medicine, 44*(6), 491–500. PMID: 12085472

Pihkala, P. (2020). Eco-anxiety and environmental education. *Sustainability (Basel), 12*(23), 10149. DOI: 10.3390/su122310149

Ready, P. D. (2010). Leishmaniasis emergence and climate change. [OIE]. *Revue Scientifique et Technique (International Office of Epizootics), 29*(3), 399–412. PMID: 18819668

Reid, C. E., Brauer, M., Johnston, F. H., Jerrett, M., Balmes, J. R., & Elliott, C. T. (2016). Critical review of health impacts of wildfire smoke exposure. *Environmental Health Perspectives, 124*(9), 1334–1343. DOI: 10.1289/ehp.1409277 PMID: 27082891

Rizzoli, A., Hauffe, H. C., Carpi, G., Vourc'h, G., Neteler, M., & Rosà, R. (2011). Lyme borreliosis in Europe. *Eurosurveillance, 16*(27), 19906. DOI: 10.2807/ese.16.27.19906-en PMID: 21794218

Rocque, R. J., Beaudoin, C., Ndjaboue, R., Cameron, L., Poirier-Bergeron, L., Poulin-Rheault, R. A., & Witteman, H. O. (2021). Health effects of climate change: An overview of systematic reviews. *BMJ Open, 11*(6), e046333. DOI: 10.1136/bmjopen-2020-046333 PMID: 34108165

Romanello, M., Di Napoli, C., Drummond, P., Green, C., Kennard, H., Lampard, P., Scamman, D., Arnell, N., Ayeb-Karlsson, S., Ford, L. B., Belesova, K., Bowen, K., Cai, W., Callaghan, M., Campbell-Lendrum, D., Chambers, J., van Daalen, K. R., Dalin, C., Dasandi, N., & Costello, A. (2022). The 2022 report of the Lancet Countdown on health and climate change: Health at the mercy of fossil fuels. *Lancet, 400*(10363), 1619–1654. DOI: 10.1016/S0140-6736(22)01540-9 PMID: 36306815

Schulte, P. A., Bhattacharya, A., Butler, C. R., Chun, H. K., Jacklitsch, B., Jacobs, T., & Wagner, G. R. (2016). Advancing the framework for considering the effects of climate change on worker safety and health. *Journal of Occupational and Environmental Hygiene, 13*(11), 847–865. DOI: 10.1080/15459624.2016.1179388 PMID: 27115294

Schulte, P. A., & Chun, H. (2009). Climate change and occupational safety and health: Establishing a preliminary framework. *Journal of Occupational and Environmental Hygiene, 6*(9), 542–554. DOI: 10.1080/15459620903066008 PMID: 19551548

Semenza, J. C., Rocklöv, J., & Ebi, K. L. (2022). Climate change and cascading risks for infectious disease transmission. *The Lancet. Planetary Health, 6*(4), e350–e358. PMID: 35397223

Sorensen, C., Murray, V., Lemery, J., & Balbus, J. (2018). Climate change and women's health: Impacts and policy directions. *PLoS Medicine, 15*(7), e1002603. DOI: 10.1371/journal.pmed.1002603 PMID: 29990343

Spector, J. T., Sheffield, P. E., & Kinney, P. L. (2019). Climate change and occupational health: Current evidence and future directions. *Journal of Occupational and Environmental Medicine, 61*(4), e155–e157. PMID: 30694882

Van Hecke, O., Cameron, A., & Cruickshank, M. (2018). Seroprevalence of Borrelia burgdorferi in occupational groups: A systematic review and meta-analysis of European studies. *BMJ Open, 8*(3), e020310.

Watts, N., Amann, M., Arnell, N., Ayeb-Karlsson, S., Beagley, J., Belesova, K., & Costello, A. (2021). The 2020 report of the Lancet Countdown on health and climate change: Responding to converging crises. *Lancet, 397*(10269), 129–170. DOI: 10.1016/S0140-6736(20)32290-X PMID: 33278353

WHO. (2017). *One Health: Operational framework*. World Health Organization.

WHO & ILO. (2021). *Climate change and occupational health: Joint policy brief.* World Health Organization and International Labour Organization.

Xu, R., Yu, P., Abramson, M. J., Johnston, F. H., Samet, J. M., Bell, M. L., & Li, S. (2020). Wildfires, global climate change, and human health. *The New England Journal of Medicine, 383*(22), 2173–2181. DOI: 10.1056/NEJMsr2028985 PMID: 33034960

Zinsstag, J., Schelling, E., Waltner-Toews, D., & Tanner, M. (2011). From "one medicine" to "one health" and systemic approaches to health and well-being. *Preventive Veterinary Medicine, 101*(3-4), 148–156. DOI: 10.1016/j.prevetmed.2010.07.003 PMID: 20832879

Chapter 2
A Systems Thinking Perspective on Occupational Health and Safety and Climate Change Interactions

Nektarios Karanikas
http://orcid.org/0000-0002-9160-025X
Queensland University of Technology, Australia

Ioannis Adamopoulos
http://orcid.org/0000-0002-4942-7123
Hellenic Open University, Greece & University of West Attica, Greece

Aida Vafae Eslahi
http://orcid.org/0000-0001-5399-0552
Qazvin University of Medical Sciences, Iran

ABSTRACT

While the interaction between climate change and Occupational Health and Safety practices manifest through the intensification of traditional workplace hazards, such as occupational heat stress, it also introduces psychosocial risks to the workforce, including climate anxiety and ecological grief. Besides, climate change measures can inadvertently expose workers to novel chemical, physical, and psychological hazards. Through a systems thinking perspective, we can identify and mitigate the unintended consequences of our interventions and ensure that the transition to a low-carbon economy does not come at the cost of worker health. Furthermore,

DOI: 10.4018/979-8-3373-8470-2.ch002

the changing climate functions as a risk multiplier that exacerbates occupational health inequities, disproportionately burdening socially, economically, or politically marginalised populations, including migrant and Indigenous workers. Ultimately, comprehensive policy frameworks and "Just Transition" principles must be continually developed to guarantee procedural, distributive, and intergenerational justice across all global labour sectors.

1. INTRODUCTION

Strategies for mitigating climate change include a variety of approaches from technology and nature-based solutions to policies and behaviours. The implementation of solar, wind, hydropower, geothermal, and ocean energy helps to reduce dependence on fossil fuels. Nuclear energy is also proposed as a "low-carbon solution for strategically preventing significant amounts of CO_2 from entering the atmosphere" each year (Lobus et al., 2023; Suman, 2021). The transport sector also plans to use liquid biofuels instead of fossil fuels (Kazemi Shariat Panahi et al., 2019). Advanced bioenergy solutions are those in third-generation bioenergy using microalgae biomass (Lobus et al., 2023), and seaweeds for the generation of biofuels and no competition with food crops land use (Yong et al., 2022).

Nature-Based Solutions (NbS) use natural processes to remove carbon and help with adaptation and have a "triple-win" benefit for the climate, biodiversity and society (Musa & Ariff Lim, 2025). Also, afforestation, reforestation and forest restoration are being introduced to sequester more biomass and soil carbon (Osaka et al., 2021). In agriculture, biochar (a carbonaceous material derived from pyrolysis of biomass) and humic acid are being considered as new techniques to store more soil organic carbon and to mitigate emissions (Antu et al., 2025; Lobus et al., 2023).

Coastal ecosystem sinks of carbon such as mangroves, salt marshes, and seagrass meadows were identified as very efficient carbon sinks (Lobus et al., 2023; Yong et al., 2022). The Seaweed cultivation for both mitigation, alleviation of ocean acidification and for provision of biomass to produce foods and fuel is gaining attention as there is no conflict for land with food crops (Antu et al., 2025). In the urban sector NbS is used to reduce flooding which arises from hydrological changes due to the climate change and is considered as an alternative tool for mitigation against flood events. For instance, green roof, permeable pavement, bioretention systems and constructed wetlands are implemented in cities, to reduce volumes of runoff and filtrate the pollutions from these surfaces (Sun et al., 2024).

The implementation of the Internet of Things (IoT), Artificial Intelligence (AI), robotics (Precision Agriculture) in agricultural section help farmers in application of agricultural inputs precisely, hence, saving input use and waste, reducing emissions (Musa & Ariff Lim, 2025). The Nanotechnology is gaining attention, with a set of techniques of usage of nano-fertilizer and nano-pesticide, implemented to improve nutrient absorption of plants and reduce environmental impacts due to agro-chemical applications (Quintarelli et al., 2024). Combined Pest Management (IPM) and organic agriculture were considered for reducing the impacts of chemicals that contributes to emission (Garg et al., 2024).

Heat stress and air quality problems are handled in cities. Green spaces (parks, trees) and green roof are created, in order to improve air quality and reduce urban heat island effect (Luyten et al., 2023; Sun et al., 2024). Increased surface reflectivity of cities (cool roofs) are suggested measures of adaptation in terms of increasing the reflectivity of surface materials to reduce ambient temperature. Another adaptation measure is air conditioning system; however, it has an adverse effect on energy consumption. In the transportation section mitigation is obtained through promoting cycling, walking and public transport instead of personal motorized transport and electrification of transport sector (Luyten et al., 2023).

In response to the interventions like the ones listed above, the relationship between climate change and Occupational Safety and Health (OSH) is conventionally viewed through a linear lens: the changing climate acts as an external stressor that imposes new hazards upon the workforce. Rising ambient temperatures, expanding vector-borne disease ranges, and extreme weather events are undeniably creating hostile working environments. However, this unidirectional perspective is insufficient for effective governance. It fails to account for the reciprocal nature of this relationship, specifically how additional OSH practices can contribute to mitigating the effects of climate crisis on workers, and how climate mitigation strategies can inadvertently introduce new occupational risks (Karanikas et al., 2021a; Karanikas et al., 2021b).

It's worth highlighting that a significant portion of climate-sensitive work, especially in agriculture, construction and the informal sector, is situated in low- and middle-income countries (LMIC), where the existing occupational health safeguards tend to be weaker, the heat burden may be more profound due to geographic and resource limitations, and capacity to monitor heat-related illnesses or injuries tends to be limited (Bianco et al. 2024; Ansah et al. 2021). Thus, although accounting for the least emissions to global Greenhouse Gas (GHG) output, LMICs seem to carry the heaviest load in climate-enhanced occupational threats.

This chapter contemplates the interactions between OSH and climate change through a systems thinking perspective. By moving beyond the “victim” narrative, where OSH is solely a recipient of climate impacts, the chapter will frame OSH as an active participant within the ecosystem that should balance worker protection and environmental sustainability. Adopting a One Health approach, this chapter demonstrates that OSH cannot be isolated from environmental health. To achieve the above, this chapter incorporates documentation of cross-sector occupational risks in LMICs directly and emphasise upon a need for interventions that are tailored to the specific context and that reflect broader policies to achieve equity objectives.

2. METHODOLOGICAL APPROACH

This chapter represents a systematic narrative review. For the process of identification, searches were conducted across PubMed, Scopus, and Web of Science. Search strategies consisted of a combination of controlled vocabulary, as applicable, and free text terms describing climate change, occupational safety and health (OSH) and systems approaches to health. The key topic areas that the search domains covered included: climate change, global warming, extreme heat, heat stress, wildfire smoke, air pollution, flooding, drought, vector borne disease, occupational injury, occupational illness, workplace safety, occupational exposure, mental health, psychosocial hazard, climate anxiety, moral injury, resilience, adaptation, mitigation, renewable energy, green jobs, circular economy, Prevention through Design (PtD), environmental justice, equity, migrant workers, vulnerability, just transition, and One Health.

Boolean operators (AND, OR) were used to combine terms and truncation were used to broaden searches where appropriate. Moreover, the grey literature and institutional reports such as World Health Organization (WHO), International Labour Organization (ILO), and Intergovernmental Panel on Climate Change (IPCC). Titles and abstracts were screened, and relevant full texts reviewed. The evidence was then synthesised thematically, leading conceptual domains.

Publications were included if they 1) were peer-reviewed empirical research, systematic reviews, scoping reviews and authoritative policy or institutional reports; 2) covered the links between climate change, climate related hazards, climate mitigation/adaptation policies and OSH effects; 3) included physical, psychosocial, mental health, organisational and/or equity related OSH effects; and 4) discussed intervention, prevention, adaptation, resilience, governance and policy related to workers or workplaces. All geographical areas and scientific fields were included to capture the global and multidisciplinary dimension of the problem. As the field of interest is relatively young and cross-sectoral, both qualitative and quantitative evidence were included.

Publications were excluded when 1) they had no occupational and/or workers' health component; 2) regarded exclusively environmental or ecological issues with no direct and visible relevance to worker health and safety; 3) comprised public health issues related to climate change with no relevance for the workers or work environment; 4) were classified as conference abstracts, editorial, commentary or non-substantial opinion papers (without analytic or empirical input), and 5) were not published in English.

Figure 1 shows the narrative review method and the systems-thinking synthesis. Evidence mapping shaped the three principal outcomes presented in the results section below, namely (1) climate-related OSH hazards and impacts, (2) OSH risks arising from climate change mitigation/adaptation and from prevention-by-design, and (3) equity and psychosocial issues, including issues relating to LMICs and workers.

Figure 1. Methodological framework for synthesising evidence on climate change and occupational health and safety, integrating systems thinking, One Health principles, and just transition governance. The presented figure is generated using Python version 3.14 (https://www.python.org).

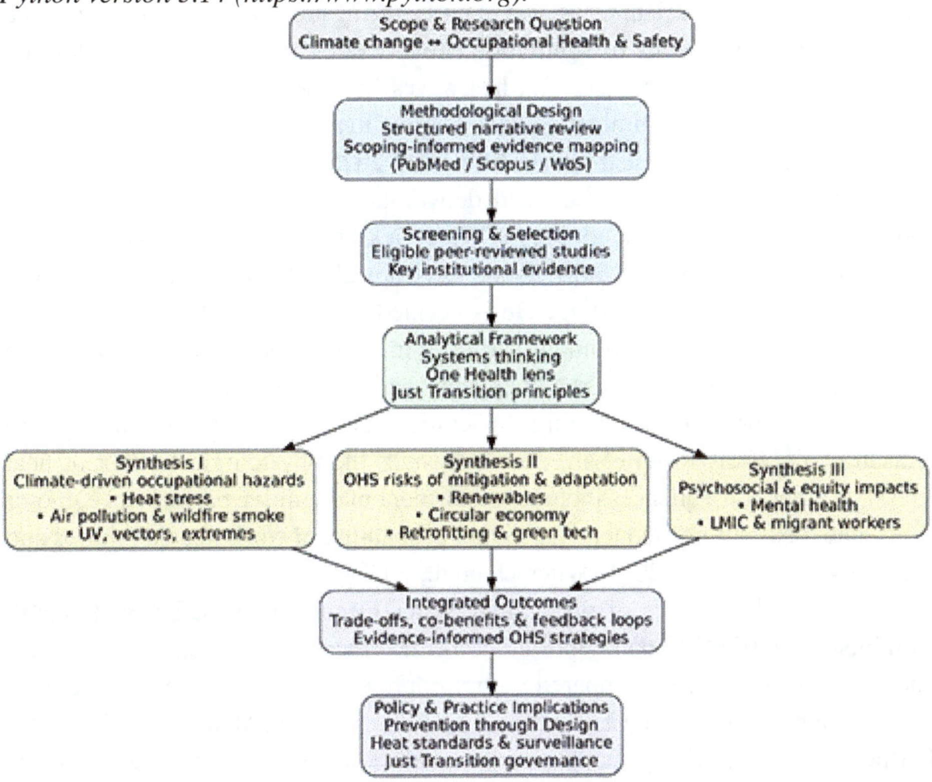

3. RESULTS

3.1 Climate Driven Occupational Hazards

3.1.1 Effects on Worker Physical Health

New OSH hazards are expected to emerge as a result of climate change, while existing hazards are being exacerbated across all work sectors and age groups. Occupational heat stress is the most frequently mentioned OSH hazard in the context of climate change, threatens outdoor workers as well as indoor workers in uncooled settings. Occupational heat stress impacts agriculture, construction, mining, and landscaping industries (Kjellstrom et al., 2016; Rossi et al., 2025) but even indoor occupations like bakers, laundries, and manufacturing are vulnerable to it if the area has no cooling and/or lack of ventilation (Levi et al., 2018; Ramadan & Mourad, 2022). As the temperature rises and heatwaves become longer, the human body becomes less capable of eliminating heat, leading to a range of problems from heat rash to life-threatening heatstroke (Habibi et al., 2025).

Combined chronic heat exposure with dehydration may cause the onset of Chronic Kidney Disease of unknown etiology (CKDu), a kidney disease of unknown etiology predominantly affecting farm laborers in tropical and hot climates (Kjellstrom et al., 2016; Levi et al., 2018). CKDu is defined as a "climate sensitive" disease associated with physically strenuous activities in hot environments without enough water intake (Habibi et al., 2025). Other short- and long- term health effects include cardiovascular problems, breathing difficulties and a weakened immune system (Ansah et al., 2021; Rossi et al., 2025). Besides the physiological effect of heat stress, its other consequences include impairment of cognitive performance, motor skills, and focus and consequently, a higher probability of work-related injuries and accidents (Habibi et al., 2021; Wuersch et al., 2023).

A nationwide analysis of the United States Occupational Safety and Health Administration (OSHA) data reported non-linear increases in occupational injury odds as heat index rose. Compared with a reference heat index of 80°F, the odds ratio for injury ranged from 1.03 at 90°F to 1.20 at 110°F (Alahmad et al., 2025). Furthermore, an intra-city spatiotemporal analysis in Greater Adelaide, Australia found elevated occupational injury and illness risk during both moderate and extreme heat, with substantial spatial variation across different areas of the city (Fatima et al., 2022). This suggests that occupational health risks from heat are not uniformly distributed across urban environments, and area-specific surveillance could inform targeted prevention planning.

Additional health hazards are connected to air quality and related respiratory impacts. Because climate change increases air pollution there is a "double burden" of combined exposure to heat and poor air quality for workers (Habibi et al., 2021). Heat promotes faster formation of the ground-level ozone which, in addition to being a respiratory irritant and potentially reducing lung capacity, can aggravate asthma (Ramadan & Mourad, 2022). Warmer weather has longer pollen seasons and produces more pollen which leads to higher chances of allergy respiratory diseases and asthma for outdoor workers (Levi et al., 2018). Fire events, which have increased in number and intensity with climate change, are another major health hazard for emergency and recovery workers, and those employed outdoors, as smoke, particulate matter and other harmful combustion products can have devastating effects on respiratory systems (Ramadan & Mourad, 2022; Tesfaye et al., 2025). Finally, more frequent drought conditions lead to increased dust at agricultural and construction sites, thus exacerbating health concerns related to inhalation (Pal et al., 2021).

A recent economic analysis estimates that, in California in 2020, at least 282 farmworkers were harmed by smoke. The same article adds that risk of injury starts increasing at smoke levels defined as 'safe' under current law (Lee, G., 2024). A comprehensive review supports the assertion that outdoor workers may be at an increased risk of exposure to wildland fire smoke, possibly resulting in increased asthma/Chronic Obstructive Pulmonary Disease (COPD) exacerbations (Weheba etal., 2024). Under California's regulations, employers are required to provide N95 masks for voluntary use when the Air Quality Index (AQI) for PM2.5 reaches 151 (Hou, 2022). The WHO's present 24-hour guideline for PM 2.5 is 15 g/m (Goshua et al., 2022).

The final group of health hazards increased by climate change that we will address here is vector-borne and biological hazards. Shifting rainfall patterns and temperature increases have broadened the geographic range and lengthened the transmission seasons of disease vectors, thus bringing biological hazards to new geographic regions and groups of workers. Increased areas of vector-borne disease (VBDs) transmission-such as the incidence of malaria, Dengue fever, Lyme disease, and Zika viruses among outdoor workers-may be experienced in the forestry, agricultural, and landscaping fields due to extended vector ranges (Levi et al., 2018; Ramadan & Mourad, 2022). Some regions may see higher rates of zoonoses such as Tick-borne encephalitis due to changes in the human-animal interface (Ramadan & Mourad, 2022). Flooding and extreme rainfall have already increased worker's risks for contracting water-borne illness (e.g., *Leptospira* and *Vibrio* species bacteria) in the fishing, sewage-treatment, and flood-rescue industries (Habibi et al., 2025; Lan et al., 2022; Levi et al., 2018). Coccidioidomycosis, a fungal disease of the soil, is predicted to become more prevalent in the arid West (Lan et al., 2022).

Higher temperatures could lead to increased evaporation of volatile substances, which can increase inhalation exposures to chemicals for agricultural and industrial workers (Ramadan & Mourad, 2022). In addition, high heat might decrease the effectiveness of Personal Protective Equipment (PPE) and increase chemical absorption through the skin because of increased sweat and skin blood flow (Lan et al., 2022). Increased ultraviolet (UV) exposure from thinning of the ozone layer and reduction of cloud cover could increase workers' risk of skin cancer, cataracts, and immune compromise (Rossi et al., 2025).

Beyond health problems, climate change could be the root cause of worker injury. More frequent and severe extreme weather events such as storms, floods, lightning, and hurricanes put workers in construction, energy, utilities, and disaster relief fields at increased risk of traumatic injury, drowning, and electrocution (Jia & Jia, 2024; Ramadan & Mourad, 2022). Workers in the fishing and marine industry are facing increasing severe conditions, as have recent reports documented instances of workers on deep sea vessels suffering injury or death due to storms and resorting to drinking salt water because of failed desalination (Yea, 2025). Extreme weather can also compromise the structural integrity of workplaces and health facilities, leading to power outages, supply chain disruptions, and unsafe working conditions for health workers and others who must remain on duty during disasters (Tesfaye et al., 2025).

3.1.2 Mitigating the Effects of Climate Change

Management techniques are applied not only to heat but also to climate-induced risks more broadly. For VBDs, appropriate interventions may involve integrated vector management, use of insect repellents and clothing and changes to working schedules to avoid mosquitoes at peak times (Jia & Jia, 2024); for air pollution they may include real-time monitoring of air quality indices, the use of respiratory protective equipment (such as N95 masks) at times of elevated pollution and wildfire smoke and dust suppression strategies (Ramadan & Mourad, 2022; Schulte et al., 2023).

Climate-induced risks, particularly those stemming from heat are generally managed following the established hierarchy of controls. The engineering controls can be air conditioning, ventilation increase in the interior environments (Amoadu et al., 2023; Habibi et al., 2024a), insulation of the buildings, cool roofs, physical shading for people required to be in outside environment (Schulte et al., 2023). Mechanisation of highly physical labour may reduce heat generated from metabolic processes (Habibi et al., 2021; Habibi et al., 2024b; Pal et al., 2021), while heatwave forecast may prompt for emergency interventions (Amoadu et al., 2023).

It has been shown that admin controls such as designated resting breaks with shaded or cooled locations, and regulated heat exposure based on temperature indices can be effective in reducing occupational heat strain (Amoadu et al., 2023; Habibi et al., 2021). Shift working of heavy manual tasks during cooler parts of the day (early morning or late evening), or in cooler seasons can be adopted (El-Sherif, 2025; Schulte et al., 2023; Wuersch et al., 2023). In addition, the provision of chilled water and electrolyte solutions and 'buddy systems' can be utilized to monitor workers for dehydration (Habibi et al., 2021; Habibi et al., 2024b; Schulte et al., 2023). The previously discussed strategies can also be supplemented with training that progressively acclimatizes new and re-acclimatizes former workers to heat over 3-7 days (Habibi et al., 2024b; Schulte et al., 2023), and training in recognizing symptoms of heat illness and in actioning procedures if heat illness symptoms are detected (Amoadu et al., 2023; Ramadan & Mourad, 2022; Rossi et al., 2025; Wuersch et al., 2023).

It has been suggested that for PPE controls to include the wearing of loose-fitting, light-coloured and breathable work clothing, along with UV-protected accessories (e.g. Sunglasses, wide-brimmed hats) (Amoadu et al., 2023; Habibi et al., 2024a; Habibi et al., 2021; Habibi et al., 2024b; Wuersch et al., 2023). Additionally, when dealing with workers who are required to wear heavy protective gear (e.g. Chemical workers and firefighters), personal cooling devices (e.g. Water-cooled vests, ice vests, phase-change material vests) have been recommended (Jia & Jia, 2024).

In addition to the risk controls listed above, there have been a number of larger scale initiatives to standardize and enhance these measures, for example 'Hothaps' (High Occupational Temperature Health and Productivity Suppression) where international collaboration has aimed to establish and validate specific strategies to combat occupational heat strain (Ansah et al., 2021; Habibi et al., 2021). The HEAT-SHIELD project, an EU-funded scheme that focused on enhancing thermal resilience in key sectors of industry (e.g. Manufacturing, construction, transportation, tourism and agriculture) utilized the development of heat-health warnings and climate-adaptation guideline development (Habibi et al., 2024a; Levi et al., 2018). National heat warning systems, along with the development of key heat metrics by public health research experts have been developed by the U.S. National Weather Service (Schulte et al., 2023).

Research examining California's 2005 heat ordinance found that following stricter enforcement in 2010, the number of heat-related deaths amongst outdoor labourers fell by 33% in relation to neighbouring states not regulated by the heat ordinance and by 51% following the changes enacted in 2015 (Dean & McCallum, 2025). Acclimatization to heat is essential; according to NIOSH, acclimatization is typically completed within 7-14 days, during which newly employed labourers only work for 20% of a typical hot shift on day one, then incrementally work for a longer duration until a full shift length is achieved (Jacklitsch et al., 2016).

The study conducted by Tustin et al., (2018) found that out of the 25 heat related illnesses observed by OSHA, 6 of these instances were fatal at a Heat Index less than 91F, suggesting that a Heat Index of 85F warrant additional safety measures. The challenges with the implementation of heat-related labour regulations are many. It was discovered in a study of piece-rate workers Washington State field labourers that there was a 6.2-fold greater chance of symptoms for heat-related illness (self-treated) among this group (Spector et al., 2015) and that workers did not increase their effort significantly above a WBGT of 29.6°C (85.3F) (Pan et al., 2021).

3.2 OHS Risks from Climate Change Mitigation and Adaptation

3.2.1 Effects on Worker Physical Health

Conversely, the move towards a low-carbon economy could generate new or new forms of dangers to workers: Green jobs" in the renewables sector involve unrecognised risks. For example, the manufacture of photovoltaic cells uses toxic chemicals such as Cadmium Telluride, Gallium Arsenide and hydrofluoric acid, and workers' safety could be affected by occupational health risks from inhalation and skin contact with these substances (Schulte et al. 2023). During the installation of photovoltaic cells, a variety of injuries, including falls at work, heat stress and electrical shock, due to the energy stored in PV systems, are also of concern to workers (Schulte et al. 2023). Wind turbine manufacturing involves epoxy resins, Styrene and glass reinforced plastics that could trigger skin irritation and respiratory problems for workers. Health and safety issues during the installation and maintenance of wind turbines at height and in confined spaces, and risk of electrocution, are significant too (Schulte et al. 2023).

There are worries that worker safety is not considered adequately (Quintarelli et al. 2024; Schulte et al. 2016) in the production of biofuels, which require the use of corrosive and volatile, as well as flammable, chemicals (Schulte & Chun 2009). Organic dust and bioaerosols released during the processing of biomass constitute occupational health risks that result in respiratory disease (Garg et al. 2024). The high flammability of hydrogen may cause risks of explosion and fire (Schulte et al. 2023). Increased occupational health risks are probable with high-risk processes such as those occurring in recycling and circular economy. In the e-recycling processes worker could be exposed to high levels of toxic substances such as metals (lead, mercury and cadmium) and flame retardants (Schulte et al, 2016). In the context of lead-acid batteries, worker safety issues concerning high levels of lead exposure have been reported (Schulte & Chun, 2009).

The composting process may also lead to an increased risk of work-related respiratory diseases, such as asthma and hypersensitivity pneumonitis, due to workers' exposure to bioaerosols and organic dusts (Lan et al, 2022). Health and safety concerns from exposure to harmful agents such as asbestos and heavy metals, in combination with workers' inadequate knowledge about Occupational Health and Safety procedures, are associated with dismantling old structures (Schulte et al, 2023). The modifications made to the buildings, in order to achieve higher energy efficiency, might lead to higher workers' exposure to indoor air contaminants. In particular, weatherisation of buildings, leading to their air tightness, has been observed to correlate with higher indoor air contaminant levels such as VOCs, radon and mould (Schulte et al, 2023).

Also, the push for "green buildings" (e.g., Leadership in Energy and Environmental Design-LEED certified) often introduces new design elements (e.g., green roofs, skylights) that may present unfamiliar fall hazards or access issues for construction and maintenance workers if safety isn't integrated into the design phase (Schulte et al., 2016). Moreover, installing insulation materials (e.g., polyurethane foam, fiberglass) carries respiratory and skin irritation risks. Retrofitting older buildings can also disturb hazardous materials like lead paint or asbestos (Quintarelli et al., 2024). In the area of agricultural innovations, changes in agricultural practices to adapt to climate change or reduce emissions can have significant OSH implications. Climate change is expanding pest habitats and extending growing seasons, which may necessitate increased use of pesticides and biocides (El-Sherif, 2025; Ramadan & Mourad, 2022). Furthermore, higher ambient temperatures can increase the volatility of these chemicals and the rate of worker respiration and skin absorption, thereby heightening toxicity risks for applicators and farmworkers (Lan et al., 2022).

The drive to use environmentally friendly chemicals can sometimes introduce worker safety risks if the substitute is not thoroughly evaluated for human toxicity. A notable example is the switch to 1-bromopropane as a "green" solvent, which was later found to cause severe neurological damage in workers (Schulte & Chun, 2009). The use of nano-fertilizers and nano-pesticides to improve efficiency introduces risks associated with engineered nanoparticles. These particles can be more toxic than their bulk counterparts due to their small size and high reactivity, potentially leading to bioaccumulation and respiratory damage (Quintarelli et al., 2024).

Seaweed farming for carbon sequestration poses maritime hazards (drowning, extreme weather, ergonomic strain) to workers (Yong et al., 2022). Protective measures can also backfirel; specialized PPE traps heat and sweat, exacerbating heat stress (Habibi et al., 2021; Van der Horst et al., 2025). Shifting work to cooler hours (e.g., early morning or night) to avoid heat stress can disrupt circadian rhythms, causing sleep deprivation, fatigue, and increased accident risk (Habibi et al., 2025; Wuersch et al., 2023). Automating dangerous tasks (e.g., pesticide spraying robots) reduces physical exposure but introduces human-machine interaction hazards, along with psychosocial risks like job insecurity and workplace surveillance (Musa & Ariff Lim, 2025; Šavrič et al., 2025).

3.2.2 Mitigating the Effects of Climate Change Measures

A primary strategy to counteract hazards in emerging "green" industries is integrating safety into the design phase of technologies and infrastructure. The PtD initiative is crucial for the rapidly growing renewable energy and green construction sectors (Schulte et al., 2016; Schulte & Chun, 2009). To mitigate risks associated with "green buildings" (e.g., falls from maintaining green roofs, indoor air quality in sealed buildings), OSH standards are being integrated into sustainability rating systems like LEED. This includes "Life Cycle Safety" credits that require architects to design safe access for maintenance workers (e.g., window cleaners) and ensure indoor air quality during and after construction (Schulte et al., 2016).

To address high-risk activities in wind energy (e.g., working at heights, confined spaces in towers), organisations like the American Society of Safety Engineers have initiated standards specifically for the construction and demolition of wind turbines. These standards address mechanical assembly, medium voltage electrical safety, and working in exposed environments (Schulte & Chun, 2009). In recycling, controls include advanced engineering controls (e.g., ventilation and dust suppression), strict hygiene protocols, and medical surveillance to monitor for heavy metal toxicity (e.g., lead, cadmium) and respiratory conditions (Schulte et al., 2016; Schulte & Chun, 2009). Also, to counteract the risk of a chemical deemed environmentally safe being toxic to workers (e.g., the use of 1-bromopropane), OSH practitioners are encouraged to actively participate in the selection of "green" materials to ensure they are also worker-safe (Schulte & Chun, 2009).

To manage the PPE paradox of heat vs. protection, the use of personal cooling systems is recommended. These include phase change material (PCM) vests, liquid cooling garments (LCGs), and air-ventilated clothing, which help regulate core body temperature without compromising chemical protection (Habibi et al., 2024a; Habibi et al., 2024b). OSH guidelines recommend modifying safe work durations based on clothing type. Workers wearing semi-permeable or impermeable suits require more frequent rest breaks in shaded or cooled areas than those in standard work clothes (Habibi et al., 2024b; Hernández-Duarte et al., 2024; Schulte & Chun, 2009).

To counteract health effects in "tight" energy-efficient buildings (e.g., radon, mould, chemical accumulation), more frequent maintenance of heating, ventilation, and air conditioning (HVAC) systems and air filtration can prevent microbial growth (Schulte & Chun, 2009). Specific testing and mitigation systems (e.g., sub-slab depressurization) are required to prevent radon accumulation, a leading cause of lung cancer (Schulte & Chun, 2009).

3.3 Psychosocial and Equity Impacts

In addition to physical health problems, the psychological impact of climate change on the workforce is profound and growing, yet under-recognised for a large range of psychosocial risks for workers. A growing body of evidence shows relationships between both acute climate events (e.g., hurricanes, wildfires, and floods), and slow onset changes (e.g., drought, heat stress, and ecosystem degradation), and anxiety, depression, and symptoms of Post-Traumatic Stress Disorder (PTSD), burnout, substance use, and suicidality (Charlson et al., 2021; Cianconi et al., 2020; Palinkas & Wong, 2020; WHO, 2022).

Overall, workplace safety, mental health and productivity are negatively affected by climate change in numerous ways. Employees in climate-sensitive occupations-such as agriculture, construction and aquaculture-were found to be more likely to suffer from depression, anxiety, mental distress and suicidality due to pressures such as increased temperatures, long periods of drought and so on. The impact of climate change stressors includes lack of security and autonomy, sense of isolation, perceived lack of control and damage to occupational identity (Tang et al., 2025). Workers in climate-sensitive industries like agriculture face "ecological grief", anxiety, and depression due to drought, crop failure, and economic uncertainty (Levi et al., 2018; Pal et al., 2021; Tang & Ho, 2025). Farmers, in particular, show increased rates of suicide linked to prolonged drought and financial strain (Pal et al., 2021; Tang & Ho, 2025).

Also, emergency responders and healthcare workers experience high rates of burnout, Post-Traumatic Stress Disorders, and compassion fatigue from responding to increasingly frequent and severe climate disasters (Tesfaye et al., 2025; Tsakonas et al., 2024). Climate disasters, like floods or extreme weather, can lead not only to physical harm, but also exhaustion and considerable psychosocial distress for health workers which require psychosocial support and emergency readiness (Tsakonas et al., 2024). Besides, rising temperatures have been linked to increased aggression and workplace violence, affecting security personnel, law enforcement, and public-facing workers (Ramadan & Mourad, 2022). Thus, the psychosocial health risks seem to be concentrated along pre-existent lines of social inequality and not necessarily being individual susceptibilities. (Quesada et al., 2011; White et al., 2023).

3.3.1 Climate Anxiety and Moral Injury

Climate anxiety and moral injury can impact workers in labour sectors that are closely related to climate change either through activities that contribute to greenhouse gas releases or through pursuing mitigation measures. Climate anxiety is a generalised state, a state of anticipation and grief, uncertainty for the future of ecosystems, and a state of moral injury when individuals are involved in actions which they consider to have adverse, environmentally damaging, consequences (Gao et al., 2025; Schwartz et al., 2022).

Young workers seem to be especially vulnerable because of high climate awareness paradoxically combined with weak institutional power (Adamopoulos et al., 2025; Ma et al., 2022). Research suggests that collective action, regardless of it being in the form of a workplace mobilisation or campaign for environmental protection, is a form of compensatory mechanism for these harms because it provides agency, solidarity and intersubjectivity (Palinkas & Wong, 2020; Schwartz et al., 2022).

3.3.2 System Stressors: Cascading and Cumulative Risks

Beyond individual psychological responses, climate change presents cascading threshing systemic stressors that affect work environments and systems. Extreme weather events disrupt critical infrastructure, increase the demands of the occupations and reduce productivity; concomitant supply chain disruptions increase the load pressures; economic instability contributes to job insecurity and builds organisational conflict (Bianco et al., 2024; Palinkas & Wong, 2020).

The shift away from fossil fuel industries (e.g., coal mining) can lead to job insecurity, economic displacement, and mental health crises in affected communities if not managed as a "just transition" (Schulte et al., 2023; Schulte & Chun, 2009; ILO, 2015). Also, there is a concern that "green jobs" are not automatically safe jobs. Rapid growth in green industries may outpace the development of specific safety standards and training, leaving workers vulnerable to unrecognized hazards (Schulte et al., 2016; Schulte & Chun, 2009).

These dynamics are particularly harmful for immigrants, women workers and Indigenous communities, in which individuals regularly face barriers in accessing mental health care and social support (Lebel et al., 2022; White et al., 2023). Slow onset climate stressors where factors such as protracted drought, landscape degradation, or cultural displacement can be anticipated, are linked to "slow burn" psychological sequelae such as solastalgia, chronic grief, frustration, and community conflict. These harms are resounding on households and across generations (Burrows et al., 2024; Middleton et al., 2020).

3.3.3 Resilience Strategies and Organisational Psychosocial Risk Management

Mitigation of the psychosocial harms from climate-related change requires a paradigm shift away from individual-level coping responses toward system-level and organisational responses. Literature commonly refers to several effective strategies, including integration of mental health literacy and first-aid programmes into OSH; ensuring social cohesion, the creation of peer-support networks and connectedness in our communities; the use of flexible work arrangement for extreme heat or other weather events; embedding psychosocial risk management in wider OSH regulatory frameworks.

Nonetheless, interventions need to be context specific and culturally based. For example, multilingual support is needed in migrant settings while land-based and culturally-rooted mental health support is a key component of serving Indigenous communities (Cunsolo & Ellis, 2018; Lebel et al., 2022). In the light of the fast-evolving climate change, novel research designs must include adaptive platform trials to assess the effectiveness of occupation-focused mental health interventions (Adamopoulos et al., 2026; Gao et al., 2025).

3.3.4 Climate Justice in Occupational Health

Besides the physical, psychosocial and mental health effects linked to climate change, the latter can also function as an accelerator of existing occupational health inequities and worsen disparities already present in global political, economic and labour systems. Drawing on political ecology, environmental justice, and critical public health, climate-related occupational risks have to be also understood as the products of unequal socio technical arrangements rather than as neutral environmental conditions (Adamopoulos et al., 2025).

Hazards including extreme heat, reduced air quality and changing patterns of vector borne disease visit points of intersection with power structures: whose labour is rewarded, whose environments are safeguarded, and who suffer at work. This additional viewpoint frames climate change not only as an environmental problem but also reflecting mechanisms that reinforce and deepen patterns of exposure. Workers who are socially, economically or politically marginalized not least migrants, Indigenous peoples, racialized populations and those working in precarious forms of employment disproportionally burdened (Bianco et al., 2024; White et al., 2023). As part of a strategy to track the ways in which climate change has become a "risk multiplier" within occupational environments, it is important to identify the dynamics of these changes.

Regarding LMICs, the concentration of climate-sensitive work, especially in agriculture, construction, and the informal economy is likely to exacerbate existing occupational health inequalities. Workers in these settings face weaker regulatory enforcement, limited access to cooling and hydration, and fragmented health surveillance (Bianco et al., 2024; Ansah et al., 2021). They are also exposed to a higher level of ambient heat, longer hours and a lack of time in which to rest or acclimatise. These workforce populations are also less likely to have access to or afford adaptive solutions such as heat early warning systems and protective personal equipment.

3.3.5 The Intersectional Burden Socioeconomic Status, Migration and Race

Intersectionality theory brings a powerful analytic framework to the study of the interplay between multiple social identities and structural forces that produce differential vulnerability to climate-related occupational hazards (Sharpe & Davison, 2022). Pregnant workers exposed to extreme heat face risks of adverse pregnancy outcomes, including stillbirth and low birth weight (Ansah et al., 2021; Habibi et al., 2025). Furthermore, aging workforces and those with pre-existing cardiovascular or respiratory conditions have reduced physiological capacity to thermoregulate, increasing their susceptibility to heat stress and air pollution (Habibi et al., 2025; Kjellstrom et al., 2016).

International migrant workers often face the "3D" jobs (dirty, dangerous, difficult) with limited legal protections, language barriers, and less access to healthcare, making them highly susceptible to heat strain and exploitation. They may also lack the ability to self-pace or take breaks due to piece-rate payment structures (Van der Horst et al., 2025). Migrant agricultural workers are paradigmatic of the above. Taken together with structural constraints posed by immigration status, linguistic barriers, and restricted access to healthcare systems, extended exposure to extreme heat, limited access to opportunities for hydration or rest, and inadequate employer-provided housing are present when discussing the impacts of climate change on migrants (Middleton et al., 2020; Sharpe & Davison, 2022).

Racialized and Indigenous workers face similar inequities. In the Circumpolar North, rapid ecological change has disrupted Indigenous livelihoods and land-based traditions, causing psychosocial strain and occupational dislocation. These harms stem as much from ongoing colonial processes as from environmental change (Lebel et al., 2022; Whyte, 2017). Across regions, racism, colonialism, and economic exploitation have historically denied opportunities to these communities, exacerbating cycles of occupational vulnerability (Middleton et al., 2020; White et al., 2023).

3.3.6 Procedural and Distributive Justice Global Inequities in OSH Governance

Environmental justice frameworks emphasise the importance of both procedural justices to ensure that responsibilities are shared across the groups concerned, and distributive justice to instil fairness regarding impact and resource allocation. However, both dimensions are significantly uneven in global OSH structures (Böhm et al., 2012; McCauley & Heffron, 2018). Global efforts to shift to low carbon technologies in high income countries are often based on resource extraction, recycling, and waste processing activities subcontracted to LMICs and unregulated labour sectors, in which regulatory structures and labour protections are least stringent (Bianco et al., 2024; White et al., 2023).

The above mirrors the larger processes of ecologically unequal exchange whereby richer countries externalize their environmental and occupational risk in more vulnerable regions of the world (Hornborg, 2019). Such a division of labour between countries and work sectors reflects a broader trend toward ecological unequal exchange (Hornborg, 2019), a phenomenon described in both ILO analyses on the informal economy (ILO, 2018) and growth of digital labour platforms (ILO, 2021). Domestically, OSH systems are often closed to undocumented workers, gig workers, and those working in the informal economy, populations that are overrepresented among migrants and racialized populations). This creates a stratified labour system whereby climate related harms have been normalized for some populations while others remain comparably insulated (Adamopoulos et al., 2025; Sharpe & Davison, 2022).

3.3.7 Intergenerational Justice: Effects on future Workforces

Intergenerational justice is another vital but underdeveloped aspect of climate-related occupational health. Decisions made today about land use, extraction of resources, workplace standards and regulations on the environment determine the structural conditions experienced by working people in the future (Gardiner, 2011; White et al., 2023). Children who are growing up in communities with increased heat exposure, elevated risk of VBDs, or are in economic precarity related to parent working conditions may experience adverse development trajectories (Middleton et al., 2020)

The potential for intergenerational transmission of occupational vulnerability is especially apparent in communities that are struggling with the multiple impacts of colonisation, insecurity from migration and economic marginalisation (Adamopoulos et al., 2026; Lebel et al., 2022). Robust climate adaptation strategies must, therefore, not drive risks to youths, informal workforce or future labour forces.

4. DISCUSSION

In this chapter the interplay between climate change and OSH at a system level has been identified. While on one hand climate change exacerbates classic OSH hazards such as heat, airborne agents and the risk of infection, new OSH hazards and psychosocial stressors arise from implementing adaptive measures (Schulte et al., 2023). There is a clear inequity in exposure to OSH hazards between the migrant, informal, and vulnerable groups (Van der Horst et al., 2025; White et al., 2023).

This finding corroborates previous reviews highlighting the heat-related morbidity among outdoor workers (Kjellstrom et al., 2016; Levi et al., 2018).

This chapter provides a new level of detail by mapping out the two-way relationship between climate actions and OSH hazards - a fact, previously excluded in syntheses predominantly focusing on climate's impact on health (e.g. Ramadan & Mourad, 2022; Amoadu et al., 2023). Additionally, it includes a broader conception of occupational health by taking both physical and psychosocial outcomes and occupational exposures into account. Our review suggests that moving towards equity requires more than technological solutions, as well as coordinated policies. Investments will need to be made locally to facilitate surveillance and improve workers' education, as well as developing just transition frameworks for LMICs.

Indeed, it is positive that governments and organisations have started managing climate change risks through legislative and standard-setting mechanisms. For instance, many regions utilise the Wet Bulb Globe Temperature (WBGT) index to set threshold limits for work. When these limits are exceeded, work must be stopped or modified (El-Sherif, 2025; Habibi et al., 2021; Pal et al., 2021; Schulte et al., 2016). ISO 7243:2017 (Ergonomics of the thermal environment) provides methodologies for assessing heat stress which inform national regulations (Schulte et al., 2016). Also, Gulf countries (e.g., Bahrain, Qatar, United Arab Emirates, Saudi Arabia) prohibit outdoor work during the hottest hours of the day (typically 12:00 PM to 4:00 PM) during summer months (El-Sherif, 2025; Van der Horst et al., 2025).

In the United States, California, Washington, and Oregon have enacted specific heat illness prevention standards requiring water, rest, and shade (El-Sherif, 2025; Schulte et al., 2023; Wuersch et al., 2023). Moreover, programs such as the "New South Wales Farm Blueprint" in Australia provide funding for mental health services, counselling, and community well-being initiatives to support agricultural workers facing drought and climate-related financial stress (Tang & Ho, 2025). It is also promising that climate risks are increasingly being incorporated into general OSH management systems, requiring employers to conduct risk assessments for extreme weather and thermal conditions (Jia & Jia, 2024; Schulte et al., 2016). Despite these efforts there is a significant lack in the implementation and utilization of them. Even in areas where a protocol exists (water, shade and rest) it is not uniformly applied, the existence of piece-rate pay promotes workers to circumvent rest periods and often migrant and informal workers lack protection because of a lack in legal cover, linguistic differences and insecure work conditions (Van der Horst et al., 2025; Wuersch et al., 2023).

For effective management of industrial transition organizations should adopt foresight techniques and look ahead for risks posed by new technology (e.g. Hydrogen, nanotechnology and geoengineering) allowing the anticipation of safety regulations instead of waiting to deal with injured parties (Schulte et al., 2023). As industries transition away from fossil fuels, Just Transition policies must be continuously evolved, helping displaced workers in dealing with potential psychological strain (e.g. Lack of motivation, anxiety about not finding employment) and safety hazards associated with new job tasks they retrain for, in the burgeoning green economy (Schulte et al., 2023).

To handle the expanded range of vectors and increased pesticide use due to climate adaptation in agriculture, integrated pest management (IPM) can minimise chemical dependence through biological and physical pest control methods (Ramadan & Mourad, 2022). Importantly, a transdisciplinary strategy which integrates human, animal, and environmental health monitoring can better predict and control zoonotic and VBDs that are spreading to new regions, thereby protecting outdoor workers (Schulte et al., 2023).

Nonetheless, the results above suggest several practical areas for action. First, OSH risk assessment must account for future climate and emerging green technology use, shifting from merely hazard control to PtD (Schulte et al., 2016). Second, just transition frameworks (ILO, 2015) must be put in place at organisational and sectoral levels and address the needs for retraining and security in employment. Third, in LMICs surveillance of health-related climate impact needs strengthening, so that occupational health burden is captured (Ansah et al., 2021; Bianco et al., 2024).

The strengths of this review chapter lie in its three main dimensions: systems thinking is utilised allowing the identification of linkages and unintended consequences which is not found in linear analyses. Grey literature, including institutional reports (WHO, ILO, IPCC), has been incorporated in this review to increase policy relevance. Moreover, equity is central; using intergenerational, procedural and intersectional justice is what sets this chapter apart from other technical OSH analyses in climate change contexts.

Still, this chapter has four main limitations. First, as a narrative review using a non-systematic search strategy with possible risk-of-bias assessments, a degree of heterogeneity is included, meaning it is not possible to perform quantitative comparisons between the included studies and reports. Second, excluding non-English literature possibly left out evidence generated in non-English speaking countries, mainly LMICs. Third, the rapidly developing green economy is not necessarily reflected in peer-reviewed literature. Despite these limitations, the inter-disciplinary, systems perspective adopted herein allows for integration of different types of findings from OSH, environment health and systems health in a useful overview of an emerging field.

Further research efforts should focus on: 1) longitudinal studies looking at health outcomes under increasingly scaling green technologies; 2) intervention studies on different heat mitigation methods and acclimatisation programs especially in LMICs and informal sector workplaces; 3) worker-driven, collaborative research with migrant, Indigenous and other marginal workers, on climate and OSH policies; and 4) developing indicators to assess the interactions between OSH and green jobs for climate resilience.

5. CONCLUSIONS

The relationship between climate change and OSH is complex and reciprocal, requiring a systems thinking and One Health perspective to manage effectively. Climate change directly threatens worker physical health by exacerbating existing hazards, notably heat stress, poor air quality, and VBDs across various industries. Conversely, measures implemented to mitigate and adapt to climate change, such as the transition to renewable energy, green building initiatives, and circular economies, introduce new occupational risks, including exposure to toxic chemicals and unfamiliar safety hazards. Furthermore, standard OSH controls can produce unintended trade-offs; for instance, the use of specialised personal protective equipment can increase a worker's susceptibility to heat strain.

In addition to physical threats, climate change imposes profound psychosocial and mental health stressors on the workforce. Workers face acute trauma following extreme weather events, alongside slow-onset challenges such as ecological grief, job insecurity within transitioning industries, and climate anxiety. Addressing these impacts necessitates a shift away from individual coping mechanisms towards comprehensive organisational support and system-level interventions.

Crucially, climate change functions as a risk multiplier that deepens existing occupational health inequities. Workers with marginalised socioeconomic statuses, migrant backgrounds, or racialized identities bear an intersectional burden, frequently facing compounding disadvantages, hazardous labour conditions, and limited regulatory protection. A systems-equity approach is therefore essential to address global disparities in procedural, distributive, and intergenerational justice, ensuring risks are not disproportionately externalised to vulnerable communities or future workforces.

Although current policies and standards provide a preliminary framework for managing these overlapping crises, their implementation remains inconsistent. Future OSH practice must integrate proactive foresight techniques, transdisciplinary monitoring, and PtD principles to anticipate and neutralise emerging risks before they manifest. Ultimately, as global economies decarbonise, robust "Just Transition" policies are imperative to guarantee that environmental sustainability is not achieved at the expense of worker health, safety, and equity.

REFERENCES

Adamopoulos, I., Syrou, N., Mpourazanis, G., Constantinidis, T. C., & Dounias, G. (2025). The Association of the Global Climate Crisis with Environmental Risks and the Impact of Heat Stress on Occupational Safety, Health, and Hygiene. *Medical Sciences Forum, 33*(1), 2. DOI: 10.3390/msf2025033002

Adamopoulos, I. P., Syrou, N. F., Lamnisos, D., Valamontes, A., Karantonis, J. T., Tsirkas, P., & Dounias, G. (2026). Classifying and mitigating occupational risks for public health inspectors in the context of the global climate crisis. *European Journal of Sustainable Development Research, 10*(1), em0336. Advance online publication. DOI: 10.29333/ejosdr/17273

Alahmad, B., Kessler, W., Alwadi, Y., Schwartz, J., Wagner, G. R., & Michaels, D. (2025). A nationwide analysis of heat and workplace injuries in the United States. *Environmental Health : A Global Access Science Source, 24*(1), 65. DOI: 10.1186/s12940-025-01231-1 PMID: 41047384

Amoadu, M., Ansah, E. W., Sarfo, J. O., & Hormenu, T. (2023). Impact of climate change and heat stress on workers' health and productivity: A scoping review. *The Journal of Climate Change and Health, 12*, 100249. Advance online publication. DOI: 10.1016/j.joclim.2023.100249

Ansah, E. W., Ankomah-Appiah, E., Amoadu, M., & Sarfo, J. O. (2021). Climate change, health and safety of workers in developing economies: A scoping review. *The Journal of Climate Change and Health, 3*, 100034. Advance online publication. DOI: 10.1016/j.joclim.2021.100034

Antu, U. B., Roy, T. K., Kulsum, T. I., Mitu, P. R., Ismail, Z., Arifin, M., Datta, M., Hossain, S. A., Islam, M. S., Mahiddin, N. A., Al Bakky, A., Hossin, S., Islam, S., & Idris, A. M. (2025). Role of humic acid for climate change adaptation measures to boost up sustainable agriculture and soil health: A potential review. *International Journal of Biological Macromolecules, 313*, 144043. DOI: 10.1016/j.ijbiomac.2025.144043 PMID: 40350124

Bianco, G., Espinoza-Chávez, R. M., Ashigbie, P. G., Junio, H., Borhani, C., Miles-Richardson, S., & Spector, J. (2024). Projected impact of climate change on human health in low- and middle-income countries: A systematic review. *BMJ Global Health, 8*(Suppl 3), e015550. Advance online publication. DOI: 10.1136/bmjgh-2024-015550 PMID: 39357915

Böhm, S., Misoczky, M. C., & Moog, S. (2012). Greening Capitalism? A Marxist Critique of Carbon Markets. *Organization Studies*, *33*(11), 1617–1638. DOI: 10.1177/0170840612463326

Burrows, K., Denckla, C. A., Hahn, J., Schiff, J. E., Okuzono, S. S., Randriamady, H., Mita, C., Kubzansky, L. D., Koenen, K. C., & Lowe, S. R. (2024). A systematic review of the effects of chronic, slow-onset climate change on mental health. *Nature Mental Health*, *2*(2), 228–243. DOI: 10.1038/s44220-023-00170-5 PMID: 41098556

Charlson, F., Ali, S., Benmarhnia, T., Pearl, M., Massazza, A., Augustinavicius, J., & Scott, J. G. (2021). Climate Change and Mental Health: A Scoping Review. *International Journal of Environmental Research and Public Health*, *18*(9), 4486. Advance online publication. DOI: 10.3390/ijerph18094486 PMID: 33922573

Cianconi, P., Betrò, S., & Janiri, L. (2020). The Impact of Climate Change on Mental Health: A Systematic Descriptive Review. *Frontiers in Psychiatry, Volume 11 - 2020*. DOI: 10.3389/fpsyt.2020.00074

Cunsolo, A., & Ellis, N. R. (2018). Ecological grief as a mental health response to climate change-related loss. *Nature Climate Change*, *8*(4), 275–281. DOI: 10.1038/s41558-018-0092-2

Dean, A., & McCallum, J. (2025). California's Heat Standard And Heat-Related Deaths Among Outdoor Workers: Article examines California's heat standard and heat-related deaths among outdoor workers. *Health Affairs (Project Hope)*, *44*(12), 1490–1496. DOI: 10.1377/hlthaff.2025.00096 PMID: 41329892

El-Sherif, G. (2025). Occupational and Environmental Health Problems Related to Climate Change in Egypt. *Egyptian Journal of Occupational Medicine*, *49*(2), 1–16. DOI: 10.21608/ejom.2025.371549.1366

Fatima, S. H., Rothmore, P., Giles, L. C., & Bi, P. (2022). Outdoor ambient temperatures and occupational injuries and illnesses: Are there risk differences in various regions within a city? *The Science of the Total Environment*, *826*, 153945. DOI: 10.1016/j.scitotenv.2022.153945 PMID: 35189241

Gao, C. X., Fava, N., Browne, V., Patrick, R., Gunasiri, H., & Menssink, J. M. (2025). Climate Change and Youth: Fast-Tracking Mental Health Solutions. *Journal of Applied Youth Studies*. Advance online publication. DOI: 10.1007/s43151-025-00174-1

Gardiner, S. M. (2011). *A perfect moral storm: the ethical tragedy of climate change*. Oxford University Press., DOI: 10.1093/acprof:oso/9780195379440.001.0001

Garg, S., Rumjit, N. P., & Arora, P. (2024). Addressing climate change impacts through sustainable agricultural solutions: A review. *Environment, Development and Sustainability*. Advance online publication. DOI: 10.1007/s10668-024-05728-6

Goshua, A., Akdis, C. A., & Nadeau, K. C. (2022). World Health Organization global air quality guideline recommendations: Executive summary. *Allergy*, *77*(7), 1955–1960. DOI: 10.1111/all.15224 PMID: 35060140

Habibi, P., Heydari, A., Dehghan, H., Moradi, A., & Moradi, G. (2024a). Climate Change and Occupational Heat Strain Among Women Workers: A Systematic Review. *Indian Journal of Occupational and Environmental Medicine*, *28*(1), 4–17. DOI: 10.4103/ijoem.ijoem_320_21 PMID: 38783874

Habibi, P., Moradi, G., Dehghan, H., Moradi, A., & Heydari, A. (2021). The impacts of climate change on occupational heat strain in outdoor workers: A systematic review. *Urban Climate*, *36*, 100770. Advance online publication. DOI: 10.1016/j.uclim.2021.100770

Habibi, P., Razmjouei, J., Badzohreh, A., & Heydari, A. (2025). Climate change and chronic kidney disease (CKD) among outdoor workers: A systematic review. *International Journal of Biometeorology*, *69*(6), 1207–1236. DOI: 10.1007/s00484-025-02896-6 PMID: 40186676

Habibi, P., Razmjouei, J., Moradi, A., Mahdavi, F., Fallah-Aliabadi, S., & Heydari, A. (2024b). Climate change and heat stress resilient outdoor workers: Findings from systematic literature review. *BMC Public Health*, *24*(1), 1711. DOI: 10.1186/s12889-024-19212-3 PMID: 38926816

Hernández-Duarte, W. A., Guerrero-Calderon, D. P., Duarte-Gómez, W. J., Fonseca-Paipa, A. N., Rodríguez-Muñoz, M., Tellez-Morales, M. A., & Garnica, L. C. C. (2024). Health conditions in land transport workers and climate change. Exploratory systematic review. *Revista Brasileira de Medicina do Trabalho : Publicação Oficial da Associaação Nacional de Medicina do Trabalho-ANAMT*, *22*(2), e20241268. DOI: 10.47626/1679-4435-2024-1268 PMID: 39371277

Hornborg, A. (2019). *Nature, Society, and Justice in the Anthropocene: Unraveling the Money-Energy-Technology Complex*. Cambridge University Press., DOI: 10.1017/9781108554985

Hou, J. (2022). Wildfire Smoke Exposure in California: Public Health Impacts and Mitigation Strategies.

International Labour Office. (2021). [The role of digital labour platforms in transforming the world of work. International Labour Organisation] [ILO]. *World Employment and Social Outlook*, •••, 2021.

International Labour Organization. (2015). Guidelines for a just transition towards environmentally sustainable economies and societies for all.

International Labour Organization. (2018). *Women and men in the informal economy: A statistical picture* (3rd ed.).

Jacklitsch, B. L., Williams, W. J., Musolin, K., Coca, A., Kim, J. H., & Turner, N. (2016). Criteria for a recommended standard: occupational exposure to heat and hot environments.

Jia, J. A., & Jia, N. I. (2024). Impacts of Climate Change on Occupational Health and Safety: A Comprehensive Review. *Asian Journal of Advanced Research and Reports*, *18*(10), 136–154. DOI: 10.9734/ajarr/2024/v18i10761

Karanikas, N., Foster, C., Beltran Hernandez, A., Harvey, A., Targal, O., & Horswill, N. (2021a). Conventional and Alternative Aviation Fuels: Occupational Exposure and Health Effects. *Journal of Chemical Health and Safety*, *28*(3), 159–170. DOI: 10.1021/acs.chas.0c00120

Karanikas, N., Steele, S., Bruschi, K., Robertson, C., Kass, J., Popovich, A., & MacFadyen, C. (2021b). Occupational health hazards and risks in the wind industry. *Energy Reports*, *7*, 3750–3759. DOI: 10.1016/j.egyr.2021.06.066

Kazemi Shariat Panahi, H., Dehhaghi, M., Kinder, J. E., & Ezeji, T. C. (2019). A review on green liquid fuels for the transportation sector: A prospect of microbial solutions to climate change. *Biofuel Research Journal*, *6*(3), 995–1024. DOI: 10.18331/BRJ2019.6.3.2

Kjellstrom, T., Briggs, D., Freyberg, C., Lemke, B., Otto, M., & Hyatt, O. (2016). Heat, Human Performance, and Occupational Health: A Key Issue for the Assessment of Global Climate Change Impacts. *Annual Review of Public Health*, *37*(1), 97–112. DOI: 10.1146/annurev-publhealth-032315-021740 PMID: 26989826

Lan, T., Goh, Y. M., Jensen, O., & Asmone, A. S. (2022). The impact of climate change on workplace safety and health hazard in facilities management: An in-depth review. *Safety Science*, *151*, 105745. Advance online publication. DOI: 10.1016/j.ssci.2022.105745

Lebel, L., Paquin, V., Kenny, T. A., Fletcher, C., Nadeau, L., Chachamovich, E., & Lemire, M. (2022). Climate change and Indigenous mental health in the Circumpolar North: A systematic review to inform clinical practice. *Transcultural Psychiatry, 59*(3), 312–336. DOI: 10.1177/13634615211066698 PMID: 34989262

Lee, G. (2024). Impact of Smoke from Wildfire and Agricultural Burning on Farmworker Health and Behavior (Doctoral dissertation, University of California, Davis).

Levi, M., Kjellstrom, T., & Baldasseroni, A. (2018). Impact of climate change on occupational health and productivity: A systematic literature review focusing on workplace heat. *La Medicina del Lavoro, 109*(3), 163–179. DOI: 10.23749/mdl.v109i3.6851 PMID: 29943748

Lobus, N. V., Knyazeva, M. A., Popova, A. F., & Kulikovskiy, M. S. (2023). Carbon Footprint Reduction and Climate Change Mitigation: A Review of the Approaches, Technologies, and Implementation Challenges. *C, 9*(4), 120. Advance online publication. DOI: 10.3390/c9040120

Luyten, A., Winkler, M. S., Ammann, P., & Dietler, D. (2023). Health impact studies of climate change adaptation and mitigation measures – A scoping review. *The Journal of Climate Change and Health, 9*, 100186. Advance online publication. DOI: 10.1016/j.joclim.2022.100186

Ma, T., Moore, J., & Cleary, A. (2022). Climate change impacts on the mental health and wellbeing of young people: A scoping review of risk and protective factors. *Social Science & Medicine, 301*, 114888. DOI: 10.1016/j.socscimed.2022.114888 PMID: 35367905

McCauley, D., & Heffron, R. (2018). Just transition: Integrating climate, energy and environmental justice. *Energy Policy, 119*, 1–7. DOI: 10.1016/j.enpol.2018.04.014

Middleton, J., Cunsolo, A., Jones-Bitton, A., Wright, C. J., & Harper, S. L. (2020). Indigenous mental health in a changing climate: A systematic scoping review of the global literature. *Environmental Research Letters, 15*(5), 053001. DOI: 10.1088/1748-9326/ab68a9

Musa, S. F. P. D., & Ariff Lim, S. (2025). Revitalising agriculture through climate change mitigation: A systematic literature review on smart technologies and sustainable practices. *International Journal of Climate Change Strategies and Management, 17*(1), 481–499. DOI: 10.1108/IJCCSM-05-2024-0071

Osaka, S., Bellamy, R., & Castree, N. (2021). Framing "nature-based" solutions to climate change. *Wiley Interdisciplinary Reviews: Climate Change, 12*(5), e729. Advance online publication. DOI: 10.1002/wcc.729

Pal, G., Patel, T., & Banik, T. (2021). Effect of climate change associated hazards on agricultural workers and approaches for assessing heat stress and its mitigation strategies–review of some research significances. *International Journal of Current Microbiology and Applied Sciences*, *10*(02), 2947–2975. DOI: 10.20546/ijcmas.2021.1002.325

Palinkas, L. A., & Wong, M. (2020). Global climate change and mental health. *Current Opinion in Psychology*, *32*, 12–16. DOI: 10.1016/j.copsyc.2019.06.023 PMID: 31349129

Pan, Q., Sumner, D. A., Mitchell, D. C., & Schenker, M. (2021). Compensation incentives and heat exposure affect farm worker effort. *PLoS One*, *16*(11), e0259459. DOI: 10.1371/journal.pone.0259459 PMID: 34727122

Quesada, J., Hart, L. K., & Bourgois, P. (2011). Structural Vulnerability and Health: Latino Migrant Laborers in the United States. *Medical Anthropology*, *30*(4), 339–362. DOI: 10.1080/01459740.2011.576725 PMID: 21777121

Quintarelli, V., Ben Hassine, M., Radicetti, E., Stazi, S. R., Bratti, A., Allevato, E., Mancinelli, R., Jamal, A., Ahsan, M., Mirzaei, M., & Borgatti, D. (2024). Advances in nanotechnology for sustainable agriculture: A review of climate change mitigation. *Sustainability (Basel)*, *16*(21), 9280. DOI: 10.3390/su16219280

Ramadan, M., & Mourad, B. (2022). Impact of climate change on occupational health and its possible preventive strategies. *Egyptian Journal of Occupational Medicine*, *46*(3), 47–62. DOI: 10.21608/ejom.2022.133585.1272

Rossi, M. F., Leone, R., & Moscato, U. (2025). Climate Change and Occupational Risks in Outdoor Workers: A Systematic Review of the Health Effects of Extreme Temperatures. *Atmosphere (Basel)*, *16*(7), 839. Advance online publication. DOI: 10.3390/atmos16070839

Šavrič, J., Ladkin, A., Cooper, C., & Turnšek, M. (2025). Climate change and tourism work: A perspective paper. *Tourism Review*. Advance online publication. DOI: 10.1108/TR-10-2024-0956

Schulte, P., Jacklitsch, B., Bhattacharya, A., Chun, H., Edwards, N., Elliott, K., Flynn, M., Guerin, R., Hodson, L., Lincoln, J., MacMahon, K. L., Pendergrass, S., Siven, J., & Vietas, J. (2023). Updated assessment of occupational safety and health hazards of climate change. *Journal of Occupational and Environmental Hygiene*, *20*(5-6), 183–206. DOI: 10.1080/15459624.2023.2205468 PMID: 37104117

Schulte, P. A., Bhattacharya, A., Butler, C., Chun, H., Jacklitsch, B., Jacobs, T., Kiefer, M., Lincoln, J., Pendergrass, S., Shire, J., Watson, J., & Wagner, G. R. (2016). Advancing the framework for considering the effects of climate change on worker safety and health. *Journal of Occupational and Environmental Hygiene*, *13*(11), 847–865. DOI: 10.1080/15459624.2016.1179388 PMID: 27115294

Schulte, P. A., & Chun, H. (2009). Climate change and occupational safety and health: Establishing a preliminary framework. *Journal of Occupational and Environmental Hygiene*, *6*(9), 542–554. DOI: 10.1080/15459620903066008 PMID: 19551548

Schwartz, S. E. O., Benoit, L., Clayton, S., Parnes, M. F., Swenson, L., & Lowe, S. R. (2022). Climate change anxiety and mental health: Environmental activism as buffer. *Current Psychology*, •••, 1–14. DOI: 10.1007/s12144-022-02735-6 PMID: 35250241

Sharpe, I., & Davison, C. M. (2022). A Scoping Review of Climate Change, Climate-Related Disasters, and Mental Disorders among Children in Low- and Middle-Income Countries. *International Journal of Environmental Research and Public Health*, *19*(5), 2896. Advance online publication. DOI: 10.3390/ijerph19052896 PMID: 35270593

Spector, J. T., Krenz, J., & Blank, K. N. (2015). Risk factors for heat-related illness in Washington crop workers. *Journal of Agromedicine*, *20*(3), 349–359. DOI: 10.1080/1059924X.2015.1047107 PMID: 26237726

Suman, A. (2021). Role of renewable energy technologies in climate change adaptation and mitigation: A brief review from Nepal. *Renewable & Sustainable Energy Reviews*, *151*, 111524. Advance online publication. DOI: 10.1016/j.rser.2021.111524

Sun, C., Rao, Q., Chen, B., Liu, X., Adnan Ikram, R. M., Li, J., Wang, M., & Zhang, D. (2024). Mechanisms and Applications of Nature-Based Solutions for Stormwater Control in the Context of Climate Change: A Review. *Atmosphere (Basel)*, *15*(4), 403. Advance online publication. DOI: 10.3390/atmos15040403

Tang, W. S. W., & Ho, C. S. H. (2025). A systematic review on the impact of climate change on occupational mental health: A focus on vulnerable industries. *Social Psychiatry and Psychiatric Epidemiology*, *60*(10), 2275–2287. DOI: 10.1007/s00127-025-02936-x PMID: 40473944

Tesfaye, A. H., Prior, J., & McIntyre, E. (2025). Impact of climate change on health workers: A scoping review. *Journal of Public Health (Berlin)*. Advance online publication. DOI: 10.1007/s10389-025-02418-z

Tsakonas, K., Badyal, S., Takaro, T., & Buse, C. G. (2024). Rapid review of the impacts of climate change on the health system workforce and implications for action. *The Journal of Climate Change and Health, 19*, 100337. DOI: 10.1016/j.joclim.2024.100337 PMID: 41647318

Tustin, A. W., Lamson, G. E., Jacklitsch, B. L., Thomas, R. J., Arbury, S. B., Cannon, D. L., Gonzales, R. G., & Hodgson, M. J. (2018). Evaluation of occupational exposure limits for heat stress in outdoor workers—United States, 2011–2016. *MMWR. Morbidity and Mortality Weekly Report, 67*(26), 67. DOI: 10.15585/mmwr.mm6726a1 PMID: 29975679

Van der Horst, L., Bouaddi, O., Williams, S., Jago, M., Lau, K., Furber, B., Zaini, A. Z., El-Ghitany, E. M., Loganathan, T., Flouris, A. D., Testa, D. J., Zimmerman, C., & Hargreaves, S. (2025). International migrant workers, heat exposure, and climate change: A systematic review of health risks and protective interventions. *BMC Global and Public Health, 3*(1), 107. DOI: 10.1186/s44263-025-00224-z PMID: 41320785

Weheba, A., Vertigan, A., Abdelsayad, A., & Tarlo, S. M. (2024). Respiratory diseases associated with wildfire exposure in outdoor workers. *The Journal of Allergy and Clinical Immunology. In Practice, 12*(8), 1989–1996. DOI: 10.1016/j.jaip.2024.03.033 PMID: 38548173

White, B. P., Breakey, S., Brown, M. J., Smith, J. R., Tarbet, A., Nicholas, P. K., & Ros, A. M. V. (2023). Mental Health Impacts of Climate Change Among Vulnerable Populations Globally: An Integrative Review. *Annals of Global Health, 89*(1), 66. DOI: 10.5334/aogh.4105 PMID: 37810609

Whyte, K. (2017). Indigenous Climate Change Studies: Indigenizing Futures, Decolonizing the Anthropocene. *English Language Notes, 55*(1-2), 153–162. DOI: 10.1215/00138282-55.1-2.153

World Health Organization. (2022). *Mental health and climate change: policy brief.* World Health Organization.

Wuersch, L., Neher, A., Marino, F. E., Bamberry, L., & Pope, R. (2023). Impacts of Climate Change on Work Health and Safety in Australia: A Scoping Literature Review. *International Journal of Environmental Research and Public Health, 20*(21), 7004. Advance online publication. DOI: 10.3390/ijerph20217004 PMID: 37947561

Yea, S. (2025). Worker Protections at Sea: Climate change and life aboard distant water fishing vessels. *Anti-Trafficking Review*, (25), 153–157. DOI: 10.14197/atr.201225259

Yong, W. T. L., Thien, V. Y., Rupert, R., & Rodrigues, K. F. (2022). Seaweed: A potential climate change solution. *Renewable & Sustainable Energy Reviews, 159*, 112222. Advance online publication. DOI: 10.1016/j.rser.2022.112222

Chapter 3
Climate Change, Agrochemicals, and Occupational Health: A One Health Assessment for Agricultural Workers

Muhammad Zia Ul Haq
http://orcid.org/0000-0001-9914-3367
Department of Agronomy, University of Agriculture, Faisalabad, Pakistan

Safdar Ali
http://orcid.org/0000-0003-4588-3006
Department of Agronomy, University of Agriculture, Faisalabad, Pakistan

Rana Nadeem Abbas
Department of Agronomy, University of Agriculture, Faisalabad, Pakistan

Umair Gull
Department of Agronomy, University of Agriculture, Faisalabad, Pakistan

Rizwan Maqbool
Department of Agronomy, University of Agriculture, Faisalabad, Pakistan

Rukhsar Saleem
http://orcid.org/0000-0002-2738-9832
Department of Agronomy, University of Agriculture, Faisalabad, Pakistan

ABSTRACT

The present chapter examines the intersection of climate change, agrochemical exposure, and occupational health through a One Health framework. Agricultural workers face escalating risks as rising temperatures and elevated carbon dioxide levels drive herbicide resistance and increased pesticide application. These climatic shifts simultaneously enhance human physiological vulnerability, where heat stress accelerates dermal absorption and impairs the renal clearance of toxins. Furthermore, socio-economic factors—including substandard housing and piece-rate pay—limit recovery and exacerbate health disparities among migrant populations. The au-

DOI: 10.4018/979-8-3373-8470-2.ch003

thors advocate for integrated mitigation strategies, such as climate-smart personal protective equipment (PPE), updated occupational exposure limits (OELs), and an agroecological transition to protect the interconnected health of humans, animals, and the environment.

INTRODUCTION

Human health and food systems rely heavily on agricultural labor to produce crops and other foods. Agricultural laborers worldwide face the possibility of exposure to agrochemicals through their work. Farmworkers make up a large part of our nation's agricultural industry and economy. Agrochemicals kill many insects but can potentially affect other insects and ecosystems. There are direct effects of pesticides on insect health and how insects help pollinate the plants we eat (Gándara et al., 2024). The effects of climate change on insect populations can also reduce the food we grow, directly impacting food security. Climate change affects insect behavior by changing when they appear in the spring or how far north they travel. Many pests will thrive in a warmer climate and require more pesticide spraying. Agrochemicals like pesticides will need to adjust to the changing climate. These changes can affect farmworkers and how they are exposed. Gándara et al. (2024) found that agrochemical exposure led to lower reproductive success in insect communities. If insect populations continue to decline, they can no longer pollinate our crops. What happens when insects that help grow our food begin to decline in numbers? Food security is threatened as a result of climate change. One way we can reduce the risks associated with climate change and agricultural health is by using tools like Climate Change - House of Safety (CC-HoS). CC-HoS evaluates direct health risks from climate change. CC-HoS also helped users discover new hazards in occupations where workers spend a significant amount of time outdoors, improving agricultural health (Braglia et al., 2023). Educating farm workers on pesticide use and knowledge of climate change can help us prevent many risks.

Climate change can be described as a "risk multiplier". Increased temperatures can lead to more pests thriving, which can increase the amount of pesticide used on crops. Many crops will also experience a lack of rainfall and increased severe weather. Changes in temperature can affect a worker's ability to work outside for extended periods without overheating. This will put more agricultural workers at risk for heat-related illness. Climate change can affect how pests act, when farmers apply agrochemicals, and how effective they are at pest control. All of these things can affect agricultural workers' health.

CLIMATE-AMPLIFIED OCCUPATIONAL HAZARDS

Climate change is not only alarming at a global scale, but certain populations are at an even greater risk than others. Outdoors workers, especially those who work in agriculture, are on the front line when it comes to facing various occupational risks (Khayat et al., 2022; Ferrari et al.). Agriculture workers face risks from rising temperatures, shifting ecology, and deteriorating air quality. These occupational hazards will be explained further in this chapter, along with a review of current evidence.

Direct Heat Stress- Pathophysiology of Heat Illnesses and the Increasing Prevalence of Chronic Kidney Disease of Unknown Etiology (CKDu) in Tropical Regions

Heat illnesses kill and negatively affect > 1 million farmworkers in the United States alone each year, making agricultural workers one of the most vulnerable populations to suffer from extreme heat effects. As the climate continues to change, this number will rise (Tigchelaar et al., 2020). Heat stress from occupational exposure to heat will vary depending on agricultural worker sensitivity and vulnerability to changing temperatures and their community's ability to adapt and be resilient to increasing temperatures (Nunfam et al., 2018). Research has shown that working people worldwide are exposed to adverse weather conditions and climate—most of the time for longer durations and at higher intensities than the rest of the population—and these exposures will continue to increase under future climate change (Ferrari et al., 2023).

Rising temperatures worldwide will have major impacts on agricultural workers. Reports have shown that rising temperatures are directly correlated with higher occupational health risk, as when agricultural workers are exposed to heat, their ability to work efficiently decreases (Amini et al., 2021). Countries that are still developing are faced with higher levels of decreased human resources productivity in agriculture.

Heat illnesses range from mild cases such as heat cramps to severe cases such as heat stroke. Heat illnesses occur when someone is doing work that their body can not keep cool. Heat protection standards/guidelines need to be created with local environmental conditions and the physical demands of the job in mind (Khayat et al., 2022).

Studies have found that there have been few to no adaptations made to decrease the growing heat risk for United States agricultural workers (Tigchelaar et al., 2020). Since crop workers face a large amount of heat exposure and health risk that is growing each year, we need to assess how effective workplace adaptations are at reducing that risk (Tigchelaar et al., 2020).

One of the most serious effects of chronic exposure to heat is the development of CKDu. CKDu is a disease that leads to kidney failure and affects countless agricultural workers in areas such as Central America and Sri Lanka. CKDu has been categorized as an epidemic in tropical regions. Most people who are diagnosed with CKDu have no other past medical history of kidney disease. Although the exact cause of CKDu is unknown, farmers who are repeatedly affected by dehydration and heat tend to have subclinical AKI, which leads to CKDu.

Literature currently shows that over the past decade, studies focusing on social aspects and impacts of occupational heat stress, along with workers' adaptation strategies, have been inadequate, especially in Europe and Africa (Nunfam et al., 2018). Africa is impacted by higher negative occupational health outcomes than Europe due to lower adaptive capacity, rising poverty levels, and less developed technologies to help decrease temperatures and change (Nunfam et al., 2018). There have been studies done on the effects of excess heat exposure on workers' health and safety, productivity, and the environmental conditions they work in, along with what adaptations they use.

Policy makers should take steps to ensure labor protections and occupational health and safety for these agricultural communities. There is a need to conduct evidence-based research studies on climate change and UHI in relation to outdoor workers to allow us to understand and raise awareness of occupational hazards. This can help establish risk awareness and help workers find coping strategies. Awareness can be created among farm workers, managers, and other stakeholders (Moda et al., 2019).

The first step to coming up with adaptations for our outdoor workers is to ensure that they have Occupational health and safety protocols that can handle the risks that climate change will bring to our workers. Going forward, we need to identify how climate change will affect these workers so we can better protect them (Braglia et al., 2023).

Changing Vector Ecology: Expanding Geographic Range of Lyme Disease, West Nile Virus, and Emerging Zoonoses

Climate change poses a major threat to crop yields, farmworkers' health and livelihoods, and animal welfare. One thing that should be done to help decrease the long-term effects is to create and start adapting to new strategies that can help fight climate change. Climate change will allow agricultural areas that have never had issues with pest/disease-carrying mosquitoes to start seeing the negative effects of these stressors (Richards et al., 2025). Agricultural workers are at risk because they work long hours outside in fields that can be populated with disease-carrying insects and animals. Since one health takes into consideration our health, animal health, and our environment. Research priorities to address and adapt to climate change in agricultural areas need to be identified and mapped with the health of farmworkers and livestock animals in mind (Richards et al., 2025). Due to rising temperatures, Lyme disease will start to see higher cases north of its current boundary in America. Also, globally, we are seeing Lyme disease being found at higher altitudes. Workers in these new areas may not know about ticks and the behaviors needed to prevent tick bites.

There are many ways that climate change can affect mosquito-borne illnesses. Some of these ways include: changes in precipitation, length of seasons, and humidity levels. All of these factors can play a role in the mosquito life cycle and survival, as well as the spread of competent vector species. Farmworkers who work outside during dawn and dusk are at higher risk of exposure. Agriculture is a large animal-human interface. As the climate continues to change, wildlife migration patterns will shift, and we will see more exposure to wildlife and large mammals. These large mammals can carry and spread disease to livestock and humans.

Atmosphere Impairment: Wildfire Smoke (PM2.5), Ozone, and Agricultural Dust

Farmers/farm workers are exposed to agricultural dust on a regular basis. Agricultural dust has been linked to decreased lung function, shortness of breath, and other adverse health effects (Crawford et al., 2021). Agricultural dust can contain many harmful things, such as chemicals, bacteria, fungi, and viruses, which all cause harm to farmworkers when exposed.

Climate change has led to longer wildfire seasons and more intense fires. Some of these fires have been recorded as the worst in history. Farmworkers who live and work in areas affected by wildfires are at risk for exposure to wildfire smoke and PM2.5. Agricultural work is strenuous work, and farmworkers are outside, breathing in high amounts of polluted air while exercising.

As the climate continues to warm, ground-level ozone levels will rise. Ozone levels are typically higher in the summer months when farmworkers are working the longest hours outside. Agricultural workers are not only exposed to PM2.5 while working in fields; they are also exposed to ozone. Agricultural workers are susceptible to more than one climate-amplified hazard at a time. Exposure to multiple things at once can be more harmful than just one exposure. There has not been enough research on studies that involve multiple exposures.

Heat stress, vector-borne illness range expansions, and deteriorating air quality pose serious threats to agricultural worker safety and wellness. Each of these climate-amplified occupational hazards will only continue to worsen as our climate changes (Khayat et al., 2022; Nunfam et al., 2018; Moda et al., 2019). Therefore, it is critical that we gain a better understanding of these risks and start adapting to them in order to better protect our agricultural workers (Ferrari et al., 2023). Adaptation strategies need to consider climate change impacts not only on crop yield but also on farmworkers and agricultural animals (Richards et al., 2025). There are many research gaps that need to be filled. More studies need to be done on occupational heat stress, especially in Africa (Nunfam et al., 2018). Researchers need to start coming up with methods to study the health effects of multiple climate-amplified hazards at one time (Moda et al., 2019).

THE CHEMICAL-CLIMATE NEXUS: ALTERED TOXICOLOGICAL PROFILES (REDUX)

The nexus of climate change and agrochemical exposure is a field of inquiry whose importance is only beginning to be understood in the context of the occupational health landscape. The various components of anthropogenic climate change will exacerbate numerous risks that face agricultural workers in diverse geographical settings around the globe. In this chapter, we highlight three key vectors by which the evolving atmospheric landscape is reshaping pesticide toxicology. First, we discuss the ways that the physiology of weeds will change under increasing CO_2 concentrations to become more competitive with their surrounding crop plants. This will force the use of additional herbicide applications, potentially heightening worker exposure. Second, increasing temperatures are associated with increased volatilization of herbicides and insecticides from plants to the atmosphere, further increasing exposure risk. Third, and finally, human physiological vulnerability to the neurotoxic effects of pesticides and many other contaminants is increased when the human body is in a state of heat stress. We discuss each of these three vectors in detail in the following sections.

The Global Herbicide Resistance Crisis

Herbicide-resistant weed populations are a growing problem in agriculture worldwide. Glyphosate is one of the most widely used herbicides in the world and is considered the most important herbicide for weed control (Broster et al., 2022; Lubis & Priwiratama, 2025). Glyphosate-resistant crops have been widely adopted, and glyphosate is now the primary method of weed control in many agricultural systems (Lubis & Priwiratama, 2025). Glyphosate is highly effective in controlling many weed species, including annual grasses, broadleaf weeds, and perennials with deep roots (Lubis & Priwiratama, 2025). The heavy reliance on glyphosate, however, has created intense selection pressure for glyphosate-resistant weed biotypes.

Glyphosate is often used in both the production phase of annual crops and in the fallow phase. This has led to an increase in the evolution and spread of glyphosate resistance in many weed species (Werth et al., 2021). In cotton production in Australia, glyphosate has been the mainstay of weed control for the past several decades. As a result, glyphosate resistance has evolved in several key weeds (Werth et al., 2021). This has forced growers to reduce their reliance on glyphosate and has led to the re-emergence of tactics that were abandoned years ago, such as the use of residual herbicides and mechanical weed control (Werth et al., 2021). Glyphosate resistance has been documented in eight countries across the globe. At the current rate of spread, it is predicted that glyphosate resistance will be almost universal unless standard management practices are overhauled (Hicks et al., 2018).

Survey data from different regions of southeastern Australia provide a useful lens into the distribution of resistant ryegrass populations. Resistance in Lolium rigidum, although glyphosate is widely used in farming systems throughout the area surveyed, remained rare in most regions outside of the South Australian Southeast region, where the weed was resistant in 27% of surveyed fields (Broster et al., 2022). Resistance to selective herbicides in other weed species is similar to or even greater than that in glyphosate, when resistance data from the same fields at the same point in time were collected (Broster et al., 2023). Glyphosate is applied not only in summer fallow for weed control, but also after pre-sowing cultivation as a pre-emergent weed control strategy, usually followed by the application of selective pre-emergent herbicides and subsequent crop planting (Broster et al., 2023). Augmented atmospheric CO_2 Levels in the coming century will also strongly influence weed competitive strength in relation to their crop counterparts.

Crop-Mimicry in Weeds

Increased atmospheric CO_2 concentrations will give rise to enhanced rates of photosynthesis in weed species under enriched CO_2 conditions. These weeds will not be able to photosynthesize under these conditions as effectively as under lower CO_2 levels and will be outcompeted by other species. The weeds that do best under elevated CO_2 conditions, however, are those that closely resemble the crops with which they are competing in their physiological response to CO_2. In a way, weeds can be said to "mimic" the crop when they respond to CO_2 in much the same way. This mimicking can negate the effects that elevated CO_2 might otherwise have on the crop in terms of a competitive advantage over the weed.

Reduction in Herbicide Efficacy

Plants growing under elevated CO_2 are often observed to have thicker leaves, as well as changes in stomatal density and cuticular wax composition. This can have important implications for herbicide application, as it may change how much herbicide is able to enter the plant and how easily it can be translocated once it is inside. Morphological changes in plants due to elevated CO_2 may lead to a reduction in the efficacy of foliar-applied herbicides. The enhanced vigor in many weeds, however, is also likely to speed up metabolic detoxification mechanisms that allow plants to break down herbicide molecules before they reach toxic levels. Overall, this increase in weed competitiveness and herbicide resistance will have major implications for herbicide application and effectiveness.

Occupational Health Implications of Resistance Management

The reliance on glyphosate for weed control has significant implications for agricultural workers. The evolution of glyphosate resistance has led to an increase in the use of alternative herbicides and management strategies, many of which may be more toxic or require more applications than glyphosate. The re-emergence of tactics that have been abandoned in the past, such as mechanical weed control, may also increase worker exposure to herbicides. Tank-mix combinations of glufosinate with indaziflam have been found to control glyphosate-resistant Leptochloa virgata populations for up to 90 days in Mexican citrus production systems, due to the residual activity of the second compound (Cruz et al., 2019). The mixture of glyphosate with a contact or residual herbicide may also be an effective strategy for weed control and the avoidance of selection for resistance (Cruz et al., 2019). The widespread introduction of both dicamba- and 2,4-D-resistant crops has altered the approach to weed management (Gage et al., 2019). Past mistakes in the stewardship of glyphosate-resistant cropping systems are directly responsible for the situation with which growers now find themselves (Gage et al., 2019). We must be careful not to repeat those mistakes with newer technologies (Gage et al., 2019). From an ecological standpoint, considering weeds as enemies to be vanquished through ever-more-powerful technology is a fundamentally flawed strategy; the emergence of multiple herbicide-resistant genotypes and the rapidly dwindling list of chemical options should be an urgent message that it is time to think differently about weeds and weed resistance (Bagavathiannan & Davis, 2018).

Human Physiological Vulnerability

Cutaneous vasodilation results in increased blood flow to the skin. This not only brings more heat to the skin's surface to aid in thermoregulation but also creates a concentration gradient that favors the diffusion of absorbed substances from the skin into the bloodstream. The increased blood flow effectively clears absorbed pesticide molecules from the dermal layer, maintaining the concentration gradient and promoting continued absorption.

Sweating is another thermoregulatory response that can affect the skin's barrier function. The presence of sweat can solubilize some pesticide formulations, increasing their availability for absorption. Additionally, sweating hydrates the stratum corneum, the outermost layer of the skin, which can increase its permeability to both hydrophilic and lipophilic substances. The occlusive effect of sweat, especially when clothing is damp against the skin, can further enhance pesticide absorption by increasing contact time. Reduced renal function due to dehydration can also impact the clearance of pesticides. The kidneys are responsible for filtering blood and excreting waste products, including metabolites of pesticides, in the urine. Impaired kidney function can lead to a decreased glomerular filtration rate, which can reduce the body's ability to excrete pesticides and their metabolites.

It should also be noted that physiological factors affected by various conditions can have a significant effect on ADME processes and, by extension, chemical exposure profiles (Jo et al., 2021). Altered hepatic or renal clearance rates have also been observed in patients with impaired function (Jo et al., 2021). This has important ramifications for the safety profiles of drugs that these patients may take (Jo et al., 2021). This is an observation that is based on pharmaceuticals, but the physiology underlying these processes is the same as that of pesticide toxicokinetics.

The nexus of chemical and climate stressors is an area of occupational health that is only beginning to be appreciated. The interplay between herbicide resistance, volatilization, and human physiology under changing climatic conditions poses a multiplicative risk to agricultural workers. Tackling these challenges will require a concerted effort from the agricultural, environmental, and public health communities, with a particular focus on the unique vulnerabilities of workers who find themselves at the intersection of these hazards.

CLIMATE CHANGE, AGROCHEMICALS & OCCUPATIONAL HEALTH: SOCIETAL FACTORS CREATING VULNERABILITY

No discussion around climate change or agricultural chemical exposure is complete without recognition of the socio-economic forces that create agricultural worker vulnerability. The workers most likely to experience heat illness while working outdoors are agricultural workers, with research finding that agricultural workers from 2000-2010 were two times more likely to develop heat-related illness compared to those working in other industries (Mizelle et al., 2022). This section explores three major socio-economic factors that impact vulnerability to heat and chemicals. The following subsections cover increased vulnerability in migrant and seasonal workers, the impact of substandard housing on workers' ability to recover from daily heat and agrochemical loads, and economic pressures pushing workers to take on unnecessary risk.

Migrant and Seasonal Workforce

Population Vulnerability and Overrepresentation in Agriculture and Construction

At-risk populations are often overrepresented in agriculture and construction due to factors inherent to the demographics of these industries. Agriculture has repeatedly been identified as both the industrial sector most likely to see heat-related illness (Heinzerling et al., 2020) and death (Heinzerling et al., 2020). When examining occupational heat-related deaths in North Carolina by industry and race/ethnicity, researchers found that the Hispanic population is both overrepresented in agriculture compared to their presence in the overall population and experiences the highest number of occupational heat-related deaths in agriculture (McClure et al., 2024). Researchers examining occupational heat-related injury among agricultural workers found that Latino/a workers had significantly higher HRI mortality than non-Latino/a workers (Mizelle et al., 2022).

Issues such as language and acculturation barriers increase migrant agricultural workers' risk of occupational heat-related injury (Hansen et al., 2020). Several studies outside of the agricultural sphere have shown that farmworkers experience many roadblocks to taking proper heat preventative measures or seeking treatment if heat-injured, with language being a common barrier (Khayat et al., 2022). Specifically, language deficits can contribute to basic misunderstandings about the dangers associated with heat exposure and how to stay safe.

One study done in Italy found that foreign agricultural and construction workers perceived heat differently than native workers. Specifically, foreign workers reported that heat did not affect their work performance as much as their native counterparts believed it would (Hansen et al., 2020). These findings support the idea that foreign-born workers are at increased risk for heat illness due to cultural/language barriers limiting their knowledge about heat illness (Hansen et al., 2020). However, cultural and ethnic factors are still understudied contributors to occupational heat-related vulnerability (Hansen et al., 2020).

Earlier studies on construction worker safety identified many of the same issues that Latino/a workers face today, including English-language deficiencies, foreign-born status, cultural differences, and a desire to "look like they were working hard" as contributors to inequities between groups of workers (Richey et al., 2022). More recent literature has expanded its focus to the occupational environment surrounding these workers. Authors began to discuss ideas such as workers not understanding their labor laws and protections, employers believing Latino/a workers will simply accept more dangerous work assignments, and poor communication between supervisors and coworkers as likely causes of disparities (Richey et al., 2022).

In addition to language barriers, migrant farmworkers face a host of other structural issues that make it difficult to follow heat safety guidelines. Researchers have found that migrant farmworkers often lack access to standard heat preventative tools, including prevention training, breaks, shady areas, and water to drink (Khayat et al., 2022). Workers also report having less control over their work conditions and a poor work safety climate, which, paired with language and cultural differences, further increases vulnerability (Khayat et al., 2022). There seems to be a basic knowledge gap where workers know what to do but are unable to enact that knowledge because of cultural differences. This gap has real-world consequences, as research examining core body temperature and heart rate indicative of heat strain found that workers who experienced heat strain were at significant risk of acute kidney injury (AKI) (Butler-Dawson et al., 2019).

An important and understudied vulnerability facing agricultural workers is climate migration. Climate migration, or the movement of persons due to climate-related events, will push individuals into the agricultural workforce as their own land is less suited for farming or home life. Workers who are pushed to southern states where jobs in agriculture and construction are prevalent are at special risk for illness due to heat and humidity (Richey et al., 2022). Workers already vulnerable to climate-change-related migration are at risk of being exposed to the same high temperatures at work.

Housing and Recovery

To understand how housing affects agricultural workers, it is important to understand heat acclimatization. One study examining serum creatinine levels as a marker for AKI found that volunteers who were not heat acclimatized had a high incidence of AKI during exercise heat stress protocols, but experienced near 0 incidence after one week of heat acclimatization (Omassoli et al., 2019). Heat acclimatization is of particular concern for agricultural workers and is included as part of the NIOSH recommendation to reduce heat illness (Moyce et al., 2019). Although not directly stated in this study, recovery from exercise heat stress would have occurred outside of the workplace in participants' housing.

Another study examining AKI and workload in agricultural workers found an association between higher workload and AKI, supporting the theory that consecutive days of working without proper recovery can result in injuries that build up on each other (Moyce et al., 2019). This theory would suggest that if workers do not properly recover from heat stress on days off work, repeated exposures that would be otherwise manageable could become injurious.

In one study, agricultural workers were asked to perform a work shift and then had their kidney function monitored (Butler-Dawson et al., 2019). Researchers found that higher temperatures led to AKI, but did not examine whether workers were able to return to normal kidney function after the work shift was completed. If workers are not able to recover from this daily heat exposure due to poor housing conditions that do not allow for proper cooling, they may be at risk for chronic kidney disease (CKD) as days of subclinical AKI add up (Omassoli et al., 2019).

Much like air pollution, recovery from a workday does not take place solely at work for agricultural workers. Agricultural workers often live in housing with little to no air conditioning or the ability to cool down during non-working hours (McClure et al., 2024). This means that not only are workers experiencing dangerous heat exposures during the workday, but they are also continuing to be subjected to high temperatures during what should be their recovery period.

Research has examined how a lack of nighttime cooling increases health risk for agricultural workers by preventing the body from returning to baseline after a heat exposure. Lab-based studies have shown that while participants acclimatized to physical work in heat, there was less of a rise in serum creatinine after repeated exposures, suggesting the kidneys' ability to regulate against further damage (Moyce et al., 2019). Without proper recovery time to allow for this acclimatization to occur, agricultural workers are more likely to experience injuries during heat exposures.

The problem with recovery does not end with heat stress. Agricultural workers are exposed to a variety of agricultural chemicals at work that also need to recover from on days off. Without a way to metabolize and rest away from these chemicals, poor housing can affect workers' ability to acclimate to chemical exposures as well.

Economic Factors

Much of whether or not a worker decides to follow safety precautions at work is based on whether they can afford not to follow these guidelines. Piece rate work, which occurs when a worker is paid based on how much work they complete instead of hours worked, creates a strong financial disincentive to following safety procedures that require taking breaks, drinking water, or slowing work effort when the temperature is high.

Construction and agriculture are two industries that research has found to be particularly concerned about heat-related injury because of the physicality of the work and the fact that it is performed outside in hot weather (Heinzerling et al., 2020). When an employee knows that every second not working is a second they aren't getting paid, occupational hazards that require stopping work to prevent injury can cause workers to choose between pay and safety. This issue creates an environment where workers are put into dangerous positions because they cannot afford to lose work time.

Some specific safety behaviors that are difficult for farmworkers to implement at work include taking breaks to prevent heat injury. Agricultural farmworkers face many occupational health and safety hazards, including working in extreme heat outdoors (Mizelle et al., 2022). While agricultural workers know they should take breaks when it is hot out to prevent heat illness, the inability to take breaks when working under a piece-rate system makes it difficult for workers to practice this preventative behavior.

Hazards affecting vulnerable subpopulations include males, young adults ages 15-24, outdoor and indoor workers, and those who work in industries like electricity, manufacturing, utility, transportation, construction, agriculture, fishing, and forestry (Varghese et al., 2018). Increased risk of injury was seen across all these industries during hot weather months, and types of injuries included wounds, lacerations and amputations, burns, falls, cuts, fractures, slips, and trips (Varghese et al., 2018). Some of these injuries can be linked to heat-related injuries that occur when workers are not able to take breaks because of piece-rate systems.

A study examining occupational heat-related illness across California workers from 2000 to 2017 found that construction, agriculture, and public administration had the highest rates of HRI (Heinzerling et al., 2020). Agriculture had an estimated rate of 38 cases per 100,000 workers, which is higher than any of the other studies discussed above identified agriculture as the industrial sector most likely to suffer from HRI (Heinzerling et al., 2020). Another study looking at occupational heat-related mortality found very similar results, with agriculture and construction having the highest heat-related deaths (Heinzerling et al., 2020). Just because certain demographic groups are seen more frequently getting injured at work does not mean that they are the cause of their own injuries.

In the past few years, research on occupational heat-related illness has shifted from focusing on the individual to their work environment. A newer study discusses how factors such as understanding of labor laws, employer perceptions of Latino/a workers, and communication between coworkers and supervisors play a role in the increased risk that agricultural workers face on the job (Richey et al., 2022). Piece-rate pay is one systemic barrier that prevents workers from taking breaks to prevent heat illness, even if they know they should.

As mentioned earlier, Latino/a workers face many cultural pressures that put them at risk of injury. Wanting to look like they are working hard enough to keep their job is something researchers have identified as a source of disparity between migrant agricultural workers and their peers (Richey et al., 2022). When you add piece-rate into this scenario, workers are not only losing out on money to take breaks, but might feel as if they cannot take breaks to appear as if they are working hard. If employers believe that Latino/a workers are not going to speak up about heat illness because they feel they need to work hard, they may give these workers more hazardous jobs.

EXAMPLES OF HOW ONE HEALTH COULD BE OPERATIONALIZED IN AGRICULTURE

As exemplified by the following case studies, climate change, agrochemical exposure, and occupational health challenges exist in vastly intricate, interconnected webs that must be understood through combined lenses of human health, animal health, and environmental health.

Case Study #1: Simultaneous exposure to pesticides and extreme heat

When discussing climate change, occupational health concerns, and agrochemical exposure, it is impossible not to address the immense threat posed by the overlap between heat exposure and pesticide exposure.

Climate change has presented a widely undocumented threat to occupational health and safety, as workers experience exposures influenced by countless variables (Braglia et al., 2023). The extent to which climate change will impact occupational heat stress depends primarily on two factors: sensitivity and vulnerability to climate change-related heat exposures, as well as current levels of adaptation capacity and resiliency planning (Nunfam et al., 2018).

In agricultural settings like California's Central Valley and other major agricultural production areas across Southern Europe, farmworkers often experience heat exposures and pesticide exposures simultaneously. One study outlined that occupational heat stress impacts on workers and adaptation strategies of workers to rising temperatures have been socially understudied in European agriculture" (Nunfam et al., 2018). This is highly relevant because developing and implementing improvements to occupational health and safety standards to keep pace with climate change's effect on outdoor workers is one adaptation measure that must be taken to better protect farmworkers now and in the future (Braglia et al., 2023).

Synergy between pesticide and heat exposure occurs in part due to the fact that heat stress can cause increased pesticide absorption through the skin, increased metabolism (which can activate certain pesticides), and difficulty concentrating (which can lead to accidents related to pesticide exposure). There is a need for "empirical research on climate change adaptation as it relates to outdoor workers to increase awareness of occupational hazards" and to assign risk awareness among workers and others involved (Moda et al., 2019). While there have been individual cases of extreme heat exposure affecting workers' health and safety, work performance, and work environment, there has been limited data collection on this occurrence, especially under changing climatic conditions (Moda et al., 2019).

One Health issues come into play because research has shown that there are overlaps in health-related effects of climate change across humans and other animals. One study indicated that smoke from wildfires affected dairy cows' metabolism, milk production, and immune systems (Richards et al., 2025). Agricultural workers also experience increased exposure to fine particles of wildfire smoke that can have negative health consequences. This indicates how heat-related climate change impacts on agriculture affect both farmers and animals involved in agriculture.

Case Study #2: Spread of vector-borne illness to new agricultural areas

Vector-borne illness spread to new territories due to climate change can act as an occupational health threat that requires the One Health approach.

West Nile Virus (WNV) seroprevalence has been detected across several occupational classes, and there is concern over work-related infection risk due to increased WNV infection rates (Stufano et al., 2025). This occupational seroprevalence should be viewed through the lens of climate change, as WNV is a vector-borne illness that will be affected by climate change.

As global temperatures continue to rise, agricultural workers will experience exposures to vectors and vector-borne illnesses that they would not have previously faced in their work environment. Climate change will cause an increase in "frequency, intensity, and duration of some extreme weather events such as heatwaves, floods, and droughts" (Odigie et al., 2024). Extreme weather can not only increase contact between humans and vectors, but also contacts between vectors and viruses, allowing for increased chances of WNV spreading to humans and causing more outbreaks (Odigie et al., 2024). It is important to begin looking at ways in which occupation can affect the risk of work-related illness through changes in disease ecology due to climate change.

Agricultural workers can act as sentinel health populations for tracking the spread of vector-borne illness. Additionally, since this is an issue that occurs within the work environment, there are opportunities for workplace interventions. Farmworkers can be provided with personal protective equipment, changes can be made to the work environment to make it safer, and protocols can be established to better manage exposure risks.

Vectorborne illnesses do not exist in a vacuum – because they can be affected by agrochemical use (agricultural chemicals can be used to kill vectors, like mosquitoes), and agrochemical use can be affected by climate change (changes in pest pressures can cause fluctuations in agrochemical use), this occupational health issue should be considered through a One Health lens.

Case Study #3: Increased runoff due to extreme weather events

Climate change-induced extreme weather can cause problems with agrochemical runoff that affect both the environment and agricultural workers.

Determinants of health, like social factors and structural systems, influence vulnerability to flood damage from hurricanes. Interventions aimed at targeting these drivers of health at their root could reduce vulnerability disparities seen after hurricanes (Bakshi et al., 2025). Because climate change will increase the number of extreme weather events like hurricanes, exploring what drives vulnerability to flood damage in hurricane-affected regions is needed to see if these findings are consistent (Bakshi et al., 2025).

Many agricultural regions store pesticides, fertilizers, and other agrochemicals outdoors, so flood water can easily pick up these agrochemicals and run them off into streams, rivers, drinking water sources, farms, and people's homes. During hurricane clean-up or regular farm maintenance after a hurricane, agricultural workers can be exposed to pesticides through dermal absorption after coming into contact with flood water, ingestion of contaminated water or food, or inhalation of vapors.

One Health themes should be considered because pesticide contamination due to flooding can affect humans, livestock that drink the water, wildlife that drink the water, and ecosystems into which the water flows. When discussing pesticide exposure in farmworkers, it is cited that "primary health care units serving rural populations lack capacity to care for individuals with pesticide exposure" (Silvério et al., 2020). Other literature shows that there is a connection between education and knowledge of occupational risks, but farmworkers may know little about the health risks that they face at their worksite (Silvério et al., 2020; Cazé et al., 2018). Farmworkers have been noted to not regularly use personal protective equipment because they view it as uncomfortable.

Hurricanes and other natural disasters can also impair agricultural workers' ability to protect themselves from pesticides. If hurricanes destroy personal protective equipment or farmworkers are tempted to return to work on their farms before it is safe to do so, exposures can occur.

Applications of One Health for recovery after a hurricane should include monitoring of human health, animal health, and environmental health.

Discussion of How These Case Studies Relate to One Another through One Health

All three of these case studies highlight how climate change will affect the occupational health and safety of agricultural workers. It is important to identify how climate change affects agricultural workers' occupational health and safety, and adaptation is one way to try to prevent some of these negative effects (Braglia et al., 2023). Currently, not enough is being done to prevent the negative health outcomes of climate change on agricultural workers.

Countries with "lower adaptive capacity, increasing poverty and technological stagnation" will be more likely to see negative occupational health outcomes (Nunfam et al., 2018). To address this, it is important to take into consideration not only how climate change will affect agricultural workers, but also how factors like poverty, access to resources, and educational levels affect their vulnerability to these changes. Climate change will affect occupational heat stress, productivity loss, accidents, loss of working hours, and mental health issues, but research has not been done to quantify or qualify these effects (Habibi et al., 2024). In order to properly understand how climate change will affect agricultural workers and devise strategies to lessen these impacts, this research must be done.

One Health can be used to operationalize these case studies by viewing each of these problems through the lens of agricultural animal health and environmental health. Air pollution is something that affects both human health and animals, and studies have shown that air pollution effects on horses have similar effects on humans (Richards et al., 2025). By looking at how climate change and agriculture affect animals, we can understand how our animal counterparts are affected by the same things that we are, allowing us to identify vulnerabilities we may not have previously considered.

INTEGRATED STRATEGIES FOR MITIGATION AND ADAPTATION

This chapter briefly discusses four mitigation and adaptation strategies that span different issue areas but are integrally related: breathable, chemical-resistant PPE that can withstand increasingly severe heat waves, data-driven early warning systems for heat/agrochemical warnings based on indices like the Heat Index and Air Quality Index, updating occupational exposure limits (OELs) to account for concurrent heat and chemical stressors, and pursuing Integrated Pest Management and the agroecological transition more broadly to decrease dependence on harmful agrochemicals. These strategies can be linked through One Health due to the interconnectedness between environmental degradation and poor human health outcomes. As the continued increase of global pesticide use suggests, decades of IPM promotion have had a limited effect on pesticidal practices globally despite negative implications for farmer livelihoods, biodiversity conservation, and human health (Tiwarri, 2024).

Climate-smart PPE: breathable, chemical-resistant materials

Climate-smart PPE encompasses personal protective equipment that can keep agricultural workers safe from heat stress while wearing chemical-resistant gloves and masks. Regular chemical-resistant gloves and masks impede the body's ability to cool itself, putting agricultural workers at serious risk of heat stroke when applying chemicals during heat waves. The adverse human health and environmental impacts of chemical pesticide misuse and overuse continue to warrant concern globally (Schenk, 2020), so we will still need to wear gloves and masks when applying pesticides until we transition away from these harmful inputs.

Breathable, chemical-resistant materials that mitigate this issue have to balance two priorities: creating a barrier to pesticide exposure and allowing heat to escape from the body. There have already been breakthroughs in this area, as scientists have developed multiple materials with built-in phase change microcapsules, moisture-wicking membranes, and selectively permeable pores that can block chemicals but allow increased vapor transmission (Mawcha et al., 2024). Smart chemicals have also been embedded with microsensors, which can detect hazardous chemicals that have come into contact with the wearer.

It makes sense that we would utilize smart technology to keep farmworkers safe, as the agriculture industry has widely adopted data-driven strategies for farm management (Barn et al., 2020). Wearable sensor technologies could alert both agricultural workers and their supervisors of dangerous conditions, whether it be due to hazardous chemicals or heat stress.

Bioclimatic Surveillance: Early Warning Systems

As these smart PPE solutions are being developed, we can simultaneously work to implement early warning systems that alert agricultural workers to when it is too dangerous to apply pesticides outside. Certain air quality thresholds and temperatures could trigger automated work-stop orders that prevent farmers from applying chemicals when the Heat Index is too high or when air pollution is dangerous.

Integrated Pest Management (IPM) has been shown to be useful in agricultural settings, with data-driven pest management having several positive effects on crop yields and health (Wiesner et al., 2022). It stands to reason that we could leverage many of the same tools used for IPM to create a successful bioclimatic surveillance system.

Creating an effective early warning system will require extensive ground-level knowledge, as hazards can vary by region and even within the same farm. Farmer knowledge and adaptability are key to both sustainable agriculture (Rustia et al., 2022) and our hypothetical early warning system. Computer vision and AI are already being used in pest monitoring as well (Wiesner et al., 2022). We can repurpose many of the tools being used for sustainable agriculture and use them to monitor environmental conditions and trigger automated work-stop orders based on AI-detected dangerous conditions.

Policy: Updating occupational exposure limits for heat and agrochemical risk

Existing OEL standards do not take climate change into account. OELs typically consider worker exposure to a single chemical, not multiple agrochemicals often used in conjunction. OELs are "important tools to protect workers from harmful workplace exposures" and can help establish quantitative, achievable targets for exposure prevention and risk assessment (Muramoto et al., 2022). However, they need to be reconceptualized with two major considerations in mind.

First, OELS are typically defined as the time-weighted average concentration that a worker can be exposed to for 8-10 hours a day, 5 days a week, throughout their career (Deguine et al., 2021). This definition assumes that conditions will stay the same throughout their career, which is not true with rising global temperatures and heat waves. Increased heat can change the rate at which chemicals are absorbed through the skin, breathed in via the lungs, and metabolized through the body.

Second, OELs vary drastically by state and country (Muramoto et al., 2022). While there is movement towards standardized, evidence-based OELs (Muramoto et al., 2022), there is an opportunity to revise them to include risk factors related to heat stress. The EPA recently transferred responsibility for providing a scientific basis for OELs from pesticide registrants to the agency directly (Kennedy & Lekshmi, 2023). This change in regulation will hopefully allow for the creation of OELs that account for combined heat and agrochemical exposure.

According to one survey, chemical risk managers are split on whether they would ever use something other than OELs as an exposure metric (Kennedy & Lekshmi, 2023). Only about 30% of respondents reported that they would always consider using another metric (Kennedy & Lekshmi, 2023). While there is some support for changing exposure limits and how they are calculated, there is a lot of work to be done to standardize OELs across states and countries. The detail and specificity of these SBS also vary, some providing just enough information for third-party evaluation of reliability (Kennedy & Lekshmi, 2023). Standardized OELs, which take climate change into account, must be clear about how they were calculated.

The agroecological transition: Integrated Pest Management (IPM)

IPM is the practice of reducing chemical pesticide use. By using techniques like crop rotation, biopesticides, and other mechanical/physical means of pest control, we can reduce agricultural worker exposure to pesticides. This strategy directly benefits farmworker health as reduced exposure to harmful chemicals reduces both chronic and acute health risks.

Integrated Pest Management is a strategy to reduce pesticide use while still effectively controlling pests. IPM "provides a practical approach to managing insect pests with minimum risk to people, property, and the environment" while also allowing growers to improve crop quality and yields (Robusti et al., 2023). IPM centers around ecological theory and stakeholder involvement (Robusti et al., 2023). While IPM serves as a health intervention itself, it is also a key part of the agroecological transition.

IPM can improve farmworker health beyond reducing pesticide exposure. By reducing the need for pesticides, we can improve food security, mitigate climate change, and preserve biodiversity (Robusti et al., 2023). IPM allows farmers to tackle pest management with a holistic approach, thus improving crop quality and ensuring ecosystems are protected (Robusti et al., 2023). Many different organizations have applied the principles of IPM using a transdisciplinary and participatory approach, integrating worker health into their plans (Rustia et al., 2022).

IPM was created to help reduce pesticide usage. It "encompasses a combination of techniques, used in a coordinated way to keep insect populations below the levels at which they cause substantial damage" (Mawcha et al., 2024). Couple this with the definition of pesticides previously provided, and IPM serves as a direct intervention to reduce farmworker pesticide exposure. Farmers should use IPM strategies instead of, or in conjunction with, pesticides to reduce worker exposure.

One of the tenets of IPM is using biopesticides. Biopesticides "offer alternatives to synthetic chemical pesticides" and are generally less toxic to apply for agricultural workers (Schenk, 2020). By replacing synthetic chemical pesticides with biopesticides where possible, we can significantly reduce farmworker exposure to harmful chemicals.

CONCLUSION: CLIMATE-PROOF WORKPLACES

Climate change will likely worsen occupational heat stress among agricultural workers. Exposure to high heat increases heat-related occupational illnesses and injuries, which lead to negative health outcomes, decreased productivity, and increases in worker absenteeism as well as healthcare usage (Gellert et al., 2022). Climate change has been shown to impact both how much pesticides are used and volatilized, which may place workers at risk of exposure (Hutter et al., 2018). Chronic pesticide exposure has documented cytotoxic and genotoxic effects on agricultural workers, and action needs to be taken to protect these workers as well as enforce occupational exposure standards (Assis et al., 2020). Chronic exposure to pesticides for over ten years showed significant associations with cases of worker acute poisoning (Somboon et al., 2022).

Experts in toxicology should continue to communicate with climatologists and labor/environmental advocates. Framing climate change as an occupational safety issue will allow employers to recognize that they need to decrease the negative outcomes associated with exposed workers (Gellert et al., 2022). Occupational heat stress has received limited attention to its social dimensions and how workers can adapt to these changing conditions, especially in disadvantaged regions with low capacity to adapt to these exposures (Nunfam et al., 2018). There needs to be an evidence-based study to determine climate change adaptation in order to identify risk perception and coping among workers, managers, and other stakeholders involved (Moda et al., 2019). Restricting highly hazardous pesticides is one solution, but agricultural development should also move towards more organic farming.

REFERENCES

Amini, M., Ghavamabadi, L., Rangkooy, H., & Dehaghi, B. (2021). Climate change and its effects on farm workers. *Environmental Health Engineering and Management, 8*(3), 179–185. DOI: 10.34172/EHEM.2021.21

Assis, M., Barcella, R., Padilha, J., Pohl, H., & Krug, S. (2020). Health problems in agricultural workers occupationally exposed to pesticides. *Revista Brasileira de Medicina do Trabalho : Publicação Oficial da Associaação Nacional de Medicina do Trabalho-ANAMT, 18*(03), 352–363. DOI: 10.47626/1679-4435-2020-532 PMID: 33597986

Bagavathiannan, M., & Davis, A. (2018). An ecological perspective on managing weeds during the great selection for herbicide resistance. *Pest Management Science, 74*(10), 2277–2286. DOI: 10.1002/ps.4920 PMID: 29569406

Bakshi, A., Doren, A., Berens, A., Jacob, C., Omari, A., Foreman, A., & Chew, G. (2025). Sociodemographic and housing factors for flood and mold damage in the Greater New Orleans Area after Hurricane Ida. *Disaster Medicine and Public Health Preparedness, 19*, 19. DOI: 10.1017/dmp.2025.13 PMID: 40556425

Barn, P., Keefe, A., Slot, N., Jardine, K., Ziembicki, S., Telfer, J., Palmer, A. L., & Peters, C. (2020). Canada Should Move Toward Adopting Harmonized Evidence-Based OELs to Consistently and Adequately Protect Workers. *Annals of Work Exposures and Health, 65*(4), 367–372. DOI: 10.1093/annweh/wxaa110 PMID: 33336241

Braglia, M., Paco, F., Frosolini, M., Gabbrielli, R., & Marrazzini, L. (2023). A Novel Tool for Preliminary Risk Assessment of Climate Change on Workers' Health and Safety in Outdoor Worksites, 1514–1521.

Broster, J., Boutsalis, P., Gill, G., & Preston, C. (2022). The extent of herbicide resistance in *Lolium rigidum* Gaud. (annual ryegrass) across south-eastern Australia as determined from random surveys. *Australian Journal of Agricultural Research, 73*(11), 1308–1317.

Broster, J., Jalaludin, A., Widderick, M., Chambers, A., & Walsh, M. (2023). Herbicide Resistance in Summer Annual Weeds of Australia's Northern Grains Region. *Agronomy (Basel), 13*(7), 1862. DOI: 10.3390/agronomy13071862

Butler-Dawson, J., Krisher, L., Yoder, H., Dally, M., Sorensen, C., Johnson, R., Asensio, C., Cruz, A., Johnson, E. C., Carlton, E. J., Tenney, L., Asturias, E. J., & Newman, L. (2019). Evaluation of heat stress and cumulative incidence of acute kidney injury in sugarcane workers in Guatemala. *International Archives of Occupational and Environmental Health, 92*(7), 977–990. DOI: 10.1007/s00420-019-01426-3 PMID: 30997573

Cazé, A., Lacerda, A., Lüders, D., Conto, J., Marques, J., & Leroux, T. (2018). Perception of the Quality of Life of Tobacco Growers Exposed to Pesticides: Emphasis on Health, Hearing, and Working Conditions. *International Archives of Otorhinolaryngology*, 23(01), 050–059.

Crawford, M., Nordgren, T., & McCole, D. (2021). Every breath you take: Impacts of environmental dust exposure on intestinal barrier function–from the gut-lung axis to COVID-19. *American Journal of Physiology. Gastrointestinal and Liver Physiology, 320*(4), G586–G600. DOI: 10.1152/ajpgi.00423.2020 PMID: 33501887

Cruz, R., Martínez, P., Silveira, H., Cruz-Hipólito, H., Palma-Bautista, C., Vázquez-García, J., & Prado, R. (2019). Management of Glyphosate-Resistant Weeds in Mexican Citrus Groves: Chemical Alternatives and Economic Viability. *Plants, 8*(9), 325. DOI: 10.3390/plants8090325 PMID: 31487903

Deguine, J., Aubertot, J., Flor, R., Lescourret, F., Wyckhuys, K., & Ratnadass, A. (2021). Integrated pest management: Good intentions, hard realities. A review. *Agronomy for Sustainable Development, 41*(3), 38. DOI: 10.1007/s13593-021-00689-w

Ferrari, G., Leal, G., Souza, R., & Galdámez, E. (2023). Impact of climate change on occupational health and safety: A review of methodological approaches. *Work : A Journal of Prevention, Assessment, and Rehabilitation, 74*(2), 485–499. DOI: 10.3233/WOR-211303 PMID: 36314181

Gage, K., Krausz, R., & Walters, S. (2019). Emerging Challenges for Weed Management in Herbicide-Resistant Crops. *Agriculture, 9*(8), 180. DOI: 10.3390/agriculture9080180

Gándara, L., Jacoby, R., Laurent, F., Spatuzzi, M., Vlachopoulos, N., Borst, N., … & Crocker, J. (2024). Pervasive sublethal effects of agrochemicals as contributing factors to insect decline.

Gellert, G., Montgomery, S., Gellert, T., Gillett, J., Kerekes, A., & Hole, J. (2022). Employee Wellness in a Changing Climate: Environmental Heat Stress Driving Need for Targeted Health Promotion and Risk Reduction. *Journal of Ecophysiology and Occupational Health*, 60–66.

Habibi, P., Heydari, A., Dehghan, H., Moradi, A., & Moradi, G. (2024). Climate Change and Occupational Heat Strain Among Women Workers: A Systematic Review. *Indian Journal of Occupational and Environmental Medicine*, *28*(1), 4–17. DOI: 10.4103/ijoem.ijoem_320_21 PMID: 38783874

Hansen, A., Williams, S., Hanson-Easey, S., Varghese, B., Bi, P., Heyworth, J., Nitschke, M., Rowett, S., Sim, M. R., & Pisaniello, D. (2020). Using a Qualitative Phenomenological Approach to Inform the Etiology and Prevention of Occupational Heat-Related Injuries in Australia. *International Journal of Environmental Research and Public Health*, *17*(3), 846. DOI: 10.3390/ijerph17030846 PMID: 32013180

Heinzerling, A., Laws, R., Frederick, M., Jackson, R., Windham, G., Materna, B., & Harrison, R. (2020). Risk factors for occupational heat-related illness among California workers, 2000–2017. *American Journal of Industrial Medicine*, *63*(12), 1145–1154. DOI: 10.1002/ajim.23191 PMID: 33075156

Hicks, H., Comont, D., Coutts, S., Crook, L., Hull, R., Norris, K., Neve, P., Childs, D. Z., & Freckleton, R. (2018). The factors driving evolved herbicide resistance at a national scale. *Nature Ecology & Evolution*, *2*(3), 529–536. DOI: 10.1038/s41559-018-0470-1 PMID: 29434350

Hutter, H., Khan, A., Lemmerer, K., Wallner, P., Kundi, M., & Moshammer, H. (2018). Cytotoxic and Genotoxic Effects of Pesticide Exposure in Male Coffee Farmworkers of the Jarabacoa Region, Dominican Republic. *International Journal of Environmental Research and Public Health*, *15*(8), 1641. DOI: 10.3390/ijerph15081641 PMID: 30081446

Jo, H., Reddy, V., Parkinson, J., Boulton, D., & Tang, W. (2021). Model-Informed Pediatric Dose Selection for Dapagliflozin by Incorporating Developmental Changes. *CPT: Pharmacometrics & Systems Pharmacology*, *10*(2), 108–118. DOI: 10.1002/psp4.12577 PMID: 33439535

Kennedy, J., & Lekshmi, J. (2023). *Holistic Pest Management Strategies in Tropical Plant Species*. IntechOpen.

Khayat, M., Halwani, D., Hneiny, L., Alameddine, I., Haidar, M., & Habib, R. (2022). Impacts of Climate Change and Heat Stress on Farmworkers' Health: A Scoping Review. *Frontiers in Public Health*, *10*, 10. DOI: 10.3389/fpubh.2022.782811 PMID: 35211437

Lubis, A., & Priwiratama, H. (2025). Effectiveness of 356 g/l and 363 g/l glyphosate in controlling *Ageratum conyzoides, Panicum repens*, and *Cyperus kyllingia* in young mature oil palm areas. *IOP Conference Series. Earth and Environmental Science*, *1494*(1), 012006. DOI: 10.1088/1755-1315/1494/1/012006

Mawcha, K., Kinyanjui, G., Simiyu, S., Babalola, O., & Ndolo, D. (2024). *Biopesticides for Sustainable Agriculture: A Review of Their Role in Integrated Pest Management*. IntechOpen.

McClure, E., Ranapurwala, S., Nocera, M., & Richardson, D. (2024). Heat-related fatalities in North Carolina, 1999–2017. *American Journal of Industrial Medicine, 67*(6), 551–555. DOI: 10.1002/ajim.23587 PMID: 38624268

Mizelle, E., Larson, K., Bolin, L., & Kearney, G. (2022). Fluid Intake and Hydration Status Among North Carolina Farmworkers: A Mixed Methods Study. *Workplace Health & Safety, 70*(12), 532–541. DOI: 10.1177/21650799221117273 PMID: 36002982

Moda, H., Filho, W., & Minhas, A. (2019). Impacts of Climate Change on Outdoor Workers and Their Safety: Some Research Priorities. *International Journal of Environmental Research and Public Health, 16*(18), 3458. DOI: 10.3390/ijerph16183458 PMID: 31533360

Moyce, S., Armitage, T., Mitchell, D., & Schenker, M. (2019). Acute kidney injury and workload in a sample of California agricultural workers. *American Journal of Industrial Medicine, 63*(3), 258–268. DOI: 10.1002/ajim.23076 PMID: 31773783

Muramoto, J., Parr, D., Perez, J., & Wong, D. (2022). Integrated Soil Health Management for Plant Health and One Health: Lessons From Histories of Soil-borne Disease Management in California Strawberries and Arthropod Pest Management. *Frontiers in Sustainable Food Systems, 6*, 6. DOI: 10.3389/fsufs.2022.839648

Nunfam, V., Adusei-Asante, K., Etten, E., Oosthuizen, J., & Frimpong, K. (2018). Social impacts of occupational heat stress and adaptation strategies of workers: A narrative synthesis of the literature. *The Science of the Total Environment, 643*, 1542–1552. DOI: 10.1016/j.scitotenv.2018.06.255 PMID: 30189570

Odigie, A., Stufano, A., Schino, V., Zarea, A., Ndiana, L., Mrenoshki, D., Ugochukwu, I. C. I., Lovreglio, P., Greco, G., Pratelli, A., Camero, M., & Tempesta, M. (2024). West Nile Virus Infection in Occupational Settings—A Systematic Review. *Pathogens (Basel, Switzerland), 13*(2), 157. DOI: 10.3390/pathogens13020157 PMID: 38392895

Omassoli, J., Hill, N., Woods, D., Delves, S., Fallowfield, J., Brett, S., Wilson, D., Corbett, R. W., Allsopp, A. J., & Stacey, M. (2019). Variation in renal responses to exercise in the heat with progressive acclimatisation. *Journal of Science and Medicine in Sport, 22*(9), 1004–1009. DOI: 10.1016/j.jsams.2019.04.010 PMID: 31085130

Richards, C., Brumley, C., Graves, J., Parker, M., Khot, L., & Postma, J. (2025). Mapping Research Priorities for Climate Change Adaptation in Agriculture: A One Health Perspective. *Workplace Health & Safety*, *73*(9), 454–465. DOI: 10.1177/21650799251334147 PMID: 40359014

Richey, M., Golightly, Y., Marshall, S., Novicoff, W., Keil, A., Nocera, M., & Richardson, D. (2022). Trends in fatal occupational injuries in Latino/a workers relative to other groups, North Carolina 2000–2017. *American Journal of Industrial Medicine*, *65*(4), 242–247. DOI: 10.1002/ajim.23331 PMID: 35128690

Robusti, E., Androcioli, H., Hata, F., Júnior, D., & Júnior, A. (2023). Farmers' and rural extension workers' perceptions on the adoption of the integrated pest management. *Delos Desarrollo Local Sostenible*, *16*(46), 2294–2307. DOI: 10.55905/rdelosv16.n46-018

Rustia, D., Chiu, L., Lu, C., Wu, Y., Chen, S., Chung, J., Hsu, J.-C., & Lin, T. (2022). Towards intelligent and integrated pest management through an AIoT-based monitoring system. *Pest Management Science*, *78*(10), 4288–4302. DOI: 10.1002/ps.7048 PMID: 35716088

Schenk, L. (2020). A scoping survey of attitudes towards occupational exposure limits and REACH-derived no-effect levels for workers among chemical risk managers at Swedish workplaces. *International Journal of Occupational Medicine and Environmental Health*, *33*(5), 611–620. DOI: 10.13075/ijomeh.1896.01493 PMID: 32699425

Silvério, A., Martins, I., Nogueira, D., Mello, M., Loyola, E., & Graciano, M. (2020). Assessment of Primary Health Care for rural workers exposed to pesticides. *Revista de Saúde Pública*, *54*, 9. DOI: 10.11606/s1518-8787.2020054001455 PMID: 31967278

Somboon, S., Jirapongsuwan, A., Kalampakorn, S., & Tipyamongkolkul, M. (2022). Acute Pesticide Poisoning among Mixed-crop Agricultural Workers in Thailand. *Trends in Sciences*, *19*(3), 2154. DOI: 10.48048/tis.2022.2154

Stufano, A., Schino, V., Sacino, G., Ravallese, R., Ravallese, R., Benedictis, L., & Lovreglio, P. (2025). West Nile Virus Seroprevalence Among Outdoor Workers in Southern Italy: Unveiling Occupational Risks and Public Health Implications. *Viruses*, *17*(3), 310. DOI: 10.3390/v17030310 PMID: 40143241

Tigchelaar, M., Battisti, D., & Spector, J. (2020). Work adaptations are insufficient to address the growing heat risk for U.S. agricultural workers. *Environmental Research Letters*, *15*(9), 094035. DOI: 10.1088/1748-9326/ab86f4 PMID: 33133229

Vandenberg, L., Pierce, E., & Arsenault, R. (2025). Pesticides are an urgent challenge to global environmental health and planetary boundaries. *Frontiers in Toxicology*, *7*, 7. DOI: 10.3389/ftox.2025.1656297 PMID: 41114400

Varghese, B., Hansen, A., Bi, P., & Pisaniello, D. (2018). Are workers at risk of occupational injuries due to heat exposure? A comprehensive literature review. *Safety Science*, *110*, 380–392. DOI: 10.1016/j.ssci.2018.04.027

Werth, J., Thornby, D., Keenan, M., Hereward, J., & Chauhan, B. (2021). Effectiveness of glufosinate, dicamba, and clethodim on glyphosate-resistant and -susceptible populations of five key weeds in Australian cotton systems. *Weed Technology*, *35*(6), 967–973. DOI: 10.1017/wet.2021.59

Wiesner, L., Araya, S., & Barle, E. (2022). Identifying nonhazardous substances in pharmaceutical manufacturing and setting default health-based exposure limits (HBELs). *Journal of Applied Toxicology : JAT*, *42*(9), 1443–1457. DOI: 10.1002/jat.4323 PMID: 35315528

KEY TERMS AND DEFINITIONS

Agrochemicals: Chemical substances, such as pesticides and fertilizers, used in agriculture that can potentially affect ecosystems and worker health.

Chronic Kidney Disease of Unknown Etiology (CKDu): A serious kidney disease affecting agricultural workers in tropical regions, often linked to repeated dehydration and heat exposure.

Climate Migration: The movement of people driven by climate-related events, often pushing individuals into high-risk agricultural labor as their original lands become unsuitable.

Cutaneous Vasodilation: The widening of blood vessels at the skin's surface during heat stress, which increases blood flow and can promote the absorption of pesticides into the bloodstream.

Glyphosate: A widely used, broad-spectrum herbicide that has faced a crisis of growing weed resistance due to heavy reliance.

Heat Stress: A condition occurring when the body cannot cool itself sufficiently during work, leading to illnesses ranging from heat cramps to fatal heat stroke.

Integrated Pest Management (IPM): An ecological approach to managing pests that combines techniques like crop rotation and biopesticides to minimize reliance on harmful synthetic chemicals.

Occupational Exposure Limits (OELs): Regulatory standards for the maximum concentration of a hazardous substance a worker can be exposed to, which currently often fail to account for the synergistic effects of heat.

One Health: An integrated approach that recognizes the interconnectedness of human health, animal health, and the shared environment.

Piece-rate Work: A pay system where workers are compensated based on output rather than hours, often discouraging them from taking necessary safety breaks in extreme heat.

Chapter 4

Accounting-Based Occupational Health Indicators Under Climate Change: A Bibliometric Review

Ioannis Ch. Lampropoulos
http://orcid.org/0000-0002-8956-2178
University of Patras, Greece

Angeliki Tsameti
http://orcid.org/0000-0002-2741-2297
University of Thessaly, Greece

Eleftherios Aggelopoulos
http://orcid.org/0000-0002-4090-9417
University of Patras, Greece

Maria Kalogera
http://orcid.org/0000-0002-4875-6785
University of Patras, Greece

Foteini Malli
University of Thessaly, Greece

ABSTRACT

Climate change is fundamentally reshaping occupational health risks, yet standardized monitoring frameworks incorporating climate-induced risks remain underdeveloped.

DOI: 10.4018/979-8-3373-8470-2.ch004

This chapter addresses this gap through a bibliometric analysis of 1,314 Scopus articles using VOSviewer, and a proposed Performance Monitoring Framework (PMF) for occupational health under climate change. Co-occurrence analysis of 72 keywords confirmed a critical absence of indicator-based monitoring research. Co-authorship analysis revealed a fragmented global network, with the Global South significantly underrepresented despite disproportionately high climate risks. The PMF is organized around three pillars: an accounting foundation aligned with GRI 403, ESG, and IFRS S1/S2; climate risk metrics covering heat stress, air pollution, zoonoses, UV radiation, and mental health; and OHS performance indicators for early warning and outcomes. Aligned with One Health principles, the framework supports organizations and policymakers in building climate-resilient occupational health systems.

1. INTRODUCTION

Climate change is increasingly recognized as one of the most significant threats to public health worldwide, with immediate and serious consequences for workers' safety and health across all economic sectors. Rising ambient temperatures, increasing frequency and intensity of extreme weather events, expanding the geographic ranges of disease vectors, and deteriorating air quality are reshaping the occupational hazard landscape to an extent that existing monitoring systems are unable to adequately address (Humphrys, 2024; Schulte et al., 2016). Workers in outdoor areas and high-exposure sectors - such as agriculture, construction, aquaculture, and waste management - are disproportionately exposed to these emerging risks, without comprehensive and standardized protection frameworks (Rossi, Leone, & Moscato, 2025; Bhati & Sheth, 2026). These concerns are also reflected in major international policy and scientific reports, such as those of the ILO, WHO, and IPCC, which highlight the growing impact of climate change on worker health and the need for integrated monitoring and policy responses at the global level (ILO, 2019; WHO, 2021; IPCC, 2023).

The intersection of climate change and occupational health has attracted increasing scientific interest in recent decades. Heat stress is emerging as the most well-documented climate risk for workers, with systematic reviews reporting symptom prevalence ranging from 64% to 90.3% in outdoor workers and a clear association between rising temperatures and increased risk of occupational accidents (Rossi et al., 2025; Ioannou et al., 2022; Tawatsupa et al., 2013). However, the broader spectrum of climate risks - including air pollution, UV radiation, zoonoses, and mental health impacts - remains significantly underrepresented in standardized monitoring and reporting frameworks (Ferrari et al., 2025; Turner et al., 2025).

Despite the existence of heat stress assessment tools-such as the Wet Bulb Globe Temperature (WBGT), ISO 7933, and the Universal Thermal Climate Index (UTCI)-existing monitoring systems remain fragmented and focused primarily on single risk categories (Mousavi, Dehghan, & Yazdanirad, 2026; Utzet et al., 2023). At the organizational level, sustainability reporting frameworks-such as GRI 403, ESG criteria, and Integrated Reporting-do not adequately incorporate climate-induced occupational health risks, with disclosures characterized by low quality and over-reporting of narratives over measurable indicators (Mariappanadar et al., 2022; Tsalis, Stylianou, & Nikolaou, 2018). Furthermore, the One Health approach - which recognizes the interdependence of human, animal, and environmental health - remains underutilized in occupational surveillance settings, despite its clear relevance to the risks faced by workers in high-exposure environments (Zinsstag et al., 2018; Cavalli, 2025).

The problem addressed in this chapter is that climate-related occupational health risks are increasingly recognized in the literature, but they remain insufficiently integrated into organizational monitoring, accounting, and sustainability reporting systems. Existing approaches tend to examine climate hazards, occupational health outcomes, and reporting frameworks separately, resulting in fragmented evidence and limited guidance for developing measurable, accounting-based occupational health indicators under climate change. This fragmentation creates a clear research gap that can be analytically derived from the literature: while climate-related occupational hazards are extensively documented, and monitoring tools and sustainability reporting frameworks are developed in parallel, these streams remain conceptually and operationally disconnected. As a result, there is limited integration of measurable occupational health indicators into accounting-based performance and reporting systems under climate change. Addressing this gap requires a structured synthesis of the literature and the development of a conceptual framework that connects occupational health monitoring with accounting-based performance measurement and One Health principles.

In this context, this chapter has a twofold objective. First, it systematically maps the literature at the intersection of occupational health, climate change, and monitoring indicators through a bibliometric analysis of 1,314 articles from the Scopus database, using the VOSviewer tool for keyword co-occurrence and co-authorship analyses. Second, based on the findings of this analysis and the theoretical overview, it proposes a comprehensive Performance Monitoring Framework (PMF) that integrates accounting principles, occupational health indicators, and climate parameters, aligned with One Health principles and international sustainability standards.

Based on this objective, the chapter is guided by the following research questions:

RQ1: How has the literature at the intersection of occupational health, climate change, and monitoring indicators evolved over time?
RQ2: What are the dominant thematic clusters and research patterns in this interdisciplinary field?
RQ3: How can accounting principles, occupational health indicators, climate parameters, and One Health principles be integrated into a conceptual Performance Monitoring Framework for climate-related occupational health risks?

2. BACKGROUND AND CONCEPTUAL FOUNDATIONS

2.1 Climate-related occupational risks

Climate change is emerging as one of the most significant threats to occupational health and safety worldwide, exacerbating existing risks and creating new challenges for both outdoor and indoor workers (Humphrys, 2024; Ferrari et al., 2025). The literature identifies several major categories of climate-related occupational risks, including heat stress, air pollution, vector-borne diseases, UV radiation, and mental health impacts, with outdoor workers, particularly in agriculture and construction, as well as workers in the informal sector, showing disproportionately higher exposure and vulnerability (Rossi, Leone, & Moscato, 2025; Bhati & Sheth, 2026).

Heat stress is the most well-documented risk. A systematic review using PRISMA guidelines found a high prevalence of symptoms in outdoor workers, ranging from 64% to 90.3% (Rossi et al., 2025), while a meta-analysis of 38 field studies in 2,409 workers from 21 countries demonstrated a significant increase in core temperature, heart rate and a dramatic decrease in manual work capacity (Ioannou et al., 2022). A systematic review in tropical and subtropical countries confirms a positive relationship between climate change and heat stress, with dehydration, fatigue, confusion, and reduced brain function as the main mechanisms (Habibi et al., 2021). In the US, 359 workplace heatstroke deaths were recorded between 2000 and 2010, with higher rates in agriculture and construction (Gubernot, Anderson, & Hunting, 2015), while a large national cohort in Thailand demonstrated that heat stress doubles the risk of occupational accidents (Tawatsupa et al., 2013). Furthermore, the HEAT-SHIELD program documents that approximately 15% of workers chronically exposed to heat stress develop acute kidney injury, while the risk of depression, anxiety, and suicidality is highlighted, particularly for farmers due to drought-related pressures (Morris et al., 2021).

In addition to heat stress, climate change affects the transport and dispersion of atmospheric pollutants, increasing workers' exposure to air pollution, while higher temperatures alter the body's absorption of chemicals, making workers more vulnerable to gases and pesticides (Ferrari et al., 2025). For indoor workers, risks include exposure to volatile organic compounds, suspended particles, and biological aerosols, with flooding significantly worsening indoor air quality through increased microbial growth (Fromme, 2023; Rose & Birchall, 2025). At the same time, climate change is expanding the geographic range of disease vectors, increasing the risk of outdoor workers being exposed to diseases such as Lyme disease, West Nile virus, malaria, and dengue fever, although available epidemiological data on the working population remains limited (Nicholas et al., 2020; Morris et al., 2021).

Regarding ultraviolet radiation, a study on forest workers in South Africa demonstrated significant deficiencies in sun protection (Rother et al., 2019), while measurements on lawnmowers in North Carolina showed that UV levels exceed safety limits during midday and summer months, highlighting an increased risk of melanoma and eye diseases (Beck, Balanay, & Johnson, 2018). Finally, mental health impacts are an emerging but consistent area of research: a review of 10 studies found that occupational heat exposure is associated with anxiety, depression, and burnout in the vast majority of studies (Florez-Acevedo, Blancas, & Spector, 2025), while indirect impacts through extreme weather events include PTSD and increased risk of suicide, particularly in vulnerable low-income groups (Berry, Bowen, & Kjellström, 2010).

These findings also highlight the unequal distribution of climate-related occupational risks, as workers in vulnerable socio-economic conditions are disproportionately exposed to hazardous environments, raising important issues related to labour vulnerability and climate justice.

Overall, the literature reveals a clear concentration of research on heat stress as the dominant climate-related occupational risk, supported by extensive empirical and epidemiological evidence. In contrast, other risk dimensions - such as air pollution, vector-borne diseases, ultraviolet radiation, and mental health impacts - remain comparatively fragmented and less systematically integrated into occupational health monitoring frameworks. This imbalance suggests that while individual climate-related hazards are well documented, the literature lacks a comprehensive and integrated perspective that captures the multi-dimensional nature of occupational risks under climate change.

2.2 Indicators, Metrics and Monitoring Standards

The development of reliable indicators and metrics for monitoring climate risks in the workplace is a critical prerequisite for the protection of occupational health. The most widely used tool for assessing heat stress is the Wet Bulb Globe Temperature (WBGT), which combines ambient temperature, wet bulb temperature, and globe temperature, and is the basis for international standards such as ISO 7243 (Meng et al., 2021; Biabani et al., 2021). The reliability of WBGT has been confirmed in a number of work environments, while methods for its estimation from remote and low-cost data sources have been developed, facilitating its wider application (Carter et al., 2020). The recent development of a WBGT calculation application demonstrated strong correlation with traditional measurements, highlighting the potential for affordable and reliable alternative assessment tools (Angol, 2024). However, a critical review of WBGT in different climate zones found that the index tends to underestimate physiological stress in dry environments and overestimate it in humid coastal environments, suggesting its combined use with physiological monitoring and local calibration (Mousavi, Dehghan, & Yazdanirad, 2026).

In addition to the WBGT, important assessment tools include the Predicted Heat Strain (PHS) model of ISO 7933, which predicts core temperature and water loss by incorporating parameters such as metabolic rate and thermal insulation of clothing, although it has limitations in extreme working conditions (Lazaro & Momayez, 2020; Nunfam et al., 2021). The Universal Thermal Climate Index (UTCI) has proven particularly useful for mapping heat strain in outdoor agricultural workers, with spatiotemporal analysis in Slovenia showing a significant increase in days of severe heat strain during the summer season (Črepinšek, Žnidaršič, & Pogačar, 2023). An emerging approach is the use of wearable sensors to monitor physiological parameters in real time, such as heart rate, skin temperature, and sweating (Sharma, Suri, & Kant, 2022; Wibowo et al., 2024; Bustos et al., 2021). At the European level, a quantitative Job Exposure Matrix (JEM) for thermal exposure at work, harmonized with ISO 7243, 8996, and 9920 standards, has recently been developed, which provides estimates by occupation and geographical area for the period 1970–2024 (de Crom et al., 2025).

Regarding occupational health performance indicators, a systematic review of 153 preliminary indicators showed that high-quality indicators mainly focus on risk control systems and communication, while indicators related to employee competence, awareness, and participation remain relatively underrepresented (Naylor et al., 2025). To support management decision-making, the Sustainable Risk Awareness Index (SRA Index) is an innovative KPI that aggregates hazard observations, near misses, and injury data into a single indicator, contributing to the systematic reporting of work environment deviations (Franzon & Rose, 2021).

Despite these tools, gaps in the existing monitoring system remain significant. There is no available transnational monitoring system in Europe that incorporates indicators on the impacts of the climate crisis on occupational health, while the available information and indicators on the threats of climate change to workers' health remain limited and fragmented (Utzet et al., 2023). A rapid review highlights significant gaps in the evaluation and implementation of occupational health and safety plans under extreme climate events, with few studies focusing on worker-specific interventions (Tsakonas et al., 2024). A gap analysis in the construction industry showed that WBGT measurements are often performed without taking into account rest breaks, while water consumption remains below recommended levels (Idris et al., 2025). Globally, the Lancet Countdown is developing climate change and health indicators, but these are not specific to the working population (Di Napoli et al., 2022), while the WHO is developing the REACH 2035 agenda for standardized indicators of occupational exposure to climate hazards, with available data still limited (Turner et al., 2025). Epidemiological analysis of three decades of heat illness data in the US highlights critical gaps in understanding the long-term effects of heat exposure, highlighting the need to evaluate the effectiveness of interventions such as acclimatization programs and real-time heat monitoring (Mullins-Jaime et al., 2026). However, these tools remain largely disconnected from organizational performance measurement and accounting-based reporting systems.

2.3 The One Health Approach and Occupational Health Monitoring

The One Health approach recognizes the interdependence of human, animal, and environmental health, proposing a holistic, interdisciplinary framework for addressing complex public health challenges, such as zoonoses, antimicrobial resistance, and climate change (Utomo, 2026; Danasekaran, 2024; Ciotti & Bernardini, 2025; López-Casas et al., 2024). Climate change is intensifying these interdependencies by expanding the geographic range of disease vectors, altering disease dynamics in wildlife and livestock species, and increasing environmental pollution, creating new risks for workers working outdoors and in high-exposure sectors (Zinsstag et al., 2018). Specifically, the One Health approach points out that the combined cost of outbreaks of zoonotic pathogens can be significantly reduced if they are detected early in vectors or animals, before they are transmitted to the human population (Zinsstag et al., 2018).

In the occupational context, workplaces that involve contact with animals or animal environments are critical points of zoonotic transmission. Workers in animal markets, slaughterhouses, and food processing facilities are exposed to increased risks of pathogen transmission, with open markets being recognized as high-risk environments for the emergence and spread of zoonoses and antimicrobial resistance (Hasan et al., 2025; Naguib et al., 2021). A systematic review and meta-analysis demonstrated a significant seroprevalence of tick-borne diseases in occupational settings in Europe, with higher rates in foresters (26%), farmers (18%), and veterinarians (8%), highlighting the need for targeted surveillance and preventive measures (Morea et al., 2025). Workers who handle wild birds, such as veterinarians, biologists, and environmental managers, face risks of exposure to avian influenza viruses, particularly in a climatic context influenced by migratory routes from infected areas (Sánchez et al., 2020).

The environmental dimension of One Health highlights the importance of joint surveillance systems for pollutants and pathogens that simultaneously affect animals and workers. Animals act as early warning sentinels for environmental infections, as outbreaks in animals often precede human cases, offering the possibility of early intervention (Haddaji, 2022). The One Health approach is also applied to the monitoring of microplastics and other environmental pollutants that simultaneously affect ecosystems, animals, and workers, proposing the use of animals as sentinels to identify environmental risks before they affect the human population (Prata et al., 2021).

One of the few sectors with an explicit integration of One Health into occupational health is aquaculture, where risk assessment combines worker health, fish health, and environmental changes into a single framework (Cavalli, 2025). This approach is a model for extending the One Health logic to other high-exposure occupational sectors. Overall, integrating One Health principles into occupational health monitoring under climate change is a necessary prerequisite for developing comprehensive, effective early warning and prevention systems that account for interactions among humans, animals, and the environment in the workplace (Naguib et al., 2021).

2.4 Accounting Approach and Performance Monitoring Frameworks

Integrating occupational health and safety into accounting and management systems is a necessary prerequisite for effectively monitoring and assessing the risks faced by workers. Management accounting supports safety-related decision-making by providing quantitative and qualitative information on the performance, costs, and benefits of specific actions, with costing methods - such as cost per injury and total accident costs - dominating practice (Tappura et al., 2015). Despite their potential, innovative occupational health and safety accounting tools face significant technical, organisational, and cognitive barriers to their integration into business processes, limiting their systematic implementation (Battaglia et al., 2020). Recognizing climate change as an emerging occupational health risk requires the integration of risk identification, monitoring, and evaluation systems into operational management frameworks, linking health impacts - such as lost productivity and increased absenteeism - to organizational performance (Schulte et al., 2016).

In the area of performance monitoring frameworks, ISO 45001 offers a structured occupational health and safety management system aligned with performance monitoring and risk control (Darabont et al., 2018). However, the traditional focus on injury indicators provides an incomplete picture, particularly for chronic or climate-induced conditions, while companies tend to focus on their overall management approach but lag behind in reporting quantitative and qualitative data beyond conventional injury indicators (Evangelinos et al., 2018). In high-exposure sectors such as construction, the need for climate-aware performance indicators covering heat stress, air pollution, and extreme weather events is recognized as an urgent priority (Baek & Choi, 2025).

Within the sustainability standards framework, the Global Reporting Initiative (GRI) is the most widely used international sustainability reporting standard (Mariappanadar et al., 2022). GRI 403 on Occupational Health and Safety provides specific disclosure requirements for OHS management systems, hazard identification, incident reporting, and injury and illness measurements (Bowers, 2021). However, a study of 250 sustainability reports found that companies mainly disclose legal compliance information, rarely translate their commitments into measurable results, and generally characterize OHS disclosures as low (Mariappanadar et al., 2022). At the same time, research on multinational construction companies confirms that OHS reporting is typically limited to a narrow set of indicators, such as accident rates and training hours (Mavroulidis et al., 2022).

In the ESG framework, occupational health is included in the social (S) pillar, with disclosures based mainly on narrative descriptions rather than quantitative indicators, while the climate dimensions of ESG are well developed compared to occupational safety indicators that remain less standardized (Martí, Bastida, & Marimon, 2025). Research on companies in the ASEAN region demonstrated a positive impact of occupational safety on corporate performance, highlighting the strategic importance of OHS indicators in the ESG framework (Khunkaew, Wichianrak, & Suttipun, 2023). Integrated Reporting combines financial and non-financial disclosures, including human, social, and natural capital, but provides less detailed requirements for occupational health indicators than GRI 403 (Tiron-Tudor, Martin, & Fărca , 2019). Finally, SDG 8 on Decent Work shows the highest link with ESRS benchmarks, underlining the importance of integrating occupational health indicators into corporate sustainability frameworks (Lukács, Molnár, & Tóth, 2026). Overall, despite the existence of these frameworks, the quality of disclosures remains low due to the limited standardization of accounting methodologies for non-financial OHS information, with the underreporting of chronic conditions—including heat disorders—remaining a significant gap (Tsalis, Stylianou, & Nikolaou, 2018).

This chapter therefore positions accounting-based performance measurement as a key area of incremental contribution, by linking climate-related occupational health risks with measurable indicators and organizational reporting frameworks. In this sense, the proposed PMF seeks to extend existing OHS and sustainability reporting approaches by connecting worker health outcomes, climate risk parameters, and accounting-based disclosure requirements within a single conceptual structure.

3. METHODOLOGY

This chapter employs a bibliometric analysis of articles indexed in the Scopus database. The main objective of the bibliometric analysis was to map and synthesize research at the intersection of occupational health and safety, climate change, and concepts related to indicators and monitoring systems.

The Scopus database was chosen due to its broad and interdisciplinary coverage, as it indexes a large number of journals, conference proceedings and books from different scientific fields (including natural sciences, medicine and health, engineering, social sciences and business administration) (Sabando-Vera et al., 2022; Pandey, Lucey, & Kumar, 2023; Singh et al., 2024; Cherappurath et al., 2025; Dhaigude & Kamath, 2025). Furthermore, the Scopus database is recognized for its suitability for bibliometric and epistemometric analyses, providing completeness in the metadata references it provides, due to the support it offers in terms of co-occurrence analyses, co-citations, as well as the mapping of scientific networks (Sabando-Vera et al., 2022; Pandey et al., 2023; Singh et al., 2024; Borromeo et al., 2025; Morelli et al., 2026).

Nevertheless, the use of a single database represents a methodological limitation, as relevant publications indexed exclusively in other databases, such as Web of Science or PubMed, may not have been captured. Future research could therefore validate and extend the findings by incorporating additional bibliographic databases.

Literature retrieval was performed using three complementary Boolean search strings in the Scopus database, applied to the TITLE-ABS-KEY fields. These search strings were designed to capture the intersection of: (a) occupational health and safety, (b) climate change and environmental stressors, and (c) indicators, monitoring frameworks, and evaluation concepts.

To enhance transparency and reproducibility, the full Boolean search strings used in the data retrieval process are reported below:

Search String 1: TITLE-ABS-KEY ("occupational health" OR "workplace health" OR "occupational safety") AND TITLE-ABS-KEY ("climate change" OR "global warming" OR "extreme heat" OR "heat stress") AND TITLE-ABS-KEY ("indicator*" OR "performance indicator*" OR "risk indicator*" OR "monitoring" OR "assessment" OR "metrics").

Search String 2: TITLE-ABS-KEY("occupational health" OR "occupational risk*" OR "worker health") AND TITLE-ABS-KEY("climate change" OR "heat stress" OR "extreme heat" OR "air pollution") AND TITLE-ABS-KEY("indicator*" OR "performance indicator*" OR "KPI*" OR "monitoring framework" OR "risk assessment" OR "early warning").

Search String 3: TITLE-ABS-KEY("occupational health" AND "climate change") AND TITLE-ABS-KEY("risk*" OR "indicator*" OR "exposure*" OR "surveillance" OR "assessment" OR "monitor*").

The three search strings returned 444, 787, and 405 records, respectively (N = 1,636). The results were combined into a single dataset. After removing duplicate records (n = 322), a final dataset of 1,314 unique articles was obtained.

In order to ensure the completeness of the dataset and to capture the broad conceptual landscape of the research field, no restrictive inclusion or exclusion criteria were applied beyond duplicate removal. This is consistent with exploratory bibliometric mapping studies aiming to capture the full conceptual landscape of the field rather than evaluate the methodological quality of individual studies. All papers, after being extracted from the Scopus database, were imported into Zotero, which is widely used for bibliographic data management, citation organization, and identification and merging of duplicate records (Scotti et al., 2025; Vendl et al., 2022; Chen, Chan, & Chan, 2025; Wister, Fyffe, & O'Dea, 2021; Antoniou & Otter, 2024). Furthermore, Zotero was chosen due to the convenience it provides to users in preparing clean and unified datasets, which can be exported to various file formats (e.g., RIS, CSV) for use in software suitable for bibliometric analyses (Scotti et al., 2025; Vendl et al., 2022; Petrescu-Mag et al., 2025).

Following duplicate removal, no additional manual checking of titles or abstracts was applied, as the goal of the analysis was to capture a broad representation of the available studies as they were extracted from the database. Therefore, for our analysis, all 1,314 records were considered suitable for further analysis. However, it is important to note that no formal screening process (e.g., PRISMA-based filtering) or quality assessment of the included studies was conducted. This may introduce heterogeneity and potential inclusion of less relevant records, and therefore represents a limitation of the study.

The aggregate file, after being exported from Zotero in a format compatible with the analysis tool, was imported into the VOSviewer tool, which is a specialized software through which the construction and visualization of bibliometric networks is carried out, emphasizing distance mapping and the analysis of co-occurrence, co-citation and bibliometric coupling (Wang & Zhang, 2025; Kriswardhana et al., 2025; Li, Xu, & Chen, 2024; Pan et al., 2025). The tool, due to its user-friendliness and its ability to group related keywords, is widely used in bibliometric analyses by presenting maps: (a) network, (b) overlay and (c) density maps, facilitating the interpretation of knowledge structures and research patterns (Yao, Qiu, & Liang, 2026; Hariry et al., 2025; Sepúlveda-Oviedo et al., 2023; Haghighi et al., 2025). Furthermore, the VOSviewer tool ensures consistency in normalization and grouping of results, limiting interoperability issues that arise between different tools (Borromeo, Manaloto, & Antonio, 2025).

VOSviewer was used to perform keyword co-occurrence analysis and visualize thematic networks based on terms included in titles, abstracts, and author keywords, with the aim of identifying dominant thematic areas (Bai et al., 2026; Huang et al., 2024; Kumar et al., 2024; Lampropoulos et al., 2026; Leal Filho et al., 2024; Liu et al., 2023; Saheb, Sidaoui, & Schmarzo, 2024; Wen et al., 2026; Yang et al., 2025). Before creating the networks, general and demographic terms that do not provide substantial thematic content to the subject under study were removed, in order to enhance the interpretability of the thematic clusters. The selected keyword occurrence threshold and preprocessing decisions were intended to balance coverage and interpretability. Alternative parameter settings, such as different minimum occurrence thresholds or keyword selection criteria, could lead to variations in the number of included keywords and the composition of thematic clusters. Therefore, the lack of a formal sensitivity or robustness analysis is acknowledged as a methodological limitation, and future research could test the stability of the identified patterns under alternative bibliometric configurations. The thematic areas that emerge and are subsequently analyzed form the basis for the interpretation of the thematic patterns and for the development of the proposed Performance Monitoring Framework. The overall process of data collection, preparation, and analysis is summarized in Figure 1.

Figure 1. Flow diagram of the data collection and preparation process (adapted from PRISMA structure; no formal PRISMA screening applied).

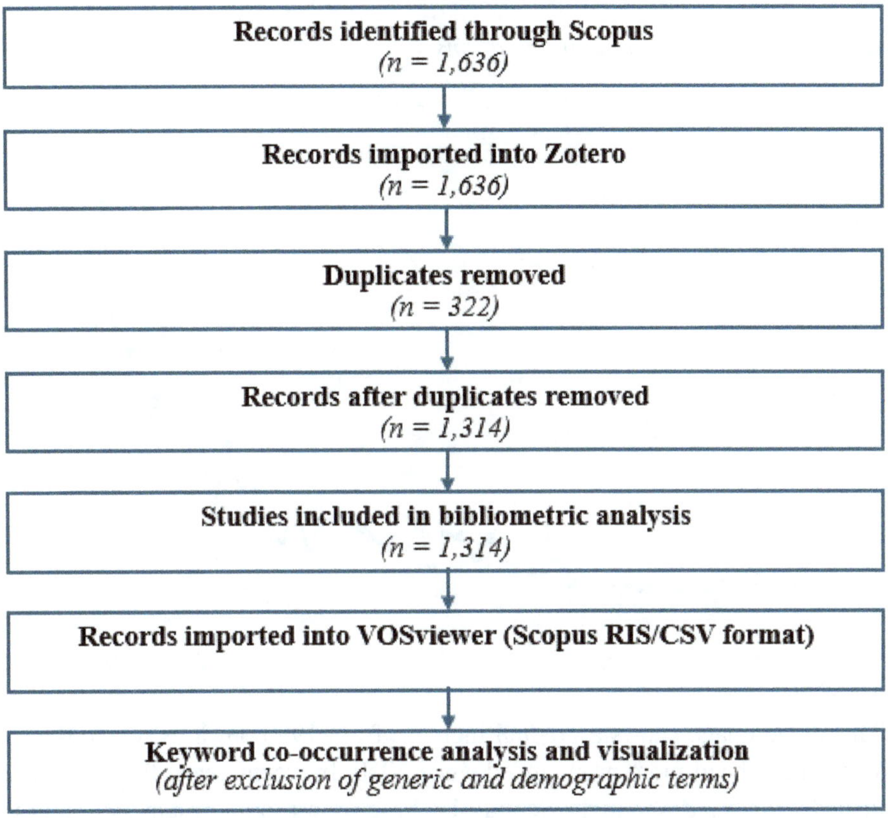

4. RESULTS

This section presents the results of the analysis of the 1,314 final articles, starting with the descriptive characteristics of the sample and then proceeding to the VOSviewer analysis.

4.1 Descriptive analysis of the results

The time distribution of the 1,314 articles extends from 1973 to the entries in the Scopus database, which existed in the system in January 2026. Figure 2 depicts the course of publications examining occupational health and safety, climate change, in combination with indicators and monitoring systems.

Figure 2. Time distribution of publications.

The above image shows a gradual increase from 1990 onwards, while the upward trend has been more pronounced since 2010. In recent years, the increase in published articles indexed in the Scopus database shows an interest in the scientific community in the issue of climate change and its impacts on occupational health and safety, seeking indicators and monitoring frameworks.

The analysis of the thematic areas showed that Medicine, Environmental Sciences, Engineering, and Social Sciences record the largest number of publications. The distribution of the thematic areas in the Scopus database is assigned to more than one area depending on the authors' affiliations and the journals, resulting in the number of entries per thematic area exceeding the number of unique articles under study.

The geographical distribution of publications shows a clear dominance of English-speaking countries. Figure 3 illustrates this distribution, with the USA leading the way with 480 publications, followed by the United Kingdom (161) and Australia (160). China ranks fourth with 147 publications, confirming its growing contribution to the field. Greece is notable with 52 publications, reflecting the increased awareness of the Greek scientific community due to the country's particular exposure to the impacts of climate change. Overall, research activity appears to be concentrated in economically developed countries, while countries in the Global South - which face disproportionately greater climate risks for workers - show limited research output. This imbalance highlights a significant geographical gap in the literature and suggests potential disparities in research capacity, funding availability, and data accessibility across regions.

Figure 3. Geographic distribution of publications by country. The bold color scale reflects the number of publications (dark brown = high production).

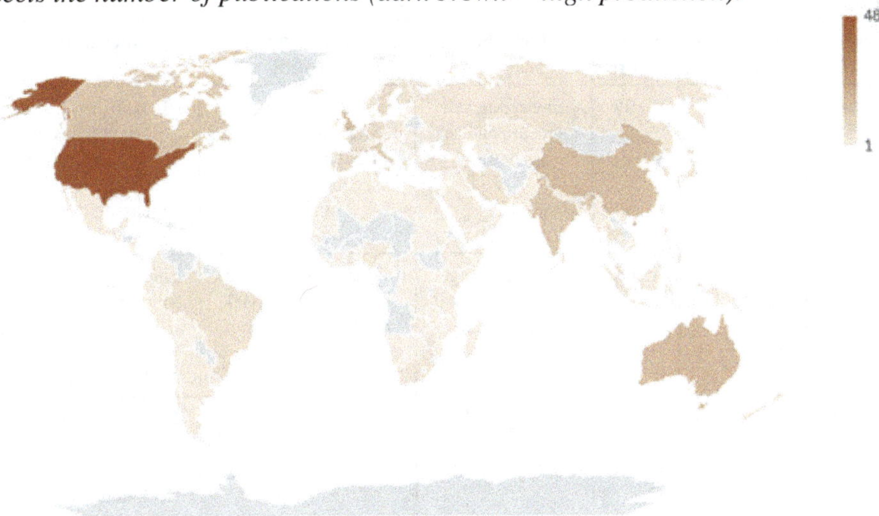

4.2 Bibliometric analysis using VOSviewer.

4.2.1 Co-occurrence analysis and thematic clusters

To visualize and interpret the thematic structures of the literature, keyword co-occurrence analysis was applied using the VOSviewer tool. Out of a total of 1,314 articles, 10,821 unique keywords were identified. The minimum occurrence threshold was set at 10 occurrences per keyword, resulting in 586 keywords meeting this criterion. However, in order to enhance the interpretability of the thematic clusters and to avoid the distortion of the results by terms without substantial thematic content, demographic descriptors (e.g. human, male, female, adult, middle aged), methodological terms (e.g. controlled study, cross-sectional study, review, article) and overly general terms (e.g. health, exposure, diseases) were excluded. After applying these exclusions, the final analysis was based on 72 keywords of high thematic relevance, which were organized by the VOSviewer algorithm into

3 thematic clusters as illustrated below. The main characteristics of the identified clusters are summarized in Table 1. The table provides a structured overview of the size, thematic focus, and interpretation of each cluster, complementing the qualitative description that follows.

Table 1. Summary of thematic clusters identified in the keyword co-occurrence analysis.

Cluster	Number of Keywords	Share (%)	Main Theme	Representative Keywords	Interpretation
Cluster 1	36	50.0%	Environmental exposure and risk assessment	air pollution, occupational exposure, particulate matter, risk assessment, industrial hygiene	Established research field focusing on chemical and atmospheric exposure and associated worker health risks
Cluster 2	35	48.6%	Climate change and occupational heat stress	climate change, heat stress, occupational health, temperature, workplace, agricultural worker	Emerging research area linking climate change with occupational health impacts, especially in high-exposure sectors
Cluster 3	1	1.4%	Industrial dimension	industry	Limited representation in the co-occurrence network, suggesting weak integration of the industrial dimension in the literature

The first and largest cluster includes 36 keywords and mainly concerns the relationship between environmental pollution and risk assessment for workers' health. The central nodes of the cluster are the terms air pollution, risk assessment, occupational exposure and occupational risks, around which concepts such as air pollutant, air pollutants, particulate matter, indoor air pollution, volatile organic compounds, benzene, dust, inhalation exposure, ambient air, air sampling, environmental exposure, environmental monitoring, health hazard, health risk, health risks, health risk assessment, health impact, cancer risk, industrial hygiene, public health and risk management are grouped. This cluster reflects a mature and established research field that focuses on the measurement and assessment of chemical and atmospheric exposure in the workplace, with an emphasis on the risks from inhalation of pollutants and toxic substances.

The second cluster includes 35 keywords and constitutes the core of the subject under study. The central nodes are the terms climate change, occupational health and heat stress, around which concepts such as heat, heat injury, heat stress disorders, high temperature, hot temperature, thermal exposure, thermal stress, environmental temperature, humidity, temperature, workplace, work environment, occupational safety, occupational hazard, occupational disease, occupational diseases, occupational accident, agricultural worker, construction industry, productivity, prevalence, epidemiology, risk factor, risk factors, prevention and control, mortality, safety and questionnaire are gathered. This cluster reflects the emerging research trend that directly links climate change to the risks of heat stress in the workplace, especially for workers in outdoor areas and in high-exposure sectors, such as agriculture and construction.

The third cluster includes only the term industry, which appears as an independent node with low connectivity to the two main clusters. This finding suggests that the industrial dimension of occupational health under climate change remains relatively underdeveloped in the existing body of research and warrants further exploration.

4.2.1.1 Network Visualization

Figure 4 presents the Network Visualization of the keyword co-occurrence analysis. It depicts the connections between the 72 keywords, with the size of each node corresponding to the frequency of occurrence and the thickness of the connecting lines reflecting the strength of co-occurrence. The two main clusters are clearly distinguishable, with Cluster 1 (red) occupying the left part of the map and Cluster 2 (green) occupying the right. The occupational health and occupational exposure nodes act as bridges between the two clusters, confirming their central role in the interconnection of the two thematic areas.

Figure 4. Network Visualization of keyword co-occurrence analysis (VOSviewer).

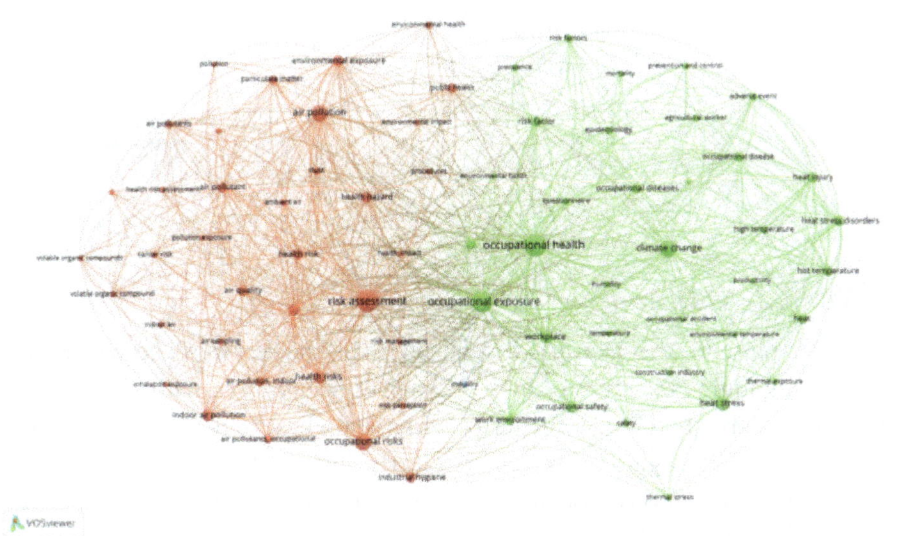

4.2.1.2 Overlay Visualization

Figure 5 presents the Overlay Visualization of the keyword co-occurrence analysis. It adds a temporal dimension to the analysis, with each node colored based on the average publication year of the articles in which the corresponding keyword appears. Dark blue-purple tones correspond to older publications (approximately 2012–2015), while light yellow-green tones correspond to more recent ones (2019–2021 and beyond). It is observed that the terms in Cluster 1, such as air pollution, risk assessment, and indoor air pollution, show an older average publication year, suggesting that this is a mature research field. In contrast, the terms in Cluster 2, such as climate change, heat stress, heat injury, and agricultural workers, show a younger average year, confirming that the connection between climate change and occupational health is an emerging and rapidly developing research field.

Figure 5. Overlay Visualization of keyword co-occurrence analysis (VOSviewer).

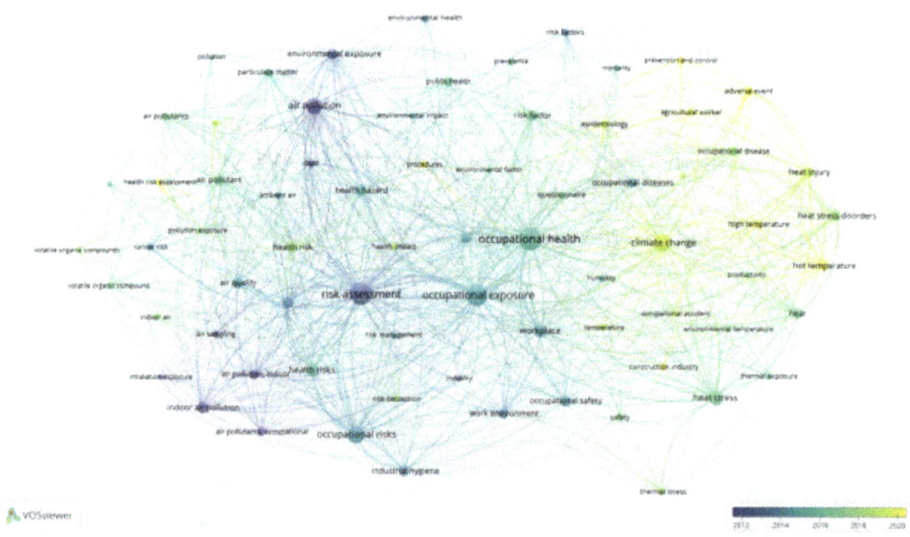

4.2.1.3 Density Visualization

Figure 6 presents the Density Visualization of the keyword co-occurrence analysis. It highlights the areas of high research concentration, with the brightest (yellow) areas corresponding to the terms with the highest frequency and strength of connections. Five high-density centers clearly stand out: occupational health, occupational exposure, risk assessment, air pollution, and heat stress and climate change. These centers indicate that research interest is mainly concentrated around the risk assessment of exposure to pollutants and heat stress under climate change, while the limited concentration around indicators and monitoring frameworks suggests a relative underrepresentation and fragmentation of research on integrated monitoring indicators and performance frameworks in this field.

Figure 6. Density Visualization of keyword co-occurrence analysis (VOSviewer).

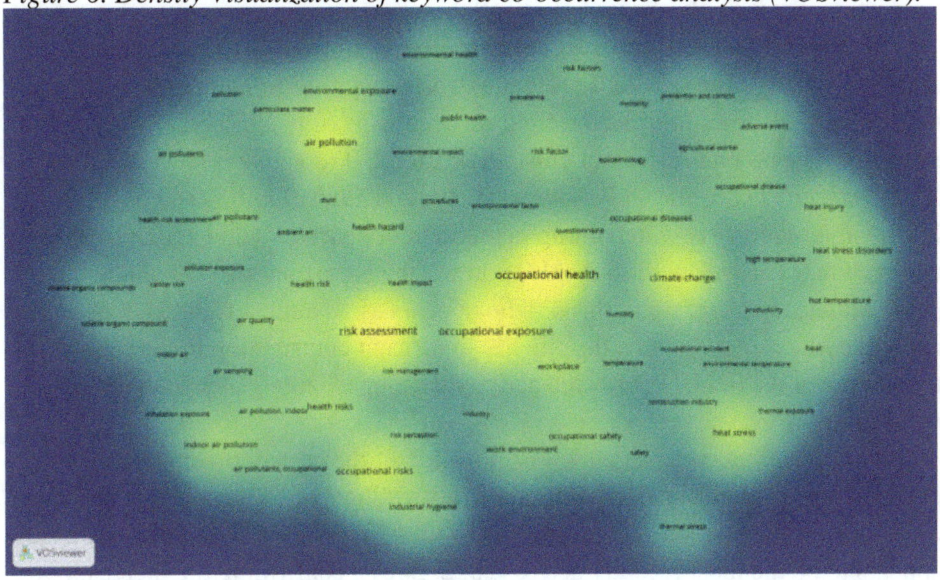

Overall, the keyword co-occurrence analysis reveals two main themes in the literature: an established focus on chemical and atmospheric exposure and risk assessment, and an emerging link between climate change and heat stress. These findings provide empirical support for the research gap identified in the Introduction, highlighting the limited integration of monitoring indicators and accounting-based performance frameworks within the existing literature.

4.2.2 Co-authorship Analysis

The co-authorship analysis was performed using the full counting method, setting a threshold of 5 publications per author. Of the 4,740 authors identified in the total of 1,314 articles, 62 met this criterion and were included in the analysis, organized into 7 thematic clusters. The relatively high number of clusters indicates a fragmented research network with limited international collaboration between individual research groups.

The author with the highest number of publications is Kjellström T. (31 publications), the main architect of the Hothaps program for assessing thermal exposure at work. This is followed by Flouris A.D. (16 publications, total link strength: 40) and Bi P. (15 publications), with Flouris showing the highest total link strength, suggesting that he is the most central collaboration node in the network.

The largest cluster (Cluster 1, 18 researchers) consists mainly of Chinese researchers (Wang, Zhang, Liu, Li, Huang), reflecting the strong output of the Chinese scientific community in the field of occupational health and environmental exposure. The second important cluster (Cluster 2, 9 researchers) includes Flouris A.D., Havenith G., Kenny G.P., Ioannou L.G., and Tetzlaff E.J., who constitute the leading international research group in the field of heat stress and work physiology under climate change conditions. Also noteworthy is the presence of Schulte P.A. (8 publications, total link strength: 1), who appears as an almost isolated node in the network, which suggests that despite his significant contribution to the theoretical foundation of the field, his research activity developed largely independently of the other groups.

Figure 7 presents the Network Visualization of the co-authorship analysis. The overlay map highlights a clear temporal stratification: the Chinese group shows an older average publication year (around 2016–2018), while the Flouris/Kenny group appears to be the most recently active (2022–2024), confirming the increasing research activity in the field of thermal stress in recent years.

Figure 7. Network Visualization of Co-authorship Analysis via VOSviewer.

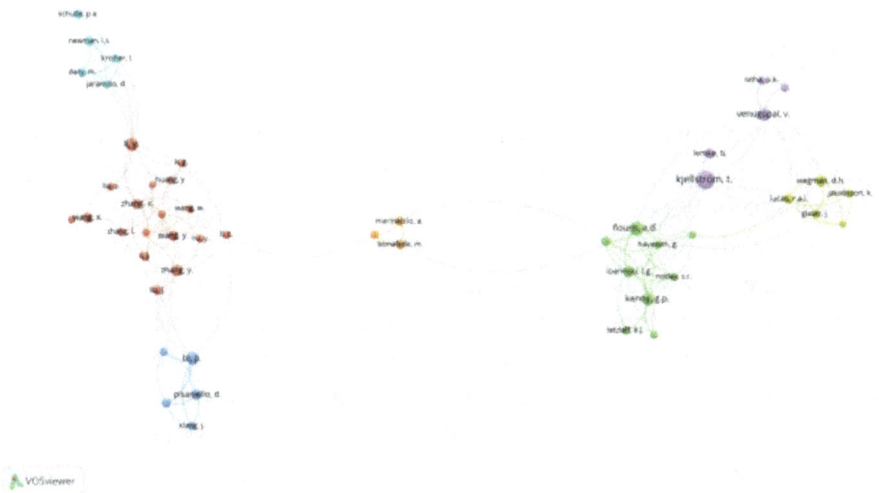

The most productive authors and their collaboration characteristics are summarized in Table 2.

Table 2. Authors with the most publications in the research field under study (Co-authorship Analysis, VOSviewer).

Author	Publications	Total Link Strength	Cluster	Thematic Focus
Kjellström, T.	31	20	5	Hothaps, work heat exposure
Flouris, A.D.	16	40	2	Heat stress, work physiology
Bi, P.	15	33	3	Occupational health, temperature
Kenny, G.P.	13	36	2	Thermoregulation, climate change
Venugopal, V.	13	22	5	Thermal exposure, tropical countries
Li, Y.	13	15	1	Environmental health, China
Zhang, Y.	11	10	1	Air pollution, China
Wegman, D.H.	11	24	4	Occupational health, occupational hazards
Ioannou, L.G.	11	32	2	Worker heat stress
Lemke, B.	10	11	5	Hothaps, work productivity

This fragmentation further reflects the interdisciplinary and still evolving nature of the research field.

5. CONCLUSION

This chapter sought to map the relationship among climate change, occupational health, and accounting performance-monitoring frameworks, combining a theoretical overview with a systematic bibliometric analysis. The findings highlight both the extent of the existing body of research and the significant gaps that remain unaddressed.

The literature review showed that climate change is exacerbating a wide range of occupational hazards, with heat stress being the most well-documented hazard, with symptom prevalence ranging from 64% to 90.3% in outdoor workers (Rossi et al., 2025), while risks from air pollution, UV radiation, zoonoses and mental health impacts are emerging (Humphrys, 2024; Ferrari et al., 2025). Despite the existence of heat stress assessment tools - such as WBGT, ISO 7933, and UTCI - existing monitoring systems remain fragmented, focused mainly on heat and insufficient to holistically capture climate hazards in the workplace (Turner et al., 2025; Utzet et al., 2023). The integration of the One Health approach is emerging as a necessary prerequisite for the development of integrated monitoring systems that account for interactions among human, animal, and environmental health in the workplace (Zinsstag et al., 2018; Cavalli, 2025). Furthermore, the analysis of existing sustainability accounting and frameworks - GRI 403, ESG, Integrated Reporting - has shown that climate-induced risks to workers remain underrepresented, with disclosures characterized by low quality and over-reporting of narratives over measurable indicators (Mariappanadar et al., 2022; Tsalis et al., 2018).

The bibliometric analysis of 1,314 articles from the Scopus database empirically confirmed these gaps. The temporal distribution of publications reveals a steadily increasing interest since 2010, with a sharp acceleration in recent years. Keyword co-occurrence analysis organized the literature into two main thematic clusters: an established line of research focusing on chemical and atmospheric exposure and risk assessment, and an emerging line directly linking climate change to heat stress and occupational health. A critical finding is the limited research concentration around terms such as "indicator", "monitoring framework", and "performance indicator", suggesting a relative and structurally embedded underrepresentation and fragmentation of research on standardized, accounting-based monitoring indicators in this field. This interpretation is consistent with the scope of bibliometric analysis, which identifies patterns of representation rather than the complete absence of research topics. Co-authorship analysis revealed a fragmented research network with limited international collaboration, while the geographical distribution of publications reveals a concentration of research activity in economically developed countries, with countries in the Global South - which face disproportionately higher climate risks - being significantly underrepresented.

Overall, the findings support the need to develop a Performance Monitoring Framework (PMF) that integrates accounting principles, occupational health indicators, and climate parameters into a coherent organizational monitoring framework, aligned with One Health principles. Such a framework can support organizations, policymakers, and researchers in developing climate-resilient occupational health systems, bridging the gap between existing scientific knowledge and practical implementation at the organizational level.

From a theoretical perspective, this study contributes by bridging the literature on occupational health, climate change, and accounting-based performance measurement, offering a structured conceptual basis for the development of integrated monitoring systems grounded in One Health principles.

Despite its conceptual nature, the proposed framework provides a structured foundation for future empirical validation and the development of standardized, accounting-based occupational health indicators under climate change.

6. SUGGESTIONS FOR FURTHER RESEARCH AND PROPOSED PERFORMANCE MONITORING FRAMEWORK (PMF)

6.1 Suggestions for Further Research

The findings of this chapter highlight a number of research gaps that need further investigation. First, research on climate risk monitoring indicators for workers remains focused almost exclusively on heat stress, while risks such as air pollution, UV radiation, zoonoses, and mental health impacts remain underrepresented in standard monitoring frameworks (Turner et al., 2025; Utzet et al., 2023). Second, the development of integrated monitoring systems that integrate data from the environment, animals, and the human population based on One Health principles is required, particularly in high-exposure work environments such as agriculture, construction, and aquaculture (Zinsstag et al., 2018; Cavalli, 2025). Third, the evaluation of the effectiveness of interventions - such as acclimatization programs, wearable sensor systems, and rest protocols - remains inadequate and requires systematic study (Idris et al., 2025; Tsakonas et al., 2024). Fourth, the geographical gap highlighted in the bibliometric analysis - with countries in the Global South being underrepresented despite disproportionately high climate risks for their workers - constitutes a critical research priority.

This imbalance may reflect structural inequalities in research capacity, funding availability, and data accessibility, which limit the production of locally grounded evidence in high-risk regions.

In addition, future research should further explore the implications of climate justice and labour vulnerability, particularly for informal sector workers, who remain underrepresented in the current literature despite facing high exposure to climate-related occupational risks. Finally, the integration of climate-induced occupational health risks into ESG, GRI, and IFRS frameworks remains inadequate and requires the development of standardized, measurable indicators that link worker health to organizational financial performance (Nyakuwanika & Panicker, 2025; Grajales-Gaviria et al., 2023).

While the present analysis provides a descriptive mapping of the research landscape, future studies could apply more advanced bibliometric or statistical techniques to further validate and extend the findings. Furthermore, future research could adopt a sector-specific approach to examine how climate-related occupational risks and monitoring indicators vary across different industries.

Future research should also examine the long-term health effects of repeated or chronic exposure to climate-related occupational hazards, including cumulative heat exposure, air pollution, and psychosocial stressors. In addition, future studies could complement bibliometric mapping with qualitative, quantitative, or case-based approaches to provide deeper empirical insights into the implementation of climate-related occupational health monitoring frameworks.

6.2 Proposed Performance Monitoring Framework (PMF)

Based on the findings of the literature review and bibliometric analysis, and drawing on established frameworks such as ISO 45001, GRI 403, ESG reporting structures, and IFRS sustainability standards, a comprehensive Performance Monitoring Framework (PMF) is proposed for measuring and reporting climate-induced occupational health risks at the organizational level. The PMF is conceptual in nature and is intended to provide a structured synthesis of existing knowledge, serving as a foundation for future empirical development. The framework is organized around three interrelated pillars: Accounting Foundation, One Health Integration, and Climate Risk Measurement.

The first pillar – Accounting Foundation – includes three functions: risk assessment, valuation, and disclosure. Accounting can support organizational adaptation to climate change by assessing vulnerability and adaptive capacity, valuing the costs and benefits of adaptation actions, and disclosing the risks associated with climate change impacts (Linnenluecke, Birt, & Griffiths, 2015). Integrating climate impacts into the balance sheet and income statement—through costs of lost productivity, absenteeism, and adaptation costs—is a necessary prerequisite for linking employee health to organizational financial performance (Stechemesser, Bergmann, & Günther, 2016). The development of accounting mechanisms for measuring environmental impacts aligned with IFRS S1 and S2 standards provides the appropriate framework for standardized reporting of climate risks (Grajales-Gaviria et al., 2023).

The proposed PMF is graphically presented in Figure 8, illustrating the three interconnected pillars, the tools and technologies supporting implementation, and the sectors of application.

Figure 8. Performance Monitoring Framework (PMF): Accounting-Based Occupational Health Indicators under Climate Change

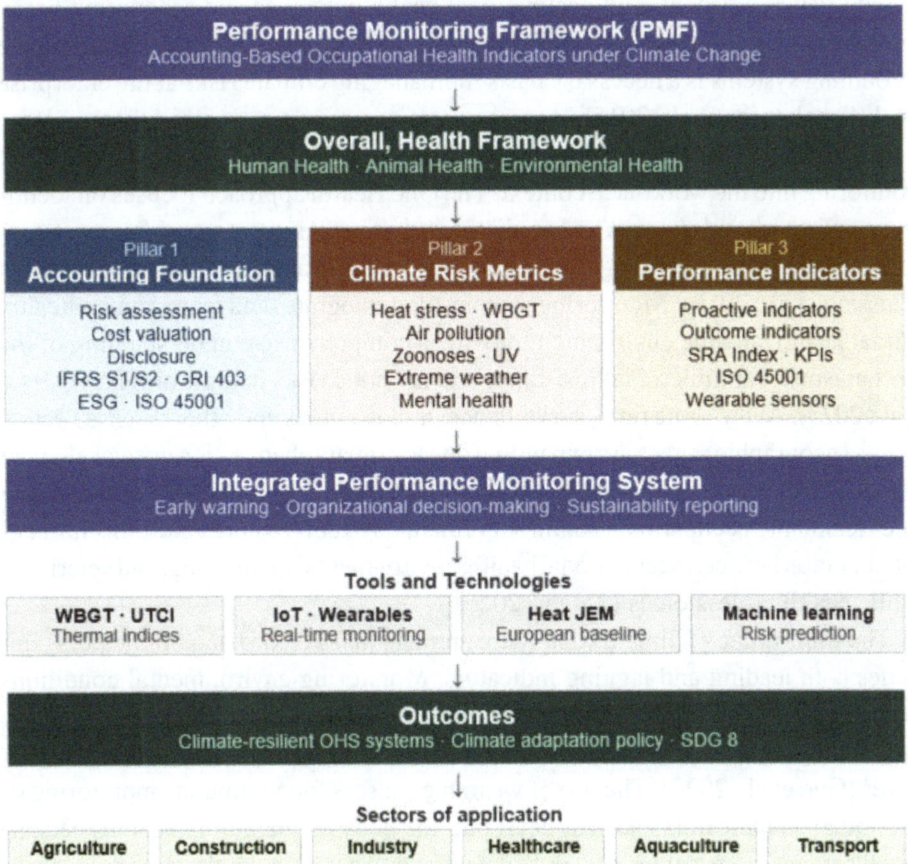

Figure 8 illustrates the proposed Performance Monitoring Framework (PMF) as an integrated system structured around three interconnected pillars: Accounting Foundation, One Health Integration, and Climate Risk Measurement. The framework highlights the interactions between these pillars, showing how accounting functions (risk assessment, valuation, and disclosure) are linked with multidisciplinary data inputs from human, animal, and environmental health, as well as with climate-related exposure indicators.

The framework also incorporates enabling tools and technologies, such as environmental monitoring systems, wearable sensors, and data analytics, which support the collection and processing of relevant information. These inputs are translated into measurable performance indicators that can be integrated into organizational decision-making and sustainability reporting frameworks (e.g., ESG, GRI, IFRS). In this way, the PMF operates as a dynamic and interconnected system that links climate risk assessment with occupational health outcomes and accounting-based performance measurement. Furthermore, the development of carbon and climate accounting systems is a necessary basis for managing climate risks at the enterprise level (Schaltegger et al., 2015).

The second pillar - One Health Integration - translates the principles of holistic monitoring into the workplace context. The One Health approach focuses on identifying environmental changes and predicting subsequent impacts, enabling upstream interventions that avoid negative impacts on agriculture, animals, and humans (Zinsstag et al., 2018). Monitoring systems that integrate data from human health, animal health, and the environment provide a comprehensive understanding of the mechanisms of antimicrobial resistance and zoonotic risks in the workplace (Desai et al., 2025). At the same time, surveillance of mosquitoes and other disease vectors - whose geographic ranges are expanding due to climate change - is a critical element of a comprehensive early warning system (Christina & Claire, 2025). Integrating One Health into occupational health surveillance systems requires multidisciplinary collaboration between occupational health, environmental monitoring, and veterinary public health professionals (Zargar, 2026).

The third pillar - Climate Risk Measurement and Performance Indicators - includes both leading and lagging indicators. Monitoring environmental conditions - temperature, humidity, radiation, and air quality - is a necessary basis for assessing and identifying occupational hazards, especially during summer periods and heat waves (Guo et al., 2023). The use of wearable sensors for continuous monitoring of physiological indicators - such as heart rate and body temperature - enables personalized preventive strategies and early detection of heat illness risks (Friedrich et al., 2025). A national study in Australia showed that the use of geographic information technologies, machine learning, and IoT to monitor local heat stress conditions is an emerging and promising direction (Fatima et al., 2025). The development of leading indicators - risk control systems, employee competence and participation - combined with outcome indicators - sickness rates, compensation, absence from work - is the foundation for reliable OHS performance measurement (Naylor et al., 2025; Pawłowska, 2015). Finally, alignment with ESG and environmental accounting frameworks - GRI 403, IFRS S1/S2, TCFD - is a necessary condition for the comparability and reliability of climate-related occupational health risk disclosures (Nyakuwanika & Panicker, 2025; Tsakonas et al., 2025).

From a practical perspective, the proposed PMF may support organizations in identifying climate-related occupational health risks, selecting measurable indicators, and linking worker health outcomes with ESG and sustainability reporting processes. For policymakers, the framework may provide an indicative structure for developing more integrated occupational health monitoring systems under climate change. However, these implications should be interpreted with caution, as the PMF is conceptual in nature and requires future empirical validation and sector-specific adaptation.

DECLARATION OF GENERATIVE AI TECHNOLOGIES IN THE WRITING PROCESS

During the preparation of this work, the authors used ChatGPT to check spelling, grammar, and language issues to improve readability. After using this tool/service, the authors reviewed and edited the content as needed and took full responsibility for the publication's content.

ACKNOWLEDGEMENTS

This paper has been financed by the funding programme "MEDICUS" of the University of Patras.

REFERENCES

Angol, B. (2024). Comparison between WBGT app prototype and WBGT monitor to assess heat stress risk in groundskeeping in an eastern North Carolina university setting.

Antoniou, E. E., & Otter, R. (2024). Phthalate exposure and neurotoxicity in children: A systematic review and meta-analysis. *International Journal of Public Health, 69*, 1606802. DOI: 10.3389/ijph.2024.1606802 PMID: 38590582

Baek, S., & Choi, E. H. (2025). Climate change health risks and workplace protective strategies for construction workers. *Scientific Reports, 16*(1), 3423. DOI: 10.1038/s41598-025-33471-z PMID: 41422155

Bai, J., Johar, S. B., Mohammad, N. B., & Xia, J. (2026). Traditional village protection and sustainable development: Knowledge graph construction and hotspot evolution. *Sustainable Futures : An Applied Journal of Technology, Environment and Society, 11*, 101577. DOI: 10.1016/j.sftr.2025.101577

Battaglia, M., Pasetti, E., Bianchi, L., & Frey, M. (2020). Innovation in managing sustainability: A tentative integration of accounting for employee health and safety. *Sinergie Italian Journal of Management, 38*(1), 73–101.

Beck, N., Balanay, J. A. G., & Johnson, T. (2018). Assessment of occupational exposure to heat stress and solar ultraviolet radiation among groundskeepers in an eastern North Carolina university setting. *Journal of Occupational and Environmental Hygiene, 15*(2), 105–116. DOI: 10.1080/15459624.2017.1392530 PMID: 29090983

Berry, H. L., Bowen, K., & Kjellstrom, T. (2010). Climate change and mental health: A causal pathways framework. *International Journal of Public Health, 55*(2), 123–132. DOI: 10.1007/s00038-009-0112-0 PMID: 20033251

Bhati, N., & Sheth, A. (2026). Occupational heat stress and adaptation among outdoor workers: A narrative review of global evidence and policy responses. *International Journal of Biometeorology, 70*(1), 19. DOI: 10.1007/s00484-025-03101-4 PMID: 41501481

Biabani, A., Falahati, M., Alimohammadi, I., Zokaei, M., Jalilian, H., Dehghani, A., & Majareh, M. N. (2021). Validation of heat pressure assessment (HPA) method and WBGT index in Iranian South Oil Company. *Journal of Health and Safety at Work*, *11*(1), 106–116.

Borromeo, A. S., Manaloto, A. M., Vicedo, R. V., Antonio, R., Cura, J. D., & Wider, W. (2025). The rise of virtual care in nursing practice: A global bibliometric review. *Journal of Nursing Regulation*, *16*(3), 141–149. DOI: 10.1016/j.jnr.2025.08.013

Bowers, B. J. (2021). Sustainability reports based on GRI 403 occupational safety and health 2018 guidelines.

Bustos, D., Guedes, J. C., Baptista, J. S., Vaz, M. P., Costa, J. T., & Fernandes, R. J. (2021). Applicability of physiological monitoring systems within occupational groups: A systematic review. *Sensors (Basel)*, *21*(21), 7249. DOI: 10.3390/s21217249 PMID: 34770556

Carter, A. W., Zaitchik, B. F., Gohlke, J. M., Wang, S., & Richardson, M. B. (2020). Methods for estimating wet bulb globe temperature from remote and low-cost data: A comparative study in Central Alabama. *GeoHealth*, *4*(5), e2019GH000231.

Cavalli, L. S. (2025). Incorporating occupational health and safety into One Health approaches to aquaculture. *Journal of Agromedicine*, *30*(2), 214–220. DOI: 10.1080/1059924X.2024.2446256 PMID: 39720893

Chen, K. Y., Chan, H. C., & Chan, C. M. (2025). Are dexamethasone implants a safe and effective breakthrough in uveitis treatment? A systematic review and meta-analysis. *Steroids*, *223*, 109675. DOI: 10.1016/j.steroids.2025.109675 PMID: 40829733

Cherappurath, N., Thoompenthodi, M. A., Elayaraja, M., Pelasseri, M., Perumbalath, S., Muriyedath, A., & Parambath, S. P. (2025). Anterior Cruciate Ligament Injury in Football (Soccer) Players: A Bibliometric Review of Global Sports Orthopaedic Literature. *Journal of Orthopaedic Reports*, •••, 100834. DOI: 10.1016/j.jorep.2025.100834

Christina, A. L., & Claire, V. M. (2025). A general population KAP survey: Supporting the design of interventions and tiger mosquito control efforts in response to arbovirus emergence in non-endemic areas. *One Health*, *21*, 101247. DOI: 10.1016/j.onehlt.2025.101247 PMID: 41292613

Ciotti, M., & Bernardini, S. (2025). One health and laboratory medicine: Challenges and perspectives for antimicrobial resistance, emerging infections, and food/ water security. *Critical Reviews in Clinical Laboratory Sciences*, •••, 1–13. DOI: 10.1080/10408363.2025.2597189 PMID: 41395681

Črepinšek, Z., Žnidaršič, Z., & Pogačar, T. (2023). Spatio-temporal analysis of the universal thermal climate index (UTCI) for the summertime in the period 2000–2021 in Slovenia: The implication of heat stress for agricultural workers. *Agronomy (Basel), 13*(2), 331. DOI: 10.3390/agronomy13020331

Danasekaran, R. (2024). One health: A holistic approach to tackling global health issues. *Indian Journal of Community Medicine : Official Publication of Indian Association of Preventive & Social Medicine, 49*(2), 260–263. DOI: 10.4103/ijcm.ijcm_521_23 PMID: 38665439

Darabont, D. C., Bejinariu, C., Ionita, I., Bernevig-Sava, M. A., Baciu, C., & Baciu, E. R. (2018). CONSIDERATIONS ON IMPROVING OCCUPATIONAL HEALTH AND SAFETY PERFORMANCE IN COMPANIES USING ISO 45001 STANDARD. [EEMJ]. *Environmental Engineering and Management Journal, 17*(11).

de Crom, T. O., Scholten, B., Traini, E., van der Sanden, K., Kingma, B., Pekel, F., & Pronk, A. (2025). Exposure to heat at work: Development of a quantitative European job exposure matrix (heat JEM). *Scandinavian Journal of Work, Environment & Health, 52*(1), 7. PMID: 40782101

Desai, S., Patel, V., Nagare, S. D., Patil, R., & More, M. J. (2025). Microbial genomics and infectious disease forecasting. In *Host Adaptation, Virulence, and Evolution* (pp. 193-223). Academic Press. DOI: 10.1016/B978-0-443-31554-1.00012-1

Dhaigude, A. S., & Kamath, G. B. (2025). Mapping responsible artificial intelligence in business and management: Trends, influence, and emerging research directions. *Journal of Open Innovation, 11*(4), 100640. DOI: 10.1016/j.joitmc.2025.100640

Di Napoli, C., McGushin, A., Romanello, M., Ayeb-Karlsson, S., Cai, W., Chambers, J., Dasgupta, S., Escobar, L. E., Kelman, I., Kjellstrom, T., Kniveton, D., Liu, Y., Liu, Z., Lowe, R., Martinez-Urtaza, J., McMichael, C., Moradi-Lakeh, M., Murray, K. A., Rabbaniha, M., & Robinson, E. J. (2022). Tracking the impacts of climate change on human health via indicators: Lessons from the Lancet Countdown. *BMC Public Health, 22*(1), 663. DOI: 10.1186/s12889-022-13055-6 PMID: 35387618

Evangelinos, K., Fotiadis, S., Skouloudis, A., Khan, N., Konstandakopoulou, F., Nikolaou, I., & Lundy, S. (2018). Occupational health and safety disclosures in sustainability reports: An overview of trends among corporate leaders. *Corporate Social Responsibility and Environmental Management, 25*(5), 961–970. DOI: 10.1002/csr.1512

Fatima, S. H., Giles, L. C., Rothmore, P., Varghese, B. M., & Bi, P. (2025). Heatwaves and occupational injuries and illnesses risk varied at localised spatial scale: A national study in Australia. *Safety Science, 181*, 106684. DOI: 10.1016/j.ssci.2024.106684

Ferrari, G. N., Leal, G. C. L., Ossani, P. C., & Galdamez, E. V. C. (2025). Investigation of the usage of machine learning to explore the impacts of climate change on occupational health: A systematic review and research agenda. *Frontiers in Public Health, 13*, 1578558. DOI: 10.3389/fpubh.2025.1578558 PMID: 40589827

Florez-Acevedo, S., Blancas, M. T., & Spector, J. T. (2025). Occupational heat exposure & mental health outcomes: A review and framework incorporating social determinants of health to guide future research. *Current Environmental Health Reports, 12*(1), 15. DOI: 10.1007/s40572-025-00479-6 PMID: 40014259

Franzon, H., & Rose, L. M. (2021, May). The SRA Index (Sustainable Risk Awareness Index): A new KPI for Management Support. In *Congress of the International Ergonomics Association* (pp. 457–466). Springer International Publishing. DOI: 10.1007/978-3-030-74608-7_57

Friedrich, J., Schick, T. S., Mess, F., & Blaschke, S. (2025). Wearable device-based interventions in heat-exposed outdoor workers-a scoping review and an explanatory intervention model. *BMC Public Health, 25*(1), 2893. DOI: 10.1186/s12889-025-24262-2 PMID: 40846922

Gaviria, D. A. G., Martínez, E. M. C., Polo, O. C. C., & Arcila, J. O. S. (2023). Accounting mechanism to measure the environmental impact of industries in the Aburrá Valley and its application of the general disclosure requirements related to financial information on sustainability (IFRS S1) and climate-related disclosures (IFRS S2). *Revista de Gestao Social e Ambiental, 17*(5), 1–16. DOI: 10.24857/rgsa.v17n5-013

Gubernot, D. M., Anderson, G. B., & Hunting, K. L. (2015). Characterizing occupational heat-related mortality in the United States, 2000–2010: An analysis using the census of fatal occupational injuries database. *American Journal of Industrial Medicine, 58*(2), 203–211. DOI: 10.1002/ajim.22381 PMID: 25603942

Guo, Y., Zhang, Y., Yu, P., Wu, Y., Wen, B., Yang, Z., & Li, S. (2023). Strategies to reduce the health impacts of heat exposure. In *Heat Exposure and Human Health in the Context of Climate Change* (pp. 293–322). Elsevier. DOI: 10.1016/B978-0-12-819080-7.00009-4

Habibi, P., Moradi, G., Dehghan, H., Moradi, A., & Heydari, A. (2021). The impacts of climate change on occupational heat strain in outdoor workers: A systematic review. *Urban Climate*, *36*, 100770. DOI: 10.1016/j.uclim.2021.100770

Haddaji, N. (2022). Environmental contaminants and antibiotic resistance as a One Health threat. In *One Health* (pp. 231–252). Academic Press. DOI: 10.1016/B978-0-12-822794-7.00010-1

Haghighi, A., Cheraghi, M., Pocachard, J., Botta-Genoulaz, V., Jocelyn, S., & Pourzarei, H. (2025). A comprehensive review and bibliometric analysis on collaborative robotics for industry: Safety emerging as a core focus. *Frontiers in Robotics and AI*, *12*, 1605682. DOI: 10.3389/frobt.2025.1605682 PMID: 41019776

Hariry, A., Mardawati, E., Iswanto, A., Karliati, T., Zaini, L., & Adly, M. (2025). Nanocellulose-Based Adhesives for Sustainable Wood-Polymer Composites: Recent Advancement and Future Perspective. *Journal of Renewable Materials*, *13*(4), 773–798. DOI: 10.32604/jrm.2025.058359

Hasan, M. A. E., Islam, M. S., Fahim, N. A. I., Antor, M. T. H., Salam, S., Jahan, R., Masud, R. I., Bakhtiyar, Z., Jany, D. A., Rana, M. L., Punom, S. A., Rahman, A. M. M. T., & Rahman, M. T. (2025). Impact of open markets on zoonotic threats and antimicrobial resistance: A one health concern. *One Health*, *21*, 101228. DOI: 10.1016/j.onehlt.2025.101228 PMID: 41127627

Huang, Y., Gong, D., Dang, K., Zhu, L., Guo, J., Yang, W., & Wang, J. (2024). The applications of anterior segment optical coherence tomography in glaucoma: A 20-year bibliometric analysis. *PeerJ*, *12*, e18611. DOI: 10.7717/peerj.18611 PMID: 39619196

Humphrys, E. (2024). Inertia in transformed times: Work health and safety amid climate change. *The Journal of Industrial Relations*, *66*(5), 685–702. DOI: 10.1177/00221856241295497

Idris, M. A., Mena, K. D., Markham, C., & Perkison, W. B. (2025). Heat Stress Prevention and Management in the Construction Industry: A Gap Analysis. *Annals of Global Health*, *91*(1), 75. DOI: 10.5334/aogh.4833 PMID: 41179855

ILO. I. (2019). Working on a warmer planet: the impact of heat stress on labour productivity and decent work. *Geneva: International Labour Organization*.

Ioannou, L. G., Foster, J., Morris, N. B., Piil, J. F., Havenith, G., Mekjavic, I. B., Kenny, G. P., Nybo, L., & Flouris, A. D. (2022). Occupational heat strain in outdoor workers: A comprehensive review and meta-analysis. *Temperature*, *9*(1), 67–102. DOI: 10.1080/23328940.2022.2030634 PMID: 35655665

IPCC. 2023: Climate Change 2023: Synthesis Report. Contribution of Working Groups I, II and III to the Sixth Assessment Report of the Intergovernmental Panel on Climate Change [Core Writing Team, H. Lee and J. Romero (eds.)]. IPCC, Geneva, Switzerland, 184 pp., DOI: 10.59327/IPCC/AR6-9789291691647

Khunkaew, R., Wichianrak, J., & Suttipun, M. (2023). Sustainability reporting, gender diversity, firm value and corporate performance in ASEAN region. *Cogent Business & Management*, *10*(1), 2200608. DOI: 10.1080/23311975.2023.2200608

Kriswardhana, W., Toaza, B., Esztergar-Kiss, D., & Duleba, S. (2025). Analytic hierarchy process in transportation decision-making: A two-staged review on the themes and trends of two decades. *Expert Systems with Applications*, *261*, 125491. DOI: 10.1016/j.eswa.2024.125491

Kumar, M., Khan, S. Y., Liu, S., Ji, W., Shen, Y., Sohrabi, A., & Mazhar, A. R. (2024). Scientific mapping and data analysis of the research landscape in perovskite solar cell technology. *Solar Energy*, *273*, 112509. DOI: 10.1016/j.solener.2024.112509

Lampropoulos, I. C., Aggelopoulos, E., Giannopoulos, V., Tsameti, A., & Karifyllas, S. G. (2026). Performance Management and Accounting Frameworks for Assessing Corporate Sustainability in the Energy Transition Era. In *Economic, Social, and Environmental Insights on Energy Transition* (pp. 65-96). IGI Global Scientific Publishing.

Lazaro, P., & Momayez, M. (2020). Development of a modified predicted heat strain model for hot work environments. *International Journal of Mining Science and Technology*, *30*(4), 477–481. DOI: 10.1016/j.ijmst.2020.05.009

Leal Filho, W., Schmidberger, I., Sharifi, A., Vargas, V. R., Rampasso, I. S., Dibbern, T., Liakh, O., Aina, Y. A., Trevisan, L. V., Mbah, M. F., Anholon, R., & Kozlova, V. (2024). Design thinking for sustainable development: A bibliometric analysis and case study research. *Journal of Cleaner Production*, *455*, 142285. DOI: 10.1016/j.jclepro.2024.142285

Li, P. F., Xu, Y., & Chen, Q. B. (2024). Global trends and research characteristics on CO2 capture and conversion from 2000 to 2023: A bibliometric review. *Sustainable Energy Technologies and Assessments*, *67*, 103836. DOI: 10.1016/j.seta.2024.103836

Linnenluecke, M. K., Birt, J., & Griffiths, A. (2015). The role of accounting in supporting adaptation to climate change. *Accounting and Finance*, *55*(3), 607–625. DOI: 10.1111/acfi.12120

Liu, Y. G., Jiang, S. T., Zhang, L., Zheng, H., Zhang, T., Zhang, J. W., Zhao, H.-T., Sang, X.-T., Xu, Y.-Y., & Lu, X. (2023). Worldwide productivity and research trend of publications concerning tumor immune microenvironment (TIME): A bibliometric study. *European Journal of Medical Research*, *28*(1), 229. DOI: 10.1186/s40001-023-01195-3 PMID: 37430294

Lukács, B., Molnár, P., & Tóth, Á. (2026). Categorisation of SDG targets into ESG pillars based on ESRS taxonomy. *Discover Sustainability*, *7*(1), 120. DOI: 10.1007/s43621-025-02550-6

Mariappanadar, S., Maurer, I., Kramar, R., & Muller-Camen, M. (2022). Is it a sententious claim? An examination of the quality of occupational health, safety and well-being disclosures in global reporting initiative reports across industries and countries. *International Business Review*, *31*(2), 101922. DOI: 10.1016/j.ibusrev.2021.101922

Mavroulidis, M., Vouros, P., Fotiadis, S., Konstantakopoulou, F., Fountoulakis, G., Nikolaou, I., & Evangelinos, K. (2022). Occupational health and safety of multinational construction companies through evaluation of corporate social responsibility reports. *Journal of Safety Research*, *81*, 45–54. DOI: 10.1016/j.jsr.2022.01.005 PMID: 35589305

Meng, X., Xiong, H., Yang, H., & Cao, Y. (2021). Dynamic prediction of indoor wet bulb globe temperature in an industrial workshop. *Applied Thermal Engineering*, *195*, 117219. DOI: 10.1016/j.applthermaleng.2021.117219

Morea, A., Schino, V., Bartolomeo, N., Ravallese, R., Sacino, G., Ravallese, R., Iatta, R., Otranto, D., Lovreglio, P., & Stufano, A. (2025). Seroprevalence of tick-borne diseases in Europe in occupational settings: A systematic review and metanalysis. *Public Health*, *248*, 105923. DOI: 10.1016/j.puhe.2025.105923 PMID: 40907304

Morelli, G., Gurrieri, A. R., Drago, C., Gatto, A., & Magazzino, C. (2026). A Biblio-Ecological Study on Sustainable Land Planning Research: Evidence from Machine Learning Methods. *Socio-Economic Planning Sciences*, *105*, 102431. DOI: 10.1016/j.seps.2026.102431

Morris, N. B., Piil, J. F., Morabito, M., Messeri, A., Levi, M., Ioannou, L. G., Ciuha, U., Pogačar, T., Kajfež Bogataj, L., Kingma, B., Casanueva, A., Kotlarski, S., Spirig, C., Foster, J., Havenith, G., Sotto Mayor, T., Flouris, A. D., & Nybo, L. (2021). The HEAT-SHIELD project-Perspectives from an inter-sectoral approach to occupational heat stress. *Journal of Science and Medicine in Sport*, *24*(8), 747–755. DOI: 10.1016/j.jsams.2021.03.001 PMID: 33757698

Mousavi, S. M., Dehghan, H., & Yazdanirad, S. (2026). A critical systematic narrative review of the WBGT index for heat stress assessment in diverse climatic regions: Insights from Iran. *International Journal of Biometeorology*, *70*(3), 63. DOI: 10.1007/s00484-025-03093-1 PMID: 41703112

Mullins-Jaime, C., Stefancik, J., Gannon, A., Bourbai, C., & Ismyana, M. (2026). What states are most at risk? An epidemiologic analysis of state-level occupational heat-related illness and fatality trends over three decades. *Journal of Safety Research*. Advance online publication. DOI: 10.1016/j.jsr.2026.01.013

Naguib, M. M., Li, R., Ling, J., Grace, D., Nguyen-Viet, H., & Lindahl, J. F. (2021). Live and wet markets: Food access versus the risk of disease emergence. *Trends in Microbiology*, *29*(7), 573–581. DOI: 10.1016/j.tim.2021.02.007 PMID: 33712334

Naylor, C., Canvin, S., Dunk, L., Gopaldasani, V., Liebenberg, A., Lux, G., Macfarlane, F., Panyam, S., Sharp, V., Thomas, A., Uru, E., & Whitelaw, J. L. (2025). What does good health look like? Development of leading occupational health performance indicators. *Annals of Work Exposures and Health*, *69*(9), 904–916. DOI: 10.1093/annweh/wxaf050 PMID: 40973106

Nicholas, P. K., Breakey, S., & Blank, P. (2020). Roles of nurse practitioners: Health consequences of climate change in vulnerable older adults. *The Journal for Nurse Practitioners*, *16*(6), 433–437. DOI: 10.1016/j.nurpra.2020.03.011

Nunfam, V. F., Afrifa-Yamoah, E., Adusei-Asante, K., Van Etten, E. J., Frimpong, K., Mensah, I. A., & Oosthuizen, J. (2021). Construct validity and invariance assessment of the social impacts of occupational heat stress scale (SIOHSS) among Ghanaian mining workers. *The Science of the Total Environment*, *771*, 144911. DOI: 10.1016/j.scitotenv.2020.144911 PMID: 33736178

Nyakuwanika, M., & Panicker, M. (2025). The Role of Environmental Accounting in Mitigating Climate Change: ESG Disclosures and Effective Reporting-A Systematic Literature Review. *Journal of Risk and Financial Management*, *18*(9), 480. DOI: 10.3390/jrfm18090480

Pan, J., Zhang, Y., Zhao, B., Shao, W., & Chen, X. (2025). Current perspectives and trend of ferroptosis in head and neck cancer: A bibliometric analysis. *Frontiers in Oral Health, 6*, 1601962. DOI: 10.3389/froh.2025.1601962 PMID: 40860793

Pandey, D. K., Lucey, B. M., & Kumar, S. (2023). Border disputes, conflicts, war, and financial markets research: A systematic review. *Research in International Business and Finance, 65*, 101972. DOI: 10.1016/j.ribaf.2023.101972

Pawłowska, Z. (2015). Using lagging and leading indicators for the evaluation of occupational safety and health performance in industry. *International Journal of Occupational Safety and Ergonomics, 21*(3), 284–290. DOI: 10.1080/10803548.2015.1081769 PMID: 26647949

Prata, J. C., da Costa, J. P., Lopes, I., Andrady, A. L., Duarte, A. C., & Rocha-Santos, T. (2021). A One Health perspective of the impacts of microplastics on animal, human and environmental health. *The Science of the Total Environment, 777*, 146094. DOI: 10.1016/j.scitotenv.2021.146094 PMID: 33677304

Reporting, S. (2019). Corporate Reporting Practices Concerning Non-financial Aspects: A Possible Prolix?. *Integrated Reporting*, 241.

Rose, D., & Birchall, S. J. (2025). Investigating the extent and quality of health-focused climate adaptation planning: Insights from Western Canadian cities. *Environmental Science & Policy, 172*, 104192. DOI: 10.1016/j.envsci.2025.104192

Rossi, M. F., Leone, R., & Moscato, U. (2025). Climate Change and Occupational Risks in Outdoor Workers: A Systematic Review of the Health Effects of Extreme Temperatures. *Atmosphere (Basel), 16*(7), 839. DOI: 10.3390/atmos16070839

Rother, H. A., John, J., Wright, C. Y., Irlam, J., Oosthuizen, R., & Garland, R. M. (2019). Perceptions of occupational heat, sun exposure, and health risk prevention: A qualitative study of forestry workers in South Africa. *Atmosphere (Basel), 11*(1), 37. DOI: 10.3390/atmos11010037

Sabando-Vera, D., Yonfa-Medranda, M., Montalván-Burbano, N., Albors-Garrigos, J., & Parrales-Guerrero, K. (2022). Worldwide research on open innovation in SMEs. *Journal of Open Innovation, 8*(1), 20. DOI: 10.3390/joitmc8010020

Saheb, T., Sidaoui, M., & Schmarzo, B. (2024). Convergence of artificial intelligence with social media: A bibliometric & qualitative analysis. *Telematics and Informatics Reports, 14*, 100146. DOI: 10.1016/j.teler.2024.100146

Sánchez, A., García-Galán, A., García, E., Gómez-Martín, Á., de La Fe, C., Corrales, J. C., & Contreras, A. (2020). Occupational exposure to influenza virus of the wild birds. *Revista Española de Salud Pública, 94*, 202003022. PMID: 32381999

Schaltegger, S., Zvezdov, D., Etxeberria, I. A., Csutora, M., & Günther, E. (Eds.). (2015). *Corporate carbon and climate accounting*. Springer International Publishing. DOI: 10.1007/978-3-319-27718-9

Schulte, P. A., Bhattacharya, A., Butler, C. R., Chun, H. K., Jacklitsch, B., Jacobs, T., Kiefer, M., Lincoln, J., Pendergrass, S., Shire, J., Watson, J., & Wagner, G. R. (2016). Advancing the framework for considering the effects of climate change on worker safety and health. *Journal of Occupational and Environmental Hygiene*, *13*(11), 847–865. DOI: 10.1080/15459624.2016.1179388 PMID: 27115294

Scotti, K. L., Jiao, C., Gainey, M. A., Bongiovanni, E. A., & Slayton, E. R. (2025). Charting open science landscapes: A systematized review of US academic libraries' engagement in open research practices. *Journal of Academic Librarianship*, *51*(3), 103054. DOI: 10.1016/j.acalib.2025.103054

Sepúlveda-Oviedo, E. H., Travé-Massuyès, L., Subias, A., Pavlov, M., & Alonso, C. (2023). Fault diagnosis of photovoltaic systems using artificial intelligence: A bibliometric approach. *Heliyon*, *9*(11), e21491. DOI: 10.1016/j.heliyon.2023.e21491 PMID: 37954345

Sharma, M., Suri, N. M., & Kant, S. (2022). Analyzing occupational heat stress using sensor-based monitoring: A wearable approach with environmental ergonomics perspective. *International Journal of Environmental Science and Technology*, *19*(11), 11421–11434. DOI: 10.1007/s13762-021-03862-6 PMID: 35106066

Singh, A., Rejeb, A., Nangru, H., & Pathak, S. (2024). Global research trends on cyberbullying: A bibliometric study. *Computers in Human Behavior Reports*, *16*, 100499. DOI: 10.1016/j.chbr.2024.100499

Stechemesser, K., Bergmann, A., & Guenther, E. (2016). Organizational Climate Accounting-Financial Consequences of Climate Change Impacts and Climate Change Adaptation. In *Corporate carbon and climate accounting* (pp. 217–242). Springer International Publishing.

Tappura, S., Sievänen, M., Heikkilä, J., Jussila, A., & Nenonen, N. (2015). A management accounting perspective on safety. *Safety Science*, *71*, 151–159. DOI: 10.1016/j.ssci.2014.01.011

Tawatsupa, B., Yiengprugsawan, V., Kjellstrom, T., Berecki-Gisolf, J., Seubsman, S. A., & Sleigh, A. (2013). Association between heat stress and occupational injury among Thai workers: Findings of the Thai Cohort Study. *Industrial Health*, *51*(1), 34–46. DOI: 10.2486/indhealth.2012-0138 PMID: 23411755

Tsakonas, K., Badyal, S., Takaro, T., & Buse, C. G. (2024). Rapid review of the impacts of climate change on the health system workforce and implications for action. *The Journal of Climate Change and Health, 19*, 100337. DOI: 10.1016/j.joclim.2024.100337 PMID: 41647318

Tsalis, T. A., Stylianou, M. S., & Nikolaou, I. E. (2018). Evaluating the quality of corporate social responsibility reports: The case of occupational health and safety disclosures. *Safety Science, 109*, 313–323. DOI: 10.1016/j.ssci.2018.06.015

Turner, M. C., Basagaña, X., Albin, M., Broberg, K., Burdorf, A., van Daalen, K. R., Guseva Canu, I., Kolstad, H. A., Kogevinas, M., Lowe, R., Pearce, N., Pega, F., Saget, C., Schubauer-Berigan, M. K., Svensson, S., Vineis, P., & Straif, K. (2025). Occupational health in the era of climate change and the green transition: A call for research. *The Lancet Regional Health. Europe, 54*, 54. DOI: 10.1016/j.lanepe.2025.101353 PMID: 40613104

Utomo, B. (2025). Connection of Human, Animal, and Environmental Health: A One Health Perspective. *One Health Integration: Global Perspectives on Animal Health and Sustainable Agriculture*, 63-80.

Utzet, M., Ayala-Garcia, A., Benavides, F. G., & Basagaña, X. (2023). Extreme temperatures and sickness absence in the Mediterranean province of Barcelona: An occupational health issue. *Frontiers in Public Health, 11*, 1129027. DOI: 10.3389/fpubh.2023.1129027 PMID: 36891341

Vendl, C., Taylor, M. D., Bräunig, J., Gibson, M. J., Hesselson, D., Neely, G. G., & Nakagawa, S. (2022). PFAS exposure of humans, animals and the environment: Protocol of an evidence review map and bibliometric analysis. *Environment International, 158*, 106973. DOI: 10.1016/j.envint.2021.106973

Wang, Y., & Zhang, L. (2025, June). Bibliometric Analysis and Visualization of Research Trends and Hotspots in the Body Donation Field. In *Transplantation Proceedings*. Elsevier. DOI: 10.1016/j.transproceed.2025.05.015

Wen, X., Liu, J., Zhao, W., Liu, X., Liu, L., Zeng, Z., & Wei, D. (2026). Photodynamic Therapy for Colorectal Cancer: A Bibliometric and Visualization Analysis (1986–2025). *Photodiagnosis and Photodynamic Therapy, 58*, 105368. DOI: 10.1016/j.pdpdt.2026.105368 PMID: 41587715

WHO. (2021). *2021 WHO health and climate change global survey report*. World Health Organization.

Wibowo, R., Do, V., Quartucci, C., Koller, D., Daanen, H. A., Nowak, D., Bose-O'Reilly, S., & Rakete, S. (2024). Effects of heat and personal protective equipment on thermal strain in healthcare workers: Part B-application of wearable sensors to observe heat strain among healthcare workers under controlled conditions. *International Archives of Occupational and Environmental Health, 97*(1), 35–43. DOI: 10.1007/s00420-023-02022-2 PMID: 37947815

Wister, A., Fyffe, I., & O'Dea, E. (2021). Technological interventions for loneliness and social isolation among older adults: A scoping review protocol. *Systematic Reviews, 10*(1), 217. DOI: 10.1186/s13643-021-01775-6 PMID: 34362447

Yang, F., Yao, R., Ren, Y., & Guo, L. (2025). Harmony in diversity: Digital literacy research in a multidisciplinary landscape. *Computers & Education, 230*, 105265. DOI: 10.1016/j.compedu.2025.105265

Yao, D., Qiu, J., & Liang, Q. (2025). Bibliometric and Visual Analysis of Inadvertent Perioperative Hypothermia: A Last 20 Years of Research (2004 to 2024). *Journal of PeriAnesthesia Nursing*.

Zinsstag, J., Crump, L., Schelling, E., Hattendorf, J., Maidane, Y. O., Ali, K. O., Muhummed, A., Umer, A. A., Aliyi, F., Nooh, F., Abdikadir, M. I., Ali, S. M., Hartinger, S., Mäusezahl, D., de White, M. B. G., Cordon-Rosales, C., Castillo, D. A., McCracken, J., Abakar, F., & Cissé, G. (2018). Climate change and one health. *FEMS Microbiology Letters, 365*(11), fny085. DOI: 10.1093/femsle/fny085 PMID: 29790983

Chapter 5
From Pregnancy to Early Childhood:
Extreme Heat as a Maternal-Child Health Equity Imperative in the U.S. Southwest – A Narrative Review

Daryl O'Neal Traylor
http://orcid.org/0000-0002-8935-1653
Chamberlain University, USA

Eboni Elana Anderson
http://orcid.org/0000-0001-7675-4528
AT Still University School of Osteopathic Medicine in Arizona, USA

Christian McKinney
California Baptist University, USA

Aishah T. Trahan-Johnson
http://orcid.org/0009-0001-9232-3899
California Baptist University, USA

Cierra Swington
http://orcid.org/0009-0005-7044-4963
California Baptist University, USA

ABSTRACT

Extreme heat is intensifying across the U.S. Southwest and is increasingly consequential for maternal and child health. This narrative review synthesizes evidence

DOI: 10.4018/979-8-3373-8470-2.ch005

on heat exposure and outcomes across pregnancy, postpartum and lactation, and early childhood through an environmental and reproductive justice lens. We searched biomedical and interdisciplinary databases from 2000 to the present and supplemented them with gray literature on heat guidance, workplace standards, and housing and energy policies. Evidence links higher temperatures with increased risks of preterm birth and stillbirth and identifies postpartum, infant feeding, and pediatric morbidity during heat events and outages. Inequities reflect uneven neighborhood heat, energy insecurity, occupational exposure, and barriers to cooling and responsive care. We propose an interprofessional agenda that integrates clinical heat risk screening and tailored plans, heat-event care coordination with referral pathways, and longer-term structural investments in housing, energy reliability, and enforceable workplace protections

INTRODUCTION

In the United States Southwest, extreme heat has become a defining environmental hazard, shaped by an arid climate, rapid urban expansion, and climate warming that is lengthening and intensifying periods of dangerous heat exposure. Observational evidence shows that heat waves have increased in frequency and duration across most regions since the mid twentieth century (Perkins-Kirkpatrick & Lewis, 2020). Recent events underscore the regional stakes: a persistent 2023 heat wave affected large portions of the western United States, including parts of the Southwest, and reflected the capacity for heat to endure for weeks rather than days (Lopez et al., 2025). Yet heat is not experienced as a uniform "natural" exposure. Within the same metropolitan area, neighborhood thermal conditions diverge sharply in ways that reflect historic and ongoing policy choices, including patterns of segregation and disinvestment. A national analysis linking contemporary heat exposure to historical housing policy illustrates how intra-urban heat burdens map onto structural inequities (Hoffman et al., 2020).

For maternal and child health, these inequities matter because pregnancy, the postpartum period, infancy, and childhood are characterized by physiologic and social conditions that can narrow the margin for adaptation to heat. Epidemiologic syntheses indicate that higher ambient temperatures during pregnancy are associated with increased risks of preterm birth and stillbirth, with effects that, while often modest in relative terms, become consequential at population scale because exposure is common and rising (Chersich et al., 2020). US-based analyses similarly report associations between elevated apparent temperature and preterm delivery, with evidence that some groups may experience stronger associations (Basu et al., 2010). These pregnancy-linked risks intersect with the material realities that determine whether families can avoid or mitigate exposure. Heat vulnerability is partly governed by household resources and housing conditions, including access to air conditioning, vegetation, and neighborhood characteristics (Reid et al., 2009). The concept of energy insecurity further clarifies how affordability, housing quality, and unstable access to cooling can translate climate hazards into preventable health risk (Hernández, 2013).

Prior reviews have established that heat and related climate exposures are associated with adverse pregnancy outcomes, and newer pediatric syntheses show that children experience heat-related risks across emergency care, respiratory disease, heat illness, learning environments, and birth outcomes (Bekkar et al., 2020; Schapiro et al., 2024). Less developed is a synthesis that follows the pathway from pregnancy through postpartum recovery, lactation, infancy, and early childhood while asking how exposure, protection, and care are socially distributed in one of the nation's most heat-exposed regions. This chapter addresses that gap by linking outcome evidence to practical levers, clinical screening, heat-event care coordination, occupational standards, and housing and energy protections, so adaptation can be evaluated by reach and equity rather than by the presence of warnings or resources alone.

An equity-oriented approach is therefore essential. Health equity is commonly framed as the absence of systematic, socially produced disparities in health and in the conditions that shape health (Braveman & Gruskin, 2003). In the context of extreme heat, equity is not only a matter of differential biologic susceptibility; it is also a matter of who lives in the hottest neighborhoods, who works in high-heat occupations, who has reliable cooling and transportation, and who can access timely prenatal, postpartum, and pediatric care. These pathways are likely to be especially salient in the Southwest, where large urban heat islands, rural and frontier geographies, agricultural labor markets, and borderland dynamics shape daily exposure, resource access, and health care continuity. The present narrative review is organized around the premise that extreme heat operates through intersecting social and structural mechanisms that can widen maternal and child health disparities if adaptation is uneven.

Against this backdrop, this narrative review has four objectives. First, it synthesizes evidence linking heat exposure and heat stress with outcomes in pregnancy, lactation, and child health, from fetal development through early childhood. Second, it explicates equity-relevant mechanisms, differential exposure, susceptibility, and access to protection and care, providing conceptual clarity and empirical examples. Third, it reviews and critiques interventions and policies across clinical, public health, workplace, and housing or energy domains through a health equity lens, emphasizing implementation feasibility and gaps in reach. Finally, it advances an equity-forward action agenda for health systems and partners in the Southwest, outlining practical and structural priorities for adaptation. By integrating epidemiologic evidence, social theory, and policy analysis, the review aims to illuminate how extreme heat functions as a structurally patterned exposure and to identify pathways toward more just and resilient systems of maternal and child health protection.

The intended audience is deliberately interdisciplinary: maternal–child health clinicians, nurses, midwives, lactation professionals, public health practitioners, health system leaders, community-based organizations, emergency preparedness partners, and policy actors concerned with heat adaptation. The chapter is not meant to rank every available intervention. Instead, it identifies where evidence is sufficient for near-term practice, where implementation evidence is promising but incomplete, and where future research must be governed differently to serve Tribal, migrant, rural, and frontier communities.

Although this review is centered on the U.S. Southwest, the framework has relevance for other high-heat settings where reproductive and child health risks are shaped by unequal housing, work, water, energy, and care systems. The specific policy instruments will differ across South Asia, Sub-Saharan Africa, southern Europe, Australia, and other heat-exposed regions, but the analytic question is portable: who is exposed, who is physiologically or socially less protected at a given exposure, and who can realistically access cooling, rest, hydration, and care? Recent global syntheses show that escalating heat exposure threatens maternal, fetal, and neonatal health and should be prioritized in climate-health programs (Lakhoo et al., 2025). This review contributes a transferable equity framework while keeping the empirical and policy analysis grounded in one region where heat, urban development, labor exposure, and energy insecurity converge.

METHODS

This narrative review was selected to accommodate a literature base that is broad, uneven, and methodologically diverse, spanning epidemiologic studies of pregnancy and child outcomes, physiologic and thermoregulation research, occupational heat exposures, policy analyses, and qualitative or community-informed reports. A transparent search strategy was developed a priori and applied across biomedical and interdisciplinary databases, including PubMed or MEDLINE, Embase, CINAHL, and Web of Science or Scopus, with PsycINFO added to capture psychosocial and caregiving pathways when relevant. Searches combined heat terms (for example, extreme heat, heatwave, heat stress, heat-related illness, ambient temperature) with population and outcome terms related to pregnancy, postpartum and lactation, infancy, and early childhood, with the Southwestern United States operationalized as Arizona, New Mexico, Nevada, Utah, Colorado, southern California, and west Texas. We prioritized literature published from 2000 to the present and included

English-language sources, with Spanish-language screening incorporated when feasible. Although we used a structured search strategy, this review is not intended to be exhaustive. Instead, we used an "anchor evidence" approach, prioritizing (1) systematic reviews and meta-analyses and (2) large, well-designed cohort and registry studies (especially U.S. and Southwest-relevant analyses) to summarize outcome associations, and (3) equity-relevant qualitative, implementation, and policy sources (including grey literature) to characterize structural mechanisms and intervention feasibility.

Anchor sources were selected using explicit relevance and credibility criteria rather than concordance with the authors' interpretive framework. For epidemiologic associations, priority was given to systematic reviews, meta-analyses, large registry or cohort studies, and studies with clearly specified exposure windows, heat metrics, outcome definitions, and adjustment strategies. For practice, policy, and equity mechanisms, priority was given to sources that directly addressed implementation feasibility, structural determinants, or differential reach among populations relevant to the U.S. Southwest. Studies were not excluded because findings were null, mixed, or inconsistent; rather, inconsistency was retained as part of the synthesis when it reflected variation in heat metrics, exposure timing, population characteristics, or outcome definition. This approach was intended to reduce overinterpretation while still allowing synthesis across a literature base that remains heterogeneous and uneven. To further improve transparency, we assessed the narrative review against the core domains of the Scale for the Assessment of Narrative Review Articles, including justification of importance, statement of aims, description of literature search, referencing, scientific reasoning, and appropriate presentation of data (Baethge et al., 2019). Full database search strings and Boolean logic are provided in Box 1.

To make the synthesis traceable, evidence was abstracted around four recurring questions: how heat exposure was measured; which maternal, infant, or child outcomes were assessed; whether equity-relevant modifiers or structural determinants were examined; and whether the source offered actionable implementation or policy detail. This approach is consistent with narrative review standards that emphasize clear aims, transparent searching, and reasoned synthesis, while recognizing that narrative reviews do not provide pooled effect estimates or formal certainty ratings in the manner of systematic reviews (Baethge et al., 2019; Grant & Booth, 2009).

Box 1. Database search strategy: Boolean logic and example search strings

Core Boolean structure (applied across databases):
(Heat exposure terms) **AND** (Pregnancy/postpartum/lactation/infant–child terms) **AND** (Southwest geography terms)
Within each concept set, synonyms were combined with **OR**. Truncation (*) captured word variants, and quotation marks preserved phrases. Search syntax was adapted to each database's controlled vocabulary and field tags.
Heat exposure terms (OR):
extreme heat OR heat wave* OR heatwave* OR heat stress OR heat strain OR heat-related illness OR hyperthermia OR hot temperature OR ambient temperature OR apparent temperature OR heat index OR thermal stress
Pregnancy–postpartum–early childhood terms (OR):
pregnan* OR gestat* OR prenatal OR antenatal OR perinatal OR obstetric* OR maternal OR postpartum OR post-partum OR puerper* OR lactat* OR breastfeeding OR breastfeed* OR human milk OR infant feeding OR neonat* OR newborn* OR infant* OR toddler* OR child* OR pediatric*
Southwest geography terms (OR):
Arizona OR "New Mexico" OR Nevada OR Utah OR Colorado OR California OR "Southern California" OR Texas OR "West Texas" OR Southwest* OR "Southwestern United States"
Example search string (PubMed/MEDLINE; field tags illustrative)
(("Heat Stress Disorders"[Mesh] OR "Hot Temperature"[Mesh] OR "heat wave*"[tiab] OR heatwave*[tiab] OR "extreme heat"[tiab] OR "heat stress"[tiab] OR hyperthermia[tiab] OR "heat-related illness"[tiab] OR "ambient temperature"[tiab] OR "apparent temperature"[tiab] OR "heat index"[tiab])
AND (pregnan*[tiab] OR gestat*[tiab] OR prenatal[tiab] OR antenatal[tiab] OR perinatal[tiab] OR obstetric*[tiab] OR postpartum[tiab] OR puerper*[tiab] OR lactat*[tiab] OR breastfeed*[tiab] OR "infant feeding"[tiab] OR neonat*[tiab] OR infant*[tiab] OR child*[tiab] OR pediatric*[tiab])
AND ("Arizona"[tiab] OR "New Mexico"[tiab] OR Nevada[tiab] OR Utah[tiab] OR Colorado[tiab] OR California[tiab] OR Texas[tiab] OR Southwest*[tiab] OR "Southwestern United States"[tiab]))
AND ("2000/01/01"[Date - Publication]: "3000"[Date - Publication])
AND (English[lang])
Example search string (Embase; controlled vocabulary + title/abstract)
('heat stress'/exp OR 'hot temperature'/exp OR 'heat wave':ti,ab OR heatwave*:ti,ab OR 'extreme heat':ti,ab OR hyperthermia:ti,ab OR 'heat-related illness':ti,ab OR 'ambient temperature':ti,ab OR 'apparent temperature':ti,ab)
AND (pregnan*:ti,ab OR gestat*:ti,ab OR prenatal:ti,ab OR antenatal:ti,ab OR perinatal:ti,ab OR obstetric*:ti,ab OR postpartum:ti,ab OR puerper*:ti,ab OR lactat*:ti,ab OR breastfeed*:ti,ab OR 'infant feeding':ti,ab OR neonat*:ti,ab OR infant*:ti,ab OR child*:ti,ab OR pediatric*:ti,ab)
AND (arizona:ti,ab OR 'new mexico':ti,ab OR nevada:ti,ab OR utah:ti,ab OR colorado:ti,ab OR california:ti,ab OR texas:ti,ab OR southwest*:ti,ab OR 'southwestern united states':ti,ab)
AND [2000–Current]/py AND [English]/lim
Example search string (Scopus/Web of Science; TITLE-ABS-KEY / TS)
TITLE-ABS-KEY((heat OR "extreme heat" OR "heat wave*" OR heatwave* OR "heat stress" OR hyperthermia OR "heat-related illness" OR "ambient temperature" OR "apparent temperature" OR "heat index")
AND (pregnan* OR gestat* OR prenatal OR antenatal OR perinatal OR obstetric* OR postpartum OR puerper* OR lactat* OR breastfeed* OR "infant feeding" OR neonat* OR infant* OR child* OR pediatric*)
AND ("Arizona" OR "New Mexico" OR "Nevada" OR "Utah" OR "Colorado" OR "California" OR "Texas" OR Southwest* OR "Southwestern United States"))
AND PUBYEAR > 1999 AND (LIMIT-TO(LANGUAGE, "English"))

Grey-literature sources were searched to identify practice guidance, implementation considerations, and policy context that may not be represented in indexed health-science databases. These sources were not treated as equivalent to epidemiologic outcome studies. They were retained only when they met at least one of three criteria: they provided jurisdictionally relevant heat-health or occupational guidance; described a policy mechanism directly related to maternal, infant, or child heat protection; or offered implementation detail not available in peer-reviewed studies. Many federal, state, and local guidance documents were therefore used for orientation and source-tracing but were not cited in the final synthesis when their content duplicated higher-level evidence or lacked sufficient methodological detail. The RAND Corporation regulatory impact assessment by Metz et al. (2021) was classified as grey literature/report evidence. Table 1 summarizes the search yield and source-selection process for this narrative review. Of 211 records identified through database and supplementary searches, 196 records remained after duplicate removal and were screened by title and abstract. Seventy-seven full-text sources were assessed, and 56 sources were ultimately included in the synthesis, consisting primarily of peer-reviewed journal articles, with one retained grey-literature/report source that added nonduplicative regulatory or implementation evidence.

Table 1. Simplified search yield and selection summary

Step	Count (n)
Records identified (databases + supplementary sources)	211
Duplicates removed	15
Records screened (title/abstract)	196
Full-text sources assessed	77
Full-text sources not cited	21
Sources included in synthesis (total)	56
— Peer-reviewed journal sources	55
— Grey literature / report	1

Note. Grey literature was screened to support practice and policy interpretation; however, only sources that added nonduplicative implementation or regulatory evidence were retained as cited sources. Metz et al. (2021) was classified as the retained grey-literature/report source.

Ethical Considerations and Positionality

Ethical considerations included using respectful, participant-centered terminology, avoiding deficit narratives, and treating Indigenous data sovereignty and jurisdictional context as foundational rather than ancillary to interpretation. The author team brings training in community and public health and we write from professional positions that provide access to scholarly and clinical evidence but do not substitute for lived experience in the communities discussed. Our interpretation was therefore guided by environmental justice, reproductive justice, and health equity scholarship, with attention to avoiding deficit explanations of risk. Where the literature concerns Indigenous, migrant, rural, and border communities, we treat community authority, tribal sovereignty, and local knowledge as methodological concerns rather than contextual details. This aligns with current health-equity scholarship calling for reflexivity and positionality in epidemiologic and equity-focused work (Poirier et al., 2023).

Because this article is a secondary synthesis, it did not involve recruitment, intervention, or analysis of identifiable human participant data. The principal ethical risks are interpretive: overstating causality, treating communities as uniformly vulnerable, extracting Indigenous or migrant experiences without attention to governance, or citing policy documents as if they were outcome studies. The review addresses these risks by distinguishing epidemiologic, implementation, policy, and grey-literature claims; retaining mixed or null findings when relevant; and using positionality, community authority, and Indigenous data sovereignty principles as safeguards against deficit framing (Carroll et al., 2020; Poirier et al., 2023).

CONCEPTUAL FRAMEWORK

This review is organized using an intersectionality-informed environmental justice and reproductive justice framework. Intersectionality directs attention to how social positions such as race and ethnicity, gender, socioeconomic status, immigration status, disability, and occupation are experienced simultaneously, and how these intersections shape exposure and response to heat (Bowleg, 2012). Environmental justice situates these patterns in place-based inequities in environmental burdens and protections, including the built environment, housing quality, and infrastructure (Mohai et al., 2009). Reproductive justice extends the equity lens across the maternal and child continuum by emphasizing the social conditions required to decide whether and when to have children and to parent with dignity and safety (Luna & Luker, 2013). Health equity is understood as the absence of systematic, socially produced and therefore remediable disparities in health and its determinants (Braveman & Gruskin, 2003), consistent with ecosocial perspectives that link structural conditions to embodied health outcomes over the life course (Krieger, 2001). Key terms used throughout the review are defined in Box 2; these definitions anchor the equity mechanisms and analytic pathway summarized in Figure 1.

Heat exposure refers to ambient thermal conditions, typically measured using air temperature or apparent temperature indices such as the heat index. Heat stress denotes the external thermal load, whereas heat strain reflects the body's physiologic response; heat illness describes clinically recognized syndromes, including heat stroke (Bouchama & Knochel, 2002). Heat vulnerability captures the likelihood of harm given exposure, shaped by susceptibility and protective capacity. Adaptation refers to actions that reduce heat-related risk, mitigation refers briefly to reducing heat-driving emissions, and resilience describes the ability to anticipate, withstand, and recover from heat events. The analytic pathway used throughout is heat exposure leading to physiologic response and then maternal and child outcomes, with effects moderated by structural determinants and access to cooling, protection, and care.

The review's organizing logic is summarized in Figure 1, Conceptual Pathway Linking Extreme Heat to Maternal and Child Health Outcomes in the U.S. Southwest: An Intersectionality-Informed Environmental and Reproductive Justice Framework. As shown in Figure 1, the pathway begins with the intersectionality context, which informs upstream structural conditions. Upstream structural conditions then shape three equity-relevant mechanisms: differential exposure, differential susceptibility, and differential access to protection and care. Differential exposure is depicted as operating through heat exposure, which then contributes to heat stress and heat strain, followed by maternal and child health outcomes. Differential susceptibility is shown as a direct influence on heat stress and heat strain, reflecting variation in how the same heat exposure can translate into physiologic load. Differential access to protection and care operates through protection and buffering from heat exposure, which is also depicted as influencing heat stress and heat strain. In this conceptualization, heat stress and heat strain function as the proximal physiologic bridge between heat exposure and maternal and child health outcomes, while structural conditions and equity mechanisms shape where and for whom that bridge becomes consequential.

BOX 2. Key definitions used in this review

Health equity — In this review, health equity means that maternal and child health—and the conditions that make protection possible—are not reliably predicted by social position, and that systems are organized so heat does not widen remediable differences in outcomes.
Health inequity — Health inequities are patterned, socially produced, and avoidable differences in health or health-protective conditions that are therefore actionable through changes in policy, infrastructure, and institutions.
Environmental justice — Environmental justice directs attention to how environmental burdens and protections are distributed and governed, emphasizing that land use, housing, and infrastructure investment shape who lives with hotter conditions and who has practical access to cooling and relief.
Reproductive justice — Reproductive justice frames safe pregnancy, recovery, infant feeding, and parenting as dependent on social conditions and rights, highlighting how heat can narrow safe choices when housing, work, and services do not provide the material supports needed for care.
Energy insecurity — Energy insecurity is constrained access to the energy services needed for safe daily life—especially the ability to maintain tolerable indoor temperatures—due to affordability pressures, inefficient housing, equipment failures, or disruptions such as utility shutoffs.
Differential exposure — Differential exposure refers to unequal contact with heat (at home, work, and during travel) that reflects structural patterns in housing, labor, and transportation rather than individual preference.

continued on following page

BOX 2. Continued

Differential susceptibility — Differential susceptibility describes variation in harm at a similar heat load, shaped by pregnancy physiology, postpartum recovery, infant thermoregulation, comorbidities, medications, hydration constraints, and social conditions that influence physiologic reserve.
Differential access to protection and care — Differential access refers to unequal ability to obtain effective safeguards (cooling, shade, rest, safe water) and timely, respectful, responsive care because of practical barriers (transport, language, schedules), institutional barriers, or system failures during heat events (e.g., outages).
Adaptation — Adaptation includes actions that reduce heat-related risk by lowering exposure, buffering physiologic strain, or strengthening response capacity, ranging from clinical screening and coordinated outreach to structural measures such as housing retrofits, shade infrastructure, workplace standards, and utility continuity protections.
Mitigation — Mitigation refers to reducing the drivers of warming, primarily through lowering greenhouse gas emissions and related pollutants, as the long-term complement to adaptation.
Resilience — Resilience is the capacity of communities and institutions to anticipate, withstand, and recover from heat events without disproportionate harm, treated here as a systems property reflected in preparedness, reliable infrastructure, coordinated response, and accessible care—not an individual virtue.
Terminology note — We use inclusive terms (e.g., *pregnant people, postpartum people*) to reflect the diversity of those who experience pregnancy and lactation, while retaining sex-specific terms when quoting or describing prior studies for accuracy.

We use inclusive terms such as *pregnant people* and *postpartum people* to reflect the diversity of those who experience pregnancy and lactation; when we reference prior studies that use sex-specific terms (e.g., "women"), we retain the original wording when describing those findings.

Figure 1. Conceptual Pathway Linking Extreme Heat to Maternal and Child Health Outcomes in the U.S. Southwest: An Intersectionality-Informed Environmental and Reproductive Justice Framework

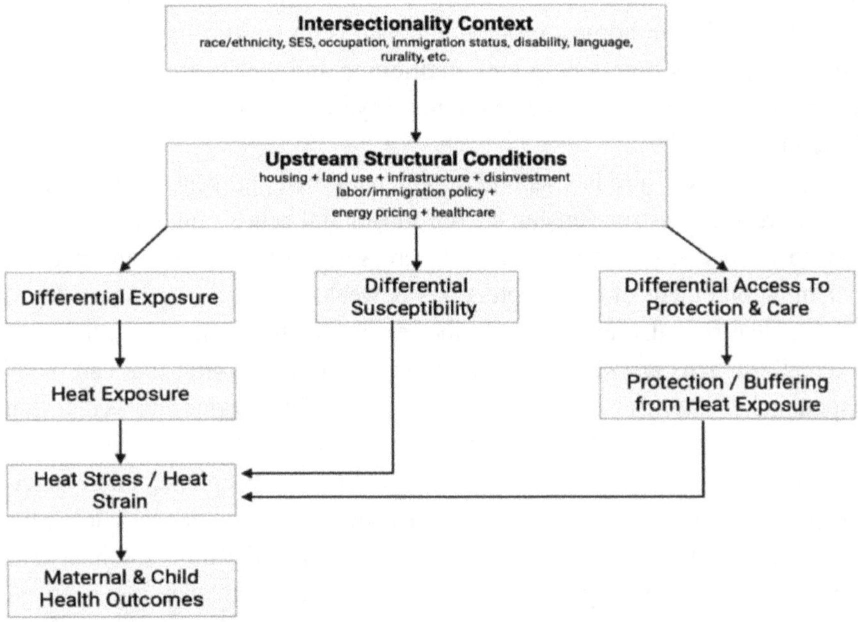

Note. *SES = socioeconomic status. Figure 1 depicts a conceptual pathway in which intersectionality context informs upstream structural conditions, which in turn shape differential exposure, differential susceptibility, and differential access to protection and care. Differential exposure operates through heat exposure and then heat stress and heat strain, leading to maternal and child health outcomes. Differential susceptibility is shown as directly influencing heat stress and heat strain. Differential access to protection and care operates through protection and buffering from heat exposure, which is shown as influencing heat stress and heat strain. Figure created by the authors.*

HEAT EXPOSURE IN THE SOUTHWEST: WHO IS EXPOSED, WHERE, AND WHY

Across the U.S. Southwest (for example, Arizona, New Mexico, Nevada, Utah, Colorado, southern California, and west Texas), heat exposure is defined not only by high daytime maxima but also by the duration of heat events, elevated nighttime minimum temperatures that limit physiologic recovery, and early season heat that arrives before acclimatization and institutional protections are fully in place (Anderson & Bell, 2011; Perkins-Kirkpatrick & Lewis, 2020). In metropolitan areas such as Phoenix, these patterns are amplified by urban heat island processes rooted in land cover, street geometry, and development decisions that shape neighborhood scale thermal conditions (Chow et al., 2012). Local exposure is further stratified by the built environment. Tree canopy and other forms of urban greening can lower near-surface temperatures and improve thermal comfort, yet canopy distribution commonly tracks disinvestment and housing vulnerability rather than need (Bowler et al., 2010). Within homes, protection depends on building quality and access to cooling, including functioning air conditioning and the ability to afford its use, which are core determinants of heat vulnerability in U.S. communities (Hernández, 2013; Reid et al., 2009).

Occupational exposure is a second, highly patterned pathway. Outdoor workers in agriculture and construction can accrue substantial heat strain during prolonged hot periods, especially when productivity pressures constrain rest and hydration, and when enforcement of heat protections is weak (Dong et al., 2019; Moyce et al., 2017). Heat risk also extends to indoor environments such as warehouses and commercial kitchens, as well as to unpaid labor and caregiving that can increase time spent in hot settings with limited autonomy. Finally, service access determines whether heat warnings translate into practical protection. Cooling centers reduce risk only when they are geographically reachable, culturally acceptable, and connected to transportation and outreach, including for people who cannot easily leave home or work (Berisha et al., 2017). Energy insecurity, utility shutoffs, and heat-related grid disruptions can abruptly convert "access to cooling" into an unmet need, compounding exposure during the very periods when cooling is most protective (Casey et al., 2020; Hernández, 2013).

HEAT STRESS IN PREGNANCY: OUTCOMES AND PLAUSIBLE MECHANISMS

Pregnancy is characterized by physiologic adjustments that can narrow the margin for thermal safety when environmental heat is intense, prolonged, or paired with limited cooling. This section follows the review outline by summarizing pregnancy-specific susceptibility, maternal and perinatal outcomes, and key measurement and equity considerations.

Physiologic Susceptibility in Pregnancy

Across gestation, metabolic heat production increases alongside plasma volume expansion and higher cardiac output, while thermoregulation relies heavily on adequate skin blood flow and evaporative cooling. When ambient heat and humidity impede heat loss, additional cardiovascular strain is required to maintain temperature homeostasis. Dehydration risk also rises with sustained sweating, and even modest volume depletion may plausibly affect uteroplacental perfusion and maternal symptoms such as dizziness, syncope, and heat exhaustion. Mechanistic syntheses emphasize several converging pathways under extreme heat, including dehydration and hemoconcentration, inflammatory and endocrine activation, and potential reductions in placental blood flow during maternal heat strain (Samuels et al., 2022).

Maternal Outcomes

Clinically, heat exposure is most directly linked to heat-related illness and acute care use, but observational studies increasingly implicate broader obstetric morbidity. In a large Southern California cohort, both short-term heatwave exposure in late gestation and higher cumulative exposure to hot days across pregnancy were associated with elevated odds of severe maternal morbidity, including cardiovascular sub-conditions (Jiao et al., 2023). Such findings are consistent with biologic plausibility: dehydration and endothelial stress may amplify hemodynamic load and contribute to complications that manifest near delivery, particularly when heat coincides with barriers to hydration, rest, and cooling. Evidence linking temperature with hypertensive disorders is also emerging in large administrative cohorts, though timing windows, temperature metrics, and residual confounding vary across studies (Xiong et al., 2020).

Perinatal Outcomes: Strength and Limits of Evidence

Among fetal and neonatal outcomes, the most consistently reported association is with preterm birth. A California study using birth registry data linked higher ambient temperatures to increased risk of preterm delivery, with effects observed across maternal subgroups (Basu et al., 2010). A broader systematic review and meta-analysis found that higher temperatures in pregnancy were associated with higher risks of preterm birth and stillbirth, and with lower birth weight, while also underscoring substantial heterogeneity in exposure definitions, lag structures, and analytic approaches (Chersich et al., 2020). The weight of evidence therefore supports an adverse signal, but effect sizes are generally modest and comparisons across settings remain limited by inconsistent exposure measurement and outcome specification.

Measurement and Equity Considerations

Exposure assessment remains a central limitation. Many studies assign temperature using weather station data linked to residential areas, which can miss neighborhood heat variation and indoor conditions, and rarely captures occupational heat (Basu et al., 2010). Yet equity-relevant effect modification is likely, because work conditions, housing quality, cooling access, chronic disease burden, and continuity of prenatal care can shape both exposure intensity and physiologic reserve. Large cohort analyses that formally test effect modification illustrate this point by documenting differential risks across sociodemographic strata and clinical characteristics (Jiao et al., 2023; Chersich et al., 2020).

HEAT STRESS DURING LACTATION AND THE POSTPARTUM PERIOD

The postpartum period places physiologic recovery alongside intensive caregiving, and extreme heat can compress the margin for safety when sleep is disrupted, hydration is difficult to maintain, or cooling is unreliable. Consistent with the review's scope, this section centers postpartum physiology and caregiving demands under heat, lactation-specific risks and barriers to breastfeeding or pumping, and infant feeding as an equity issue during heat events and power disruptions. Heat exposure during this window is often experienced in domestic and workplace settings where responsibilities are nondeferrable and where access to shade, rest, and cooling is unevenly distributed.

Postpartum heat vulnerability should be understood clinically rather than treated as a uniform extension of pregnancy risk. Recovery differs after vaginal birth, cesarean birth, hypertensive disorders, hemorrhage, infection, anemia, cardiometabolic disease, mental health conditions, and sleep disruption. These conditions can affect hydration, mobility, medication use, warning-sign recognition, and the ability to leave home for cooling or care. This is consistent with the "fourth trimester" model of postpartum care, which frames postpartum recovery as an ongoing process requiring individualized assessment of physical, psychological, and social needs rather than a single visit (American College of Obstetricians and Gynecologists [ACOG], 2018).

Lactation guidance during heat events should also be clinically precise. The available evidence does not support advising routine supplemental water for healthy exclusively breastfed infants under six months solely because of hot weather; studies summarized in a systematic review found that exclusively breastfed infants maintained normal hydration under hot conditions (Edney et al., 2022). At the same time, lactating people have increased fluid demands and may experience fatigue, perceived low supply, or difficulty feeding or pumping when heat intersects with sleep deprivation, workplace constraints, limited privacy, unsafe water, or power loss. Evidence on direct heat-related changes in human milk production or composition remains limited; small physiologic studies suggest that human milk osmolality is relatively regulated and not simply determined by maternal urinary osmolality (Soto-Méndez et al., 2016). The practice implication is not to medicalize heat exposure as inevitable lactation failure, but to protect feeding conditions: reliable cooling, safe water, rest, privacy, refrigeration, backup power, and rapid lactation support when supply concerns, mastitis symptoms, or infant hydration concerns arise.

This phrasing avoids overclaiming. It acknowledges limited direct evidence on milk composition/supply while strengthening the clinical rationale for targeted support. Edney et al. found that exclusively breastfed infants maintained normal hydration in hot conditions, and Soto-Méndez et al. found no significant association between maternal urinary osmolality and milk osmolality at a common time point. ACOG's postpartum guidance supports individualized, ongoing postpartum care rather than a single postpartum encounter.

Lactation increases daily fluid and energy needs, and heat can heighten fatigue and dehydration risk for postpartum people who are feeding on demand, traveling with infants, or working in hot environments. Evidence from hot-weather settings suggests that caregivers may introduce supplemental fluids or alter feeding practices during periods of high temperature, often reflecting perceived infant thirst rather than physiologic necessity (Edney et al., 2022; Part et al., 2022). These dynamics matter for equity because heat-related constraints on time, privacy, and material resources can narrow feasible feeding options, particularly for those in low-wage jobs or precarious work.

Breastfeeding and pumping also depend on infrastructure that becomes more fragile during heat waves. Workplace lactation support hinges on protected time, supervisor support, and access to adequate space and refrigeration, yet these conditions are least reliable in many hourly and service-sector positions (Lauer et al., 2019; Litwan et al., 2021). Heat can amplify these barriers through longer commutes, limited transit comfort, and difficulty keeping expressed milk safely stored.

Infant feeding inequities sharpen when extreme heat coincides with water insecurity or power outages. Formula preparation requires safe water, sanitation, and temperature control, and disruptions can increase contamination and spoilage risks while simultaneously undermining safe storage of expressed milk (Gribble & Berry, 2011; Hwang et al., 2021). An equity-oriented response therefore treats lactation support, reliable cooling, and targeted emergency planning as essential components of climate resilience for postpartum families (Cerceo et al., 2024).

CHILD HEALTH IMPACTS: IN UTERO THROUGH EARLY CHILDHOOD

Consistent with the scope of this review, child health impacts are considered along a continuum that begins in utero and extends through early childhood, with attention to age-specific vulnerability, co-exposures, and contextual gaps in the U.S. Southwest. Prenatal heat exposure is most plausibly understood as fetal exposure mediated through maternal heat strain. When maternal thermoregulation is challenged, compensatory cardiovascular and endocrine responses may alter uteroplacental perfusion, hydration status, and inflammatory signaling, creating biologically credible pathways through which the timing and intensity of heat exposure could influence fetal development. This life-course framing is supported by recent pediatric review evidence showing that extreme heat is linked not only to adverse birth outcomes but also to pediatric emergency department use, asthma exacerbations, heat illness, and school-related functioning, although adaptation studies remain comparatively sparse (Schapiro et al., 2024).

In the neonatal and infant period, physiologic immaturity and dependence on caregiver's shape risk in ways that are not captured by ambient temperature alone. Population-based work links warmer conditions to adverse infant outcomes, including higher infant mortality during warm seasons, with patterns that suggest multiple pathways rather than a single mechanism (Basu et al., 2015). Heat may also contribute to fatal risk in infancy through sleep-related vulnerability; a national case-crossover analysis reported associations between higher ambient temperature and sudden infant death syndrome (Jhun et al., 2017). Morbidity signals are visible in healthcare utilization as well. In large U.S. datasets, higher warm-season temperatures and extreme heat have been associated with increased pediatric emergency department visits, including for heat-related illness and dehydration (Bernstein et al., 2022; Stowell et al., 2023). Nighttime heat may be especially salient for infants because sleep and thermoregulation are tightly coupled; objective home monitoring during a UK heatwave documented measurable disruption of infant sleep at higher indoor temperatures, supporting sleep disruption as a plausible and undermeasured pathway (Berger et al., 2023).

For toddlers and young children, downstream impacts often unfold through compound environmental conditions and constraints on safe activity. Higher temperatures intensify photochemical ozone formation, and ozone is a well-established trigger for asthma exacerbations, with modeling studies projecting substantial ozone-related asthma emergency department burdens under warming scenarios (Nassikas et al., 2020). In the Southwest, wildfire smoke is an increasingly relevant co-exposure, and recent evidence suggests that extreme heat and wildfire smoke can jointly elevate hospitalization risk beyond what either hazard produces alone (Jones-Ngo et al.,

2025). Heat also reshapes daily life by curtailing outdoor play and straining childcare and school settings; for example, school-based evidence from Arizona documents frequent restriction of outdoor recess during the hottest months, with implications for physical activity and routines (Poulos & Wilson, 2025). Despite growing evidence, data gaps remain striking. Few studies quantify indoor heat in childcare or housing environments, and rural and Tribal communities remain underrepresented in the literature despite unique infrastructural and access constraints. Addressing these gaps will be essential for designing interventions that recognize how age, geography, and housing conditions intersect to shape children's lived heat exposure.

HEALTH INEQUITIES AND INTERSECTING VULNERABILITIES

Health inequities shape heat risk through three related processes: differential exposure, differential susceptibility, and differential access to protection and care. An intersectionality lens is useful because heat is experienced through overlapping social positions, including race and ethnicity, socioeconomic status, immigration status, disability, gender, and occupation, which can compound one another (Bowleg, 2012).

Differential exposure in the Southwest is driven by where people live, work, and travel. Contemporary urban heat islands track historic and ongoing segregation, with formerly redlined neighborhoods consistently hotter across U.S. cities (Hoffman et al., 2020). Housing conditions then mediate how outdoor heat becomes an indoor exposure, and uneven access to cooling technologies further widens the gradient. Occupational stratification concentrates exposure as well, from outdoor labor in agriculture and construction to indoor heat in warehouses and kitchens, often with limited control over rest or hydration. Migrant and seasonal workers, border communities, and workers facing language barriers or fear of retaliation may have fewer practical options to reduce exposure even when risk is recognized. Heat vulnerability mapping work highlights that community determinants such as housing quality, social isolation, and transportation access shape who is exposed and who can reach relief (Reid et al., 2009).

Differential susceptibility reflects the likelihood of harm given a comparable thermal load. Energy insecurity is a central mechanism: households may be unable to afford adequate cooling, may ration electricity, or may occupy dwellings that retain heat overnight, increasing physiologic strain atop baseline conditions and reproductive health vulnerabilities (Hernández, 2013). Utility shutoffs intensify this vulnerability, and disconnection policy has been shown to differentially affect vulnerable populations, including low-income households (Flaherty et al., 2020).

Differential access to protection and care concerns whether safeguards and clinical responses are reachable and responsive. In Maricopa County, a public health evaluation of cooling centers documented gaps in awareness and practical barriers to use, including reliance on informal referral networks and obstacles for visitors who were unemployed or unhoused (Berisha et al., 2017). In rural areas and some Tribal communities, distance to cooling resources and limited clinic or hospital capacity can further delay care during extreme heat. For Indigenous communities, access constraints often intersect with water insecurity and infrastructure deficits, along with jurisdictional complexity that can impede timely investment and accountability (Tanana et al., 2021; Wilson et al., 2021). Carceral settings represent an extreme form of constrained access, where exposure is shaped by institutional control over cooling and limited recourse for people experiencing symptoms (Tuholske et al., 2024). Clinical equity also matters for whether heat-related symptoms are recognized, counseled, and treated. Relationship-centered approaches grounded in critical race theory have been proposed as a practical antidote to structural racism in perinatal care (Hardeman et al., 2020).

INTERVENTIONS AND POLICY LANDSCAPE: WHAT EXISTS, WHAT WORKS, AND FOR WHOM

Consistent with the review outline's multilevel structure, the intervention and policy landscape for extreme heat spans clinical encounters, health system coordination, public health preparedness, workplace regulation, and housing and energy protections. Across these domains, the evidence base is uneven: some approaches have been evaluated for population health benefit, while others are widely implemented without strong data on reach, fidelity, or equity impacts (Randazza et al., 2023).

Because the intervention evidence base is uneven, we distinguish here between recommendations supported by evaluated evidence, recommendations supported by observational or implementation evidence, and recommendations grounded primarily in biologic plausibility, clinical judgment, or equity-based policy reasoning. This distinction is important for practice: the absence of randomized or quasi-experimental evidence should not be mistaken for lack of need, particularly when heat exposure is already producing foreseeable risk. At the same time, recommendations should be presented with enough specificity that clinicians, health systems, and policymakers can judge the evidentiary basis for action. Table 2 summarizes recommended heat-adaptation strategies by evidence basis and practical interpretation. The table is intended to clarify which approaches are supported by evaluated public health evidence, occupational or implementation research, clinical plausibility, or structural policy reasoning, while also identifying areas where maternal–child-specific outcome evaluation remains limited.

Table 2. Evidence strength and implementation status of recommended heat-adaptation strategies

Strategy	Evidence basis	Interpretation for practice
Heat-health warning systems and coordinated heat action plans	Evaluated public-health evidence, mortality-focused studies, implementation analyses	Strongest population-level evidence, but maternal–child specificity and equity reach require better evaluation
Clinical heat-risk screening and individualized heat plans	Clinical plausibility, emerging public-health guidance, limited direct trial evidence	Reasonable near-term practice, but should be embedded in workflow and linked to resources
Occupational heat protections	Occupational epidemiology, physiologic studies, regulatory guidance	Strong rationale for enforceable protections; pregnancy/lactation-specific evidence remains limited
Cooling centers and cooling referrals	Public-health practice evidence and local evaluations	Useful when accessible, but transportation, hours, trust, and eligibility barriers limit equity impact
Housing weatherization, cool roofs, and energy protections	Modeling studies, housing-health evidence, energy-insecurity literature	Equity-relevant structural interventions; more maternal–child outcome evaluation is needed
Community health worker, nurse, midwife, WIC, and home-visiting outreach	Evidence from community-based care models and climate-resilience literature	Promising implementation pathway, especially for families facing access barriers; requires staffing, training, and reimbursement

Note. Evidence basis reflects the predominant type of support identified in this narrative review and should not be interpreted as a formal evidence grade or comparative ranking of strategies. Strategies are summarized according to their relevance for maternal–child heat adaptation, implementation feasibility, and equity impact in the U.S. Southwest. Interpretation for practice emphasizes whether each approach has evaluated population-level evidence, occupational or implementation support, clinical plausibility, structural policy relevance, or a need for further maternal–child outcome evaluation. WIC = Special Supplemental Nutrition Program for Women, Infants, and Children.

At the clinical and health system level, feasible near-term actions include routine heat risk screening in prenatal and postpartum care, individualized hydration and cooling plans, and low threshold triage pathways for heat symptoms, with careful attention to occupational and housing constraints that can make standard advice impracticable. Existing heat health messaging for pregnancy often lacks specificity, omits structural barriers, or places responsibility on individuals without naming the resources required to act, underscoring the need for patient centered counseling that explicitly links clinical guidance to referrals for cooling, transportation, and utility supports (Stassen et al., 2024). Clinical pathways also benefit from alignment with evidence-based heat illness recognition and treatment guidance, which can be adapted to obstetric and pediatric settings during heat events (Lipman et al., 2019).

Throughout, we treat individual-level advice (hydration, rest, cooling) as clinically necessary but often insufficient unless paired with feasible supports: workplace autonomy to take breaks, safe housing and reliable electricity for cooling, transportation to cooling resources, and low-barrier access to responsive care.

Public health interventions include heat action plans, early warning systems, and cooling infrastructure. A classic evaluation of Philadelphia's heat watch warning system estimated meaningful mortality benefits relative to program costs, illustrating that coordinated alerts, outreach, and response can save lives when activation thresholds and communication strategies are locally calibrated (Ebi et al., 2004). In the Southwest, cooling centers are a prominent adaptation strategy; evaluation work from Maricopa County highlights both their role and their constraints, particularly for residents facing transportation barriers, limited hours, or distrust of formal services (Berisha et al., 2017). Heat action plans commonly incorporate risk communication and cooling centers, but content analyses show recurring gaps in language accessibility and in targeted, operationalized strategies for high-risk groups, which is where equity intent often fails at implementation (Randazza et al., 2023).

Workplace protections are central in a region where heat exposure is structurally concentrated in outdoor and non-climate controlled indoor jobs. Occupational heat strain is associated with measurable health risks and productivity losses, supporting enforceable standards that mandate water, rest, shade, and heat acclimatization planning (Flouris et al., 2018). Empirical evidence from California farm settings suggests that compliance with existing rules does not fully eliminate physiologic heat risk, emphasizing that enforcement, training quality, work pace, and piece rate pressures matter, and that protections must be real in practice for pregnant and lactating workers as well as for the broader workforce (Langer et al., 2021). Policy development for indoor workplaces is also advancing; a standardized regulatory impact assessment for California's proposed indoor heat regulation illustrates how anticipated benefits depend on coverage, enforcement capacity, and employer compliance, all of which can vary by industry and worker bargaining power (Metz et al., 2021). International heat-health guidance similarly treats warning systems as one component of broader heat-health action plans, linking meteorologic thresholds with communication, public health response, clinical preparedness, and evaluation rather than treating alerts as self-executing protections (World Meteorological Organization & World Health Organization, 2015).

Housing and energy policies shape who can translate warnings and clinical advice into protection. Utility disconnection policies and shutoff practices can exacerbate heat risk for medically and economically vulnerable households, making continuity of electricity for cooling an equity relevant health protection rather than a discretionary amenity (Flaherty et al., 2020). Built environment interventions such as cool roofs can reduce heat exposure, and modeling work suggests that targeting these upgrades can improve the distributive equity of urban adaptation by prioritizing neighborhoods with higher heat burden and vulnerability (Broadbent et al., 2022). Taken together, the equity critique is straightforward: interventions that rely on individual uptake or private purchase tend to benefit those with stable housing, flexible work, and financial slack, whereas policies that change workplaces, housing conditions, and service access are more likely to narrow disparities if they are enforced, accessible, and co designed with affected communities (Randazza et al., 2023).

PRACTICE AND IMPLEMENTATION: AN INTERPROFESSIONAL EQUITY-FORWARD ACTION AGENDA

Embedding heat adaptation within maternal and child health care requires translation from epidemiologic evidence into daily clinical and public health practice. Standardized, equity-sensitive screening should be incorporated into prenatal, postpartum, and pediatric encounters to identify heat risk systematically. Electronic health records can prompt clinicians to document household cooling access, energy insecurity, transportation options, and occupational exposure. Risk stratification should trigger individualized, patient-centered heat plans that address hydration, rest, medication management, and emergency thresholds. Materials must be linguistically and culturally appropriate, integrating interpreter services and trusted community channels (Karliner et al., 2007; Lipman et al., 2019). Counseling that acknowledges structural barriers, such as housing instability or power shutoffs, can reduce stigma and improve trust, consistent with relationship-centered models of perinatal care (Hardeman et al., 2020).

Implementation should be presented as a systems task, not an additional counseling burden placed on already compressed visits. Heat-risk screening will require decisions about who asks the questions, where responses are documented, how positive screens trigger referral, and how follow-up is tracked. Evidence from maternal and infant care settings shows that social risk screening and referral processes vary across clinics and that high patient volume and limited electronic health record capacity can impede implementation (Brown et al., 2023). In community health centers serving agricultural, rural, and border communities, these barriers may be compounded by staffing shortages, seasonal work schedules, transportation barriers, language needs, and limited availability of referral partners. A feasible heat-screening workflow should therefore be brief, team-based, and actionable: one or two heat-specific questions embedded into intake or rooming; standing referral pathways for cooling, utility support, and occupational concerns; patient-facing materials in locally used languages; and a defined follow-up role for community health workers, nurses, midwives, WIC staff, home visitors, or care coordinators.

At the system level, heat response protocols should embed care coordination during official heat alerts, including proactive outreach to high-risk obstetric and pediatric patients. Partnerships with local health departments and community-based organizations can create real-time referral pathways to cooling centers, energy assistance, and weatherization programs (Berisha et al., 2017; Flaherty et al., 2020). Memoranda of understanding between hospitals, utility providers, and community agencies can ensure electricity continuity for medically vulnerable households during heat emergencies (Hernández, 2013). At the population level, co-designed public health interventions, developed through participatory processes with tribal, migrant, and rural communities, can strengthen trust and relevance while addressing contextual barriers to uptake (Randazza et al., 2023). Metrics for evaluating such efforts should include both reach and equity outcomes, such as differential access across language, geography, and income strata (Braveman & Gruskin, 2003).

Implementation requires deliberate attention to feasibility, acceptability, and sustainability. Community health workers, WIC programs, and home-visiting models provide feasible entry points for integrating heat education and resource linkage. Avoiding stigmatizing narratives that individualize responsibility is essential for acceptability among marginalized populations (Hardeman et al., 2020). Sustainability depends on institutional accountability and cross-sector coordination, underpinned by routine measurement of equity impacts and implementation outcomes such as fidelity, penetration, and cost (Proctor et al., 2011). In this framing, equity is not an outcome separate from implementation but an indicator of whether adaptation functions as intended.

Interprofessional and Community-Based Implementation Roles

Heat adaptation in maternal and child health cannot rest on obstetric, pediatric, or emergency care alone. Community health nurses, midwives, lactation consultants, WIC staff, home visitors, doulas, promotoras, community health workers, pharmacists, social workers, and allied health professionals often have earlier and more sustained contact with families than acute care systems do. Their roles are especially important when heat risk is mediated by household cooling, infant feeding constraints, work schedules, transportation, language access, or fear of institutional contact.

Midwifery continuity models are relevant because they organize care around relational continuity rather than isolated encounters; recent Cochrane evidence indicates that midwife continuity of care is associated with more positive experiences and some improved birth outcomes compared with other models of care (Sandall et al., 2024). Community health worker models also provide a pragmatic route for heat preparedness because they combine navigation, health education, surveillance, referral, and trust-building functions. A scoping review of community health workers during extreme weather events found roles spanning clinical support, service navigation, community education, data collection, psychosocial support, and emergency response (Domingo et al., 2024). These models should not be romanticized: they require training, supervision, compensation, clear referral pathways, and protection for workers who may themselves be exposed to heat. A genuinely interprofessional heat plan should therefore specify which professional group performs screening, who provides counseling, who coordinates social and utility referrals, who follows up during heat alerts, and how responsibility is shared without shifting uncompensated labor onto the least-resourced workers.

RESEARCH GAPS AND PRIORITIES FOR THE SOUTHWEST

Despite growing awareness of heat-related reproductive and child health risks, major empirical and structural gaps persist in the U.S. Southwest. Exposure assessment remains limited by reliance on outdoor temperature metrics that fail to capture indoor or occupational heat conditions. Studies integrating household sensors, wearable thermometers, and microclimate modeling are needed to quantify true exposures, particularly for populations in mobile homes, informal dwellings, and non-air-conditioned workplaces (Quinn et al., 2014; Hondula et al., 2021). Incorporating time–activity and workplace assessments would illuminate cumulative thermal loads among pregnant agricultural and service-sector workers (Moyce & Schenker, 2018).

Methodologically, stronger causal designs, including quasi-experimental and natural experiments, are needed to identify effect pathways and evaluate adaptation measures. Standardizing outcome definitions and exposure windows would improve synthesis across jurisdictions (Basu et al., 2010; Craig et al., 2017). Integrative mechanistic studies linking physiologic strain, placental function, and birth outcomes could clarify biologic plausibility and inform prevention thresholds.

Equity-critical gaps among tribal, migrant, rural, and frontier communities are not merely accidental omissions in the literature; they reflect structural barriers in research funding, data systems, jurisdictional authority, language access, transportation, broadband, and the historic misuse of data from Indigenous and marginalized communities. Future research should therefore be governed differently, not simply expanded. Studies involving Tribal Nations should be developed through tribal approval processes, data-sharing agreements, community-defined benefit, and Indigenous data governance principles. The CARE principles emphasize collective benefit, authority to control, responsibility, and ethics, including the principle that Indigenous Peoples should determine data governance protocols and participate in stewardship decisions (Carroll et al., 2020). Community-based participatory research offers a complementary approach by treating affected communities as partners in question selection, study design, interpretation, and dissemination rather than as recruitment sites (Wallerstein & Duran, 2010). In practice, this requires funding community partners, compensating local advisors, returning usable findings, respecting tribal review requirements, and measuring outcomes that communities identify as meaningful, such as household cooling reliability, water access, safe transportation, clinic reachability, and continuity of culturally safe care.

LIMITATIONS

Several limitations should guide interpretation. First, this review synthesizes heterogeneous evidence across exposure metrics, geography, outcomes, and study designs; it should not be read as a meta-analysis or as a quantitative estimate of heat-attributable maternal–child risk. Second, many studies assign outdoor ambient temperature by residence, which can misclassify personal, indoor, occupational, and travel-related heat exposure (Basu et al., 2010; Hondula et al., 2021). Third, evidence for lactation, postpartum recovery, childcare settings, and rural or Tribal heat adaptation remains thinner than evidence for preterm birth and acute pediatric utilization. Fourth, policy and implementation sources clarify feasibility and reach but generally cannot establish health effects on their own. These limits do not weaken the case for adaptation; they show where near-term practice should proceed with monitoring and where future studies need stronger exposure measurement, community governance, and outcome evaluation.

CONCLUSION

In the U.S. Southwest, extreme heat functions as a structurally patterned hazard rather than a uniform environmental exposure. The burden of heat is produced and distributed through the same social and policy arrangements that shape where families live, how housing retains heat, who can afford or maintain cooling, who works in high-heat jobs, and who can access timely and responsive prenatal, postpartum, and pediatric care. For pregnant and lactating people and young children, these exposures intersect with physiologic vulnerability and caregiving demands, making heat adaptation inseparable from health equity. The practical implication of this review is that reducing heat-related maternal and child health harm requires shifting from advice that assumes individual capacity to interventions that change the conditions of exposure and the availability of protection.

Because this is a narrative synthesis emphasizing anchor evidence rather than an exhaustive systematic review, some relevant studies may not be captured; nonetheless, the conclusions are supported by convergent findings from major syntheses, large cohorts, and equity-relevant policy/practice sources. Several high-impact actions can be implemented immediately. First, routine heat risk identification should become part of standard perinatal and pediatric care, integrated into documentation workflows and paired with individualized heat plans that include clear symptom thresholds, contingency guidance for power loss, and culturally and linguistically appropriate materials. Second, health systems should operationalize heat-event care coordination in the same way they manage other predictable surges, with proactive outreach to higher-risk patients, flexible scheduling and telehealth options during extreme heat, and standing referral pathways to cooling resources, transportation support, and utility assistance. Third, public health and clinical partners should evaluate success using metrics that make equity visible, including whether interventions reach people facing the greatest barriers and whether disparities in heat-related illness, adverse birth outcomes, and acute care utilization narrow over time.

Longer-term priorities must address the structural determinants that constrain the ability to act on heat guidance. A first priority is heat-safe, affordable housing and energy protections that prevent loss of cooling during extreme events and reduce chronic indoor heat exposure, including weatherization, resilient cooling infrastructure, and safeguards against utility disconnection for medically and economically vulnerable households. A second priority is enforceable workplace protections for outdoor and non-climate-controlled indoor labor, with standards that are practical in real settings, supported by meaningful enforcement, and responsive to pregnancy and lactation. Advancing these near-term and long-term actions together reframes heat adaptation as a maternal and child health protection strategy with equity as its measure of effectiveness, not an aspirational add-on.

CONFLICT OF INTEREST STATEMENT

The authors declare that there are no conflicts of interest regarding the research, authorship, and/or publication of this manuscript.

FUNDING STATEMENT

This research received no external funding.

AUTHORSHIP AND CONTRIBUTIONS STATEMENT

Authorship for this manuscript was determined in accordance with internationally recognized authorship guidelines (e.g., ICMJE). All authors made substantial contributions to the planning/conception of the work, conduct of the research, and execution of the manuscript; participated in drafting and/or critically revising the manuscript for important intellectual content; approved the final version for publication; and agree to be accountable for all aspects of the work.

REFERENCES

American College of Obstetricians and Gynecologists. (2018). Optimizing postpartum care. *ACOG Committee Opinion No. 736. Obstetrics and Gynecology, 131*(5), e140–e150. https://www.acog.org/clinical/clinical-guidance/committee-opinion/articles/2018/05/optimizing-postpartum-care. DOI: 10.1097/AOG.0000000000002633 PMID: 29683911

Anderson, G. B., & Bell, M. L. (2011). Heat waves in the United States: Mortality risk during heat waves and effect modification by heat wave characteristics in 43 U.S. communities. *Environmental Health Perspectives, 119*(2), 210–218. DOI: 10.1289/ehp.1002313 PMID: 21084239

Baethge, C., Goldbeck-Wood, S., & Mertens, S. (2019). SANRA—A scale for the quality assessment of narrative review articles. *Research Integrity and Peer Review, 4*(1), 5. Advance online publication. DOI: 10.1186/s41073-019-0064-8 PMID: 30962953

Basu, R., Malig, B., & Ostro, B. (2010). High ambient temperature and the risk of preterm delivery. *American Journal of Epidemiology, 172*(10), 1108–1117. DOI: 10.1093/aje/kwq170 PMID: 20889619

Bekkar, B., Pacheco, S., Basu, R., & DeNicola, N. (2020). Association of air pollution and heat exposure with preterm birth, low birth weight, and stillbirth in the US: A systematic review. *JAMA Network Open, 3*(6), e208243. DOI: 10.1001/jamanetworkopen.2020.8243 PMID: 32556259

Berger, S. E., Ordway, M. R., Schoneveld, E., Lucchini, M., Thakur, S., Anders, T., Natale, L., & Barnett, N. (2023). The impact of extreme summer temperatures in the United Kingdom on infant sleep: Implications for learning and development. *Scientific Reports, 13*(1), 10061. DOI: 10.1038/s41598-023-37111-2 PMID: 37344536

Berisha, V., Hondula, D. M., Roach, M., White-Newsome, J. L., McCormick, S., Marnin, M. L., & Jolliet, O. (2017). Assessing adaptation strategies for extreme heat: A public health evaluation of cooling centers in Maricopa County, Arizona. *Weather, Climate, and Society, 9*(1), 71–80. DOI: 10.1175/WCAS-D-16-0033.1

Bernstein, A. S., Sun, S., Weinberger, K. R., Spangler, K. R., Sheffield, P. E., & Wellenius, G. A. (2022). Warm season and emergency department visits to U.S. children's hospitals. *Environmental Health Perspectives, 130*(1), 017001. DOI: 10.1289/EHP8083 PMID: 35044241

Bouchama, A., & Knochel, J. P. (2002). Heat stroke. *The New England Journal of Medicine, 346*(25), 1978–1988. DOI: 10.1056/NEJMra011089 PMID: 12075060

Bowleg, L. (2012). The problem with the phrase women and minorities: Intersectionality-an important theoretical framework for public health. *American Journal of Public Health, 102*(7), 1267–1273. DOI: 10.2105/AJPH.2012.300750 PMID: 22594719

Bowler, D. E., Buyung-Ali, L., Knight, T. M., & Pullin, A. S. (2010). Urban greening to cool towns and cities: A systematic review of the empirical evidence. *Landscape and Urban Planning*, *97*(3), 147–155. DOI: 10.1016/j.landurbplan.2010.05.006

Braveman, P., & Gruskin, S. (2003). Defining equity in health. *Journal of Epidemiology and Community Health, 57*(4), 254–258. DOI: 10.1136/jech.57.4.254 PMID: 12646539

Broadbent, A. M., Declet-Barreto, J., Krayenhoff, E. S., Harlan, S. L., & Georgescu, M. (2022). Targeted implementation of cool roofs for equitable urban adaptation to extreme heat. *The Science of the Total Environment, 811*, 151326. DOI: 10.1016/j.scitotenv.2021.151326 PMID: 34757097

Brown, J., Ahmed, N., Biel, M., Patchen, L., Rethy, J., Thomas, A., & Arem, H. (2023). Considerations in implementation of social risk factor screening and referral in maternal and infant care in Washington, DC: A qualitative study. *PLoS One, 18*(4), e0283815. Advance online publication. DOI: 10.1371/journal.pone.0283815 PMID: 37053233

Carroll, S. R., Garba, I., Figueroa-Rodríguez, O. L., Holbrook, J., Lovett, R., Materechera, S., Parsons, M., Raseroka, K., Rodriguez-Lonebear, D., Rowe, R., Sara, R., Walker, J. D., Anderson, J., & Hudson, M. (2020). The CARE principles for Indigenous data governance. *Data Science Journal, 19*, 43. https://datascience.codata.org/articles/dsj-2020-043. DOI: 10.5334/dsj-2020-043

Casey, J. A., Fukurai, M., Hernández, D., Balsari, S., & Kiang, M. V. (2020). Power outages and community health: A narrative review. *Current Environmental Health Reports, 7*(4), 371–383. DOI: 10.1007/s40572-020-00295-0 PMID: 33179170

Cerceo, E., Saxer, K., Grossman, L., Shapley-Quinn, K., & Feldman-Winter, L. (2024). The climate crisis and breastfeeding: Opportunities for resilience. *Journal of Human Lactation : Official Journal of International Lactation Consultant Association, 40*(1), 33–50. DOI: 10.1177/08903344231216726 PMID: 38158719

Chersich, M. F., Pham, M. D., Areal, A., Haghighi, M. M., Manyuchi, A., Swift, C. P., Wernecke, B., Robinson, M., Hetem, R., Boeckmann, M., & Hajat, S.Climate Change and Heat-Health Study Group. (2020). Associations between high temperatures in pregnancy and risk of preterm birth, low birth weight, and stillbirths: Systematic review and meta-analysis. *BMJ (Clinical Research Ed.)*, *371*, m3811. DOI: 10.1136/bmj.m3811 PMID: 33148618

Chow, W. T. L., Brennan, D., & Brazel, A. J. (2012). Urban heat island research in Phoenix, Arizona: Theoretical contributions and policy applications. *Bulletin of the American Meteorological Society*, *93*(4), 517–530. DOI: 10.1175/BAMS-D-11-00011.1

Craig, P., Katikireddi, S. V., Leyland, A., & Popham, F. (2017). Natural experiments: An overview of methods, approaches, and contributions to public health intervention research. *Annual Review of Public Health*, *38*(1), 429–451. DOI: 10.1146/annurev-publhealth-031816-044327 PMID: 28125392

Domingo, A., Little, M., Beggs, B., Brubacher, L. J., Lau, L. L., & Dodd, W. (2024). Examining the role of community health workers amid extreme weather events in low- and middle-income countries: A scoping review. *Public Health*, *236*, 133–143. DOI: 10.1016/j.puhe.2024.07.023 PMID: 39182471

Dong, X. S., Largay, J. A., Wang, X., Sokas, R., & Mohr, D. (2019). Heat-related deaths among construction workers in the United States. *American Journal of Industrial Medicine*, *62*(12), 1047–1057. DOI: 10.1002/ajim.23024 PMID: 31328819

Ebi, K. L., Teisberg, T. J., Kalkstein, L. S., Robinson, L., & Weiher, R. F. (2004). Heat watch warning systems save lives: Estimated costs and benefits for Philadelphia 1995–1998. *Bulletin of the American Meteorological Society*, *85*(8), 1067–1073. https://www1.udel.edu/SynClim/BAMS_Ebi_Kalkstein.pdf. DOI: 10.1175/BAMS-85-8-1067

Edney, J. M., Kovats, S., Filippi, V., & Nakstad, B. (2022). A systematic review of hot weather impacts on infant feeding practices in low- and middle-income countries. *Frontiers in Pediatrics*, *10*, 930348. DOI: 10.3389/fped.2022.930348 PMID: 36147803

Flaherty, M., Carley, S., & Konisky, D. M. (2020). Electric utility disconnection policy and vulnerable populations. *The Electricity Journal*, *33*(10), 106859. DOI: 10.1016/j.tej.2020.106859

Flouris, A. D., Dinas, P. C., Ioannou, L. G., Nybo, L., Havenith, G., Kenny, G. P., & Kjellstrom, T. (2018). Workers' health and productivity under occupational heat strain: A systematic review and meta-analysis. *The Lancet. Planetary Health*, *2*(12), e521–e531. DOI: 10.1016/S2542-5196(18)30237-7 PMID: 30526938

Grant, M. J., & Booth, A. (2009). A typology of reviews: An analysis of 14 review types and associated methodologies. *Health Information and Libraries Journal*, *26*(2), 91–108. DOI: 10.1111/j.1471-1842.2009.00848.x PMID: 19490148

Gribble, K. D., & Berry, N. J. (2011). Emergency preparedness for those who care for infants in developed country contexts. *International Breastfeeding Journal*, *6*(1), 16. DOI: 10.1186/1746-4358-6-16 PMID: 22059481

Hardeman, R. R., Karbeah, J., & Kozhimannil, K. B. (2020). Applying a critical race lens to relationship-centered care in pregnancy and childbirth: An antidote to structural racism. *Birth (Berkeley, Calif.)*, *47*(1), 3–7. DOI: 10.1111/birt.12462 PMID: 31630454

Hernández, D. (2013). Energy insecurity: A framework for understanding energy, the built environment, and health among vulnerable populations in the context of climate change. *American Journal of Public Health*, *103*(4), e32–e34. DOI: 10.2105/AJPH.2012.301179 PMID: 23409876

Hoffman, J. S., Shandas, V., & Pendleton, N. (2020). The effects of historical housing policies on resident exposure to intra-urban heat: A study of 108 US urban areas. *Climate (Basel)*, *8*(1), 12. DOI: 10.3390/cli8010012

Hondula, D. M., Kuras, E. R., Betzel, S., Drake, L., Eneboe, J., Kaml, M., Munoz, M., Sevig, M., Singh, M., Ruddell, B. L., & Harlan, S. L. (2021). Novel metrics for relating personal heat exposure to social risk factors and outdoor ambient temperature. *Environment International*, *146*, 106271. DOI: 10.1016/j.envint.2020.106271 PMID: 33395929

Hwang, C. H., Iellamo, A., & Ververs, M. (2021). Barriers and challenges of infant feeding in disasters in middle- and high-income countries. *International Breastfeeding Journal*, *16*(1), 62. DOI: 10.1186/s13006-021-00398-w PMID: 34425848

Jhun, I., Mata, D. A., Nordio, F., Lee, M., Schwartz, J., & Zanobetti, A. (2017). Ambient temperature and sudden infant death syndrome in the United States. *Epidemiology (Cambridge, Mass.)*, *28*(5), 728–734. DOI: 10.1097/EDE.0000000000000703 PMID: 28661937

Jiao, A., Sun, Y., Avila, C., Chiu, V., Slezak, J., Sacks, D. A., Abatzoglou, J. T., Molitor, J., Chen, J.-C., Benmarhnia, T., Getahun, D., & Wu, J. (2023). Analysis of heat exposure during pregnancy and severe maternal morbidity. *JAMA Network Open, 6*(9), e2332780. DOI: 10.1001/jamanetworkopen.2023.32780 PMID: 37676659

. Jones-Ngo, C. G., Schmidt, R. J., Monier, E., Ludwick, S., Al-Hamdan, M. Z., Vargo, J., & Conlon, K. C. (2025). Joint effects of wildfire smoke and extreme heat on hospitalizations in California, 2011–2020. *GeoHealth, 9*(6), e2024GH001237. DOI: 10.1029/2024GH001237

Karliner, L. S., Jacobs, E. A., Chen, A. H., & Mutha, S. (2007). Do professional interpreters improve clinical care for patients with limited English proficiency? *Health Services Research, 42*(2), 727–754. DOI: 10.1111/j.1475-6773.2006.00629.x PMID: 17362215

Krieger, N. (2001). Theories for social epidemiology in the 21st century: An ecosocial perspective. *International Journal of Epidemiology, 30*(4), 668–677. DOI: 10.1093/ije/30.4.668 PMID: 11511581

Lakhoo, D. P., Brink, N., Radebe, L., Craig, M. H., Pham, M. D., Haghighi, M. M., Wise, A., Solarin, I., Luchters, S., Maimela, G., Chersich, M. F., Chikandiwa, A., Nakstad, B., Wright, C. Y., Brimicombe, C., Technau, K.-G., Harden, L., Boeckmann, M., Strehlau, R., & Brink, N. (2025). A systematic review and meta-analysis of heat exposure impacts on maternal, fetal and neonatal health. *Nature Medicine, 31*(2), 684–694. DOI: 10.1038/s41591-024-03395-8 PMID: 39500369

Langer, C. E., Mitchell, D. C., Armitage, T. L., Moyce, S. C., Tancredi, D. J., Castro, J., Vega-Arroyo, A. J., Bennett, D. H., & Schenker, M. B. (2021). Are Cal/OSHA regulations protecting farmworkers in California from heat-related illness? *Journal of Occupational and Environmental Medicine, 63*(6), 532–539. DOI: 10.1097/JOM.0000000000002189 PMID: 33741829

Lauer, E. A., Armenti, K., Henning, M., & Sirois, L. (2019). Identifying barriers and supports to breastfeeding in the workplace experienced by mothers in the New Hampshire Special Supplemental Nutrition Program for Women, Infants, and Children utilizing the Total Worker Health framework. *International Journal of Environmental Research and Public Health, 16*(4), 529. DOI: 10.3390/ijerph16040529 PMID: 30781764

Lipman, G. S., Gaudio, F. G., Eifling, K. P., Ellis, M. A., Grissom, C. K., Morris, A., & Otten, E. M. (2019). Wilderness Medical Society Clinical Practice Guidelines for the prevention and treatment of heat illness: 2019 update. *Wilderness & Environmental Medicine, 30*(4), S33–S46. DOI: 10.1016/j.wem.2018.10.004 PMID: 31221601

Litwan, K., Tran, V., Nyhan, K., & Pérez-Escamilla, R. (2021). How do breastfeeding workplace interventions work?: A realist review. *International Journal for Equity in Health, 20*(1), 148. DOI: 10.1186/s12939-021-01490-7 PMID: 34172068

Lopez, H., Lee, S.-K., West, R., Kim, D., & Jia, L. (2025). The longest-lasting 2023 western North American heat wave was fueled by the record-warm Atlantic Ocean. *Nature Communications, 16*(1), 6544. DOI: 10.1038/s41467-025-61859-y PMID: 40670333

Luna, Z., & Luker, K. (2013). Reproductive justice. *Annual Review of Law and Social Science, 9*(1), 327–352. DOI: 10.1146/annurev-lawsocsci-102612-134037

Metz, D., Prier, S., & Miller, B. M. (2021). *Prevention of heat-related illness in indoor places of employment: Standardized regulatory impact assessment (SRIA) of the proposed California regulation for heat illness prevention in indoor places of employment (WR-A1386-1)*. RAND Corporation., DOI: 10.7249/WRA1386-1

Mohai, P., Pellow, D., & Roberts, J. T. (2009). Environmental justice. *Annual Review of Environment and Resources, 34*(1), 405–430. DOI: 10.1146/annurev-environ-082508-094348

Moyce, S., Mitchell, D., Armitage, T., Tancredi, D., Joseph, J., & Schenker, M. (2017). Heat strain, volume depletion and kidney function in California agricultural workers. *Occupational and Environmental Medicine, 74*(6), 402–409. DOI: 10.1136/oemed-2016-103848 PMID: 28093502

Nassikas, N., Spangler, K. R., Fann, N., Nolte, C. G., Dolwick, P., Spero, T. L., Sheffield, P., & Wellenius, G. A. (2020). Ozone-related asthma emergency department visits in the US in a warming climate. *Environmental Research, 183*, 109206. DOI: 10.1016/j.envres.2020.109206 PMID: 32035409

Part, C., Filippi, V., Cresswell, J. A., Ganaba, R., Hajat, S., Nakstad, B., Roos, N., Kadio, K., Chersich, M., Lusambili, A., Kouanda, S., & Kovats, S. (2022). How do high ambient temperatures affect infant feeding practices? A prospective cohort study of postpartum women in Bobo-Dioulasso, Burkina Faso. *BMJ Open, 12*(10), e061297. DOI: 10.1136/bmjopen-2022-061297 PMID: 36198451

Perkins-Kirkpatrick, S. E., & Lewis, S. C. (2020). Increasing trends in regional heatwaves. *Nature Communications, 11*(1), 3357. DOI: 10.1038/s41467-020-16970-7 PMID: 32620857

Poirier, B., Haag, D., Soares, G. H., & Jamieson, L. M. (2023). Whose values, what bias, which subjectivity?: The need for reflexivity and positionality in epidemiological health equity scholarship. *Australian and New Zealand Journal of Public Health, 47*(4), 100079. Advance online publication. DOI: 10.1016/j.anzjph.2023.100079 PMID: 37633183

Poulos, A., & Wilson, K. (2025). Effect of heat on outdoor recess in Arizona elementary schools, July–September 2023. *Public Health Reports (Washington, D.C.), 140*(5–6), 437–440. Advance online publication. DOI: 10.1177/00333549251342205 PMID: 40616399

Proctor, E., Silmere, H., Raghavan, R., Hovmand, P., Aarons, G., Bunger, A., Griffey, R., & Hensley, M. (2011). Outcomes for implementation research: Conceptual distinctions, measurement challenges, and research agenda. *Administration and Policy in Mental Health, 38*(2), 65–76. DOI: 10.1007/s10488-010-0319-7 PMID: 20957426

Quinn, A., Tamerius, J. D., Perzanowski, M., Jacobson, J. S., Goldstein, I., Acosta, L., & Shaman, J. (2014). Predicting indoor heat exposure risk during extreme heat events. *The Science of the Total Environment, 490*, 686–693. DOI: 10.1016/j.scitotenv.2014.05.039 PMID: 24893319

Randazza, J. M., Hess, J. J., Bostrom, A., Hartwell, C., Nori-Sarma, A., Spangler, K. R., Sun, Y., Weinberger, K. R., Wellenius, G. A., & Errett, N. A. (2023). Planning to reduce the health impacts of extreme heat: A content analysis of heat action plans in local United States jurisdictions. *American Journal of Public Health, 113*(5), 559–567. DOI: 10.2105/AJPH.2022.307217 PMID: 36926967

Reid, C. E., O'Neill, M. S., Gronlund, C. J., Brines, S. J., Brown, D. G., Diez-Roux, A. V., & Schwartz, J. (2009). Mapping community determinants of heat vulnerability. *Environmental Health Perspectives, 117*(11), 1730–1736. DOI: 10.1289/ehp.0900683 PMID: 20049125

Samuels, L., Nakstad, B., Roos, N., Bonell, A., Chersich, M., Havenith, G., Luchters, S., Day, L.-T., Hirst, J. E., Singh, T., Elliott-Sale, K., Hetem, R., Part, C., Sawry, S., Le Roux, J., & Kovats, S. (2022). Physiological mechanisms of the impact of heat during pregnancy and the clinical implications: Review of the evidence from an expert group meeting. *International Journal of Biometeorology, 66*(8), 1505–1513. DOI: 10.1007/s00484-022-02301-6 PMID: 35554684

Sandall, J., Fernandez Turienzo, C., Devane, D., Soltani, H., Gillespie, P., Gates, S., Jones, L. V., Shennan, A. H., & Rayment-Jones, H. (2024). Midwife continuity of care models versus other models of care for childbearing women. *Cochrane Database of Systematic Reviews*, *2024*(4), CD004667. Advance online publication. DOI: 10.1002/14651858.CD004667.pub6 PMID: 38597126

Schapiro, L. H., McShane, M. A., Marwah, H. K., Callaghan, M. E., & Neudecker, M. L. (2024). Impact of extreme heat and heatwaves on children's health: A scoping review. *The Journal of Climate Change and Health*, *19*, 100335. DOI: 10.1016/j.joclim.2024.100335 PMID: 41647315

Soto-Méndez, M. J., Maldonado, A., Burgunder, L., Scieszka, L., Gil, Á., & Solomons, N. W. (2016). Effects of maternal hydration status on the osmolality of maternal milk. *Nutrición Hospitalaria*, *33*(Suppl. 3). Advance online publication. DOI: 10.20960/nh.318 PMID: 27491583

Stassen, R. C., Benedek, R., Gabbe, B., De Witte, S., Mertens, L., & Glazer, K. B. (2024). Extreme heat and pregnancy: A content analysis of heat health risk communication in English language public health guidance. *Disaster Medicine and Public Health Preparedness*, *18*, e44. DOI: 10.1017/dmp.2024.44 PMID: 38497500

Stowell, J. D., Sun, Y., Spangler, K. R., Milando, C. W., Bernstein, A., Weinberger, K. R., Sun, S., & Wellenius, G. A. (2023). Warm-season temperatures and emergency department visits among children with health insurance. *Environmental Research, Health : ERH*, *1*(1), 015002. DOI: 10.1088/2752-5309/ac78fa PMID: 36337257

Tanana, H., Combs, J., & Hoss, A. (2021). Water is life: Law, systemic racism, and water security in Indian Country. *Health Security*, *19*(1_suppl, S1), S78–S82. DOI: 10.1089/hs.2021.0034 PMID: 33944613

Tuholske, C., Lynch, V. D., Spriggs, R., Ahn, Y., Raymond, C., Nigra, A. E., & Parks, R. M. (2024). Hazardous heat exposure among incarcerated people in the United States. *Nature Sustainability*, *7*(4), 394–398. Advance online publication. DOI: 10.1038/s41893-024-01293-y PMID: 41727418

Wallerstein, N., & Duran, B. (2010). Community-based participatory research contributions to intervention research: The intersection of science and practice to improve health equity. *American Journal of Public Health*, *100*(S1), S40–S46. DOI: 10.2105/AJPH.2009.184036 PMID: 20147663

Wilson, N. J., Montoya, T., Arseneault, R., & Curley, A. (2021). Governing water insecurity: Navigating Indigenous water rights and regulatory politics in settler colonial states. *Water International*, *46*(6), 783–801. DOI: 10.1080/02508060.2021.1928972

World Meteorological Organization & World Health Organization. (2015). *Heatwaves and health: Guidance on warning-system development*. World Meteorological Organization. https://heathealth.info/wp-content/uploads/WMO_WHO_Heat_Health_Guidance_2015.pdf

Xiong, T., Chen, P., Mu, Y., Li, X., Di, B., Li, J., Qu, Y., Tang, J., Liang, J., & Mu, D. (2020). Association between ambient temperature and hypertensive disorders in pregnancy in China. *Nature Communications*, *11*(1), 2925. Advance online publication. DOI: 10.1038/s41467-020-16775-8 PMID: 32522990

Chapter 6

Psychological Impacts of Climate Change Education Among University Students

Fredrick Kayusi
http://orcid.org/0000-0003-1481-4016
Maasai Mara University, Kenya

Michael Keari Omwenga
http://orcid.org/0009-0008-7982-1114
Pwani University, Kenya

Maad M. Mijwil
http://orcid.org/0000-0002-2884-2504
Al-Iraqia University, Iraq

Guma Ali
http://orcid.org/0000-0003-3234-6420
Muni University, Uganda

ABSTRACT

Global climate change is having a variety of effects on students' health and wellbeing. Considering the serious issues it causes; climate change is at the top of the global agenda. This study aimed to examine knowledge of climate change and its psycho-social effects among students. Quantitative design approach was used. Sample size of 291 students aged 18-24 years were involved in the study out of the 320 targeted population. Questionnaires and focus group discussion tools were used during the study. Data was analysed using the mean, median, and mode for the Likert scale

DOI: 10.4018/979-8-3373-8470-2.ch006

questions. cross-tabulations were also done to compare two different variables. Results indicated students affected psychologically were at the average of 37.5%, while climate change effect on mental health has an overall impact mean (SD) of 4.38 (0.851). The findings of this study suggest a more comprehensive approach to education that can produce resilient, well-rounded students.

INTRODUCTION

The interrelated problems of environmental degradation and mental health have become significant issues for society in recent years (Arunachalam and Velmurugan, 2018). People's mental health is more negatively impacted by higher temperatures and more regular harsh weather events. Climate change has both direct and indirect effects on mental health (Nothling et al., 2024). Hurricanes, tornadoes, and wildfires are examples of extreme weather occurrences that can result in stress, anxiety, and trauma. Global environmental degradation can intensify feelings of fear, loss, and anxiety about the future. For example, as climatic patterns change, droughts and floods become more severe, disrupting agriculture and food security and leading to social unrest and violence. High temperatures have also been linked to detrimental effects on mental health, including an increase in aggression, violence, and suicide.

The threat of climate change is one that we are all familiar with. It has been entrenched in our daily lives and manifests itself in our economics, media, and policy. It is a worldwide issue that requires consideration and prompt action if we are to prevent global warming from rising above 1.5°C in the ensuing decades (Fletcher, 2018). A group of strong-willed, resilient individuals are needed to take on the massive existential problem posed by climate change, which has a significant impact on modern society. This weight is more likely to fall on younger generations, who are also more inclined to initiate sustained action campaigns. These are the generations inheriting a planet that has changed due to climate change, endangering the entire notion of a stable future (Bethune, 2019).

Young generation are therefore under a great deal of pressure to take up environmental issues. Desperation, sadness, and anxiety related to climate change can be significant barriers to taking action. The magnitude of climate change is complicated, leading to paralyzing fear or despair. A vital aspect of the environmental movement that is frequently disregarded is affective readiness (Ray, 2018). Making individuals resilient to the psychological impact that climate change takes is just as important as working to strengthen ecosystems. Young people are ideally positioned to spearhead climate action; thus, it is especially important to support resilience and healthy coping strategies in this generation (Chawla, 2017).

By merging social and economic factors of health to create a range of mental health issues, climate change exacerbates inequality. Children, the elderly, communities of colour, low-income areas (particularly Indigenous communities), and climate migrants are among the groups most at risk for mental health issues. Additionally, research has started to record a growing set of climate-specific psychological reactions, which are characterized as distress associated with changes in the school environment that impact children' academic performance. Global warming and the advancement of artificial intelligence (AI) have had a substantial effect on mental health. The next generation will be most impacted by this change since AI and the environment have become essential components of their everyday life and education (Andr´e, et al., 2018). Further research is necessary because, despite the significance of these technologies, little is known about how they together impact mental health and academic achievement.

Disparities in mental health brought on by natural disasters raise serious ethical questions. Sub-Saharan African nations have endured decades of severe heat, drought, and flooding despite making up less than 4% of global carbon dioxide emissions (Green Climate Fund, 2025). Studies show that 50% of persons in rural Africa below the Sahara who have been uprooted by floods report suffering anxiety, PTSD, or depression. Millions of individuals have been devastated by these occurrences, which have also had a major psychological impact (Chawla, 2017). The need to address these disparities by encouraging strategies to develop and integrate mental health programmes and environmental adaptation measures in vulnerable regions is highlighted by the fact that HICs, which account for about 70% of historical carbon dioxide emissions, have stronger recovery systems and face fewer challenges.

According to the climate justice idea, HICs have a moral obligation to offer LMICs economic, technical, and policy support in order to lessen the negative effects on psychological health in these nations, given that LMICs contribute considerably to world emissions. Funding for climate adaptation and resilience has been made available through programmes like the Green Climate Fund (GCF) (Green Climate Fund, 2025). As of 2024, the GCF had allocated $13.5 billion to 243 projects in 129 nations, including some high-income nations like Australia as well as low- and middle-income nations (Green Climate Fund, 2025).

AI has shown promise in the medical domain, particularly in diagnosing and treating mental health conditions. Artificial intelligence (AI) algorithms may search a variety of data sets, including social media activity and electronic medical records, for early signs of mental illness. Through a range of specialized and easily accessible mental health services provided by artificial intelligence (AI)-powered virtual agents, this proactive strategy optimizes therapy procedures and expedites intervention, which improves individual results of diseases. By using AI algorithms to evaluate treatment-response data, AI-driven therapies including psychological education, nonjudgmental listening, and evidence-based therapy techniques may improve current methods and increase the chance of positive results (Chawla, 2017).

AI is currently a potent and unpredictable force in education that significantly enhances each student's learning outcomes (Naser et al., 2015). According to Salas-Pilco and Yang (2022), AI enhances academic performance by streamlining administrative procedures, offering focused support systems, and developing individualised learning experiences. However, in order to fully benefit from AI in education, researchers must overcome difficult problems pertaining to bias, resource limitations, and ethical dilemmas. Collaboration between academics, educators, legislators, and technology developers from many disciplines and stakeholders is necessary to effectively address these challenges (Zawacki-Richter, Marín, Bond & Gouverneur, 2019). Prior research has highlighted these difficulties and offered suggestions for their appropriate integration while assessing the long-term effects of AI technology on student performance and educational progress (Tsai, Ong & Chen, 2023).

The growing impact of climate change on mental health necessitates the use of inclusive, intersectional, culturally sensitive, and cost-effective strategies. Several tactics are essential to addressing the different vulnerabilities among affected groups and fostering an equitable and long-lasting rehabilitation process (Andr´e, et al., 2018). Important gaps can be filled by culturally sensitive initiatives including integrating Indigenous techniques and training community health workers in psychological first aid. Intersectional vulnerabilities must be addressed for interventions to be successful. Women, rural residents, those with disabilities, and indigenous communities face unique challenges during natural catastrophes. Recovery routes will be supported by culturally appropriate mental health treatments, inclusive evacuation strategies, and gender-sensitive interventions. Enhancing telemedicine, mobile clinics, and rural healthcare infrastructure can assist shorten the time it takes to receive mental health care.

By integrating data on chronic illness, poverty, and access to mental health treatment with geographical fire data, these computer models are beginning to identify which people may most urgently need mental health care. For instance, AI design that considers age, handicap, and CALD status may enhance mental health outcomes after disasters and increase the technologies' ability to resolve disparities (Ma and Huo, 2023). By incorporating mental health measures into policy initiatives, AI-powered mental health therapies are incorporated into disaster recovery plans to ensure a thorough and equitable approach (Salas-Pilco and Yang, 2022). Additionally, AI can strengthen long-term resilience and lessen short-term psychological effects, especially for underprivileged and marginalised people. As AI is created for disaster management, sustainability, ethical management, and integration must be given top importance to avoid escalating already-existing disparities.

This study investigated at how emotionally equipped future climate activists, policymakers, and scientists are to face the problem of their future by using students who are impacted by the environment attending public schools (Ma and Huo, 2023). Moods or feelings are referred to as "affective" in psychology. minimizing the effects of climate change. acts as an evaluation to determine whether climate change instruction in public schools is causing mental health issues for environmental students and whether there are strategies to assist them become more resilient (Omwenga et al., 2024). This study aims to shed light on students' individual experiences and provide educators with information about how climate change education affects the students' mental health (Diprose, 2019).

Purpose of The Study

The aim of the study was to determine the extent to which climate change has affected mental health of students and also to evaluate the student knowledge on climate change.

Objectives of The Study

i. To establish knowledge on climate change among students in higher institution.
ii. To evaluate psychosocial impacts of climate among students.
iii. To establish solutions to climate change problems on mental health

LITERATURE REVIEW

Climate Change and Its Effects

The field of climate research has been established for centuries; the first temperature and weather records date back to the 1600s, when the thermometer was invented (Eaton, 2017). Since then, climate science has expanded on these ideas to provide researchers with a more complex knowledge of the climate. Since the 1970s, as physics has allowed us to better understand the interactions between many aspects of climate, climate models have become increasingly complicated. As scientific methods and instruments improved, the volume of climate research literature increased rapidly between 1951 and 1997 (Eaton, 2017).

Environmental consciousness also became more prominent in popular culture during the 1960s. By 1975, the government had restricted or outright banned all of the chemicals listed in Rachel Carson's book Silent Spring, which brought attention to the risks associated with the use of chemicals like DDT in pest control and agriculture. According to some academics, Carson's book changed American attitudes toward the environment and sparked a period of activity by supporting the modern environmental movement (Fletcher, 2018).

Disasters like chemical burns, oil spills, and the extinction of species sparked increased agitation, which ultimately resulted in the inaugural Earth Day commemoration on April 22, 1970. The National Oceanic and Atmospheric Administration (NOAA, 2019) and the Environmental Protection Agency (EPA) were established by October 3, 1970 (NOAA, 2019). Since then, as the threat of climate change increases each, the need for environmental protection has only gotten stronger. In the contemporary era, humans are mostly to blame for climate change. With high confidence, human activity has raised global temperatures by 1.0°C, according to a 2018 report by the Intergovernmental Panel on Climate Change. If nothing is done, this increase is likely to reach 1.5°C between 2030 and 2052 (Fletcher, 2018).

Both direct and indirect impacts will result from this temperature increase. Extreme weather-related health problems are one of the most obvious direct repercussions of climate change. Other direct effects, such forced migration brought on by sea level rise, occur over a longer time span but are unmistakably related to climate change (Clayton, Manning, Krygsman, & Speiser, 2017). Both present and future generations will be impacted by long-term effects like sea level rise, which will persist for hundreds to millennia (Fletcher, 2018). Despite being less directly linked to climate change, indirect consequences are nevertheless harmful. Weakened infrastructure and unstable food systems are two examples of how climate change has a widespread, indirect impact on humans. Climate change also affects people indirectly through stress reactions caused by direct consequences and chronic stress from long-term environmental changes (Maier, Whitehead, & Walter, 2018). Climate change indirectly affects people's mental well-being, such as when they experience food insecurity or lose their house to a natural disaster (Maier et al., 2018).

Psychological Effects of Climate Change

The psychological effects of climate change are described by scholars using a wide variety of terms. "Eco-anxiety" is among the most generic words. According to Clayton et al. (2017), the American Psychological Association defines eco-anxiety as "a chronic fear of environmental doom" or "watching the slow and seemingly irrevocable impacts of climate change unfold, and worrying about the future for oneself, children, and later generations." This is how eco-anxiety, which is a concern that climate change is stealing the future, is different from an emotional reaction to feeling the direct effects of climate change since it is more existential (Pihkala, 2018).

The term "solastalgia," which was first used by Glenn Albrecht in 2005, predates the development and popularization of eco-anxiety. However, solastalgia does not just refer to eco-anxiety; rather, it indicates a more unique emotional reaction. According to Albrecht, "solastalgia is the loss or lack of comfort and the sense of isolation connected to the present state of one's home and territory" (Wallace et al., 2020). Solastalgia differs from eco-anxiety in that it relates to the agony of losing one's home in the here and now rather than a generalized worry of the future. Solastalgia is typically brought on by a firsthand experience, such as witnessing the irreversible changes brought about by climate change in one's native country. Its foundation is the idea that a person's sense of place is important to who they are and that it is emotionally upsetting to lose that feeling of place (Nolet, 2016).

Additionally, Albrecht created the word "ecoparalysis" to refer to a distinct type of ecological anguish in which an individual feels overburdened by their responsibility for an environmental problem. The concept of ecoparalysis is noteworthy because it suggests that inaction is not always the result of apathy. In actuality, ecoparalysis results from the contrary; rather than being unconcerned with the environment, people become too concerned and emotionally immobilized (Pihkala, 2018). Sally Weintrobe, a climate change psychotherapist, refers to a similar phenomenon as "environmental melancholia," which occurs when people's feelings about an environmental threat overwhelm them (NOAA, 2019).

Albrecht also created the word "ecoparalysis" to refer to a distinct type of ecological anguish in which an individual feels overburdened by their environmental concerns. Because it puts forth the notion that inaction is not always the result of apathy, ecoparalysis is noteworthy. Ecoparalysis really results from the contrary; rather than being indifferent to the environment, people become overly concerned and emotionally paralyzed (Pihkala, 2018). A similar phenomenon is described by climate change psychotherapist Sally Weintrobe, who refers to it as "environmental melancholia" when people experience overwhelming emotions in response to an environmental threat (Ray, 2021). The emotive side of climate action is illustrated by ecoparalysis and environmental melancholia, which show how inaction can impede much-needed progress when people are unable to regulate their sadness about environmental challenges (Ray, 2018).

Psychological Impacts on Students

People of all ages are affected by climate fear, guilt, and despair, but it's particularly critical to recognize the psychological effects on youth. Young people are in a unique position to make big changes in the future, but if they are brought down by their own emotional reactions to climate change, they will be doomed from the beginning (Bethune, 2019). As previously noted, psychological issues such as substance abuse and other mental health illnesses might be exacerbated by emotional distress caused by climate change (Albrecht, 2005; Clayton et al., 2017). Young people in this generation are unlikely to be inspired to tackle a global issue like climate change if they are already dealing with mental health concerns of their own (Stevenson, 2018). According to a 2019 APA study, members of Generation Z said their mental health was worse than that of previous generations. The top three stressors that Gen Z cited were mass shootings, an increase in suicide rates, and climate change. News stories on climate change induced stress, according to 58% of Gen Z (Bethune, 2019). Young people are obviously experiencing a great deal of emotional distress as a result of climate change, and they typically have a negative outlook on the future. This is made worse by the additional stressors that Gen Z reports experiencing, which leaves them unmotivated to take the required actions to alter the pessimistic mindset that so many young people have (Ojala & Bengtsson, 2019).

Additionally, the way adults handle young people's concerns affects how they feel about themselves. Research indicates that youth who possess constructive coping strategies and a feeling of agency also perceive adults to be attentive to their feelings while discussing climate change. Young people, on the other hand, who use disavowal as a coping mechanism for climate change believe that adults do not value their feelings and that they have no control over the future (Ojala & Bengtsson, 2019). In order to help young people fight climate change, older generations have a responsibility to provide them with the capacity to express and responsibly manage their emotions.

Environmental Education and Affective Preparedness Integration

Students need to be taught good coping mechanisms for their climate emotions because of the debilitating consequences that negative climate emotions can have and because of their particular ability to impact change. It is possible to characterize the failure of the intergenerational duty to prepare future generations to reduce climate change as a failure of effective student education. Starting with environmental education is a good way to give pupils the affective coping abilities and other tools they need to fight climate change (UVM, 2021).

Environmental education, according to the North American Association for Environmental Education (NAAEE), is "a process that helps people, groups, and institutions learn more about the environment, and develop skills and understanding concerning how to address global challenges" (2 NAAEE). An interdisciplinary, continuous process, environmental education encourages civic engagement and action. Affective and climate knowledge are both necessary for environmental education to foster a sense of self-efficacy and drive to address climate change (UVM, 2021).

One significant method that environmental education might alter thought patterns to reduce climate change is by fostering a sustainability mindset. According to Nolet (2016), a person who has a sustainability worldview is future-focused, solutions-focused, and has beliefs and views that promote sustainability. Given the threat posed by climate change, a sustainable worldview is necessary. According to sustainability education researcher Victor Nolet, the enormous problems we confront today are unlike anything that has ever been faced by humanity, but the ideas that guide our financial systems, educational institutions, and public policies haven't changed much since the start of the Industrial Revolution (Nolet, 2016). Environmental education helps students adopt a sustainable worldview, which is necessary in light of the changing world.

To change behavior and increase the likelihood that individuals will take action in response to climate change, however, climate knowledge alone is insufficient (Stevenson et al., 2018). In addition to physical adaptation, climate change necessitates mental adaptation. Psychology themes can be included into environmental education to address this in an interdisciplinary manner. Teachers can gain insight from psychology on how people perceive and process climate change. Additionally, it can assist environmental educators in comprehending the possible impacts of climate change on mental health and the ways in which cultivating emotional resilience and mindfulness can aid students in adapting to the changing climate (Maier et al., 2018).

METHODOLOGY

The study used correlational research design and qualitative approach, The targeted population was 320, out of which a total sample of 291 was used in the study, 18 to 24 years old students who expressed concern about climate change participated in semi-structured interviews and focus group discussion. Quantitative design approach was used and Cross-tabulations to compare two different group of variables; climate change emotions and climate change knowledge. Data was analysed using mean, median and mode for Likert Scale Questions (Kayusi et al., 2025).

RESULTS OF THE STUDY

Response Rate

A total of 320 questionnaires were sent out to respondents, and 291 (90.9%) of them were filled out and returned while 29 (9.1%) did not return the questionnaires. This was sufficient to produce results that were reliable and consistent. According to Nulty (2008), a response rate of more than 70% is suitable to yield sufficient results. Given that 50% of return rates are excellent, 60% are good, and 70% are very good for data analysis, the study's response rate of 90.9% was judged adequate (Maria & Voss, 2019). Consequently, the study's respondent percentage is sufficient *(figure 1)*.

Figure 1. Response Rate of Students to Questionnaires

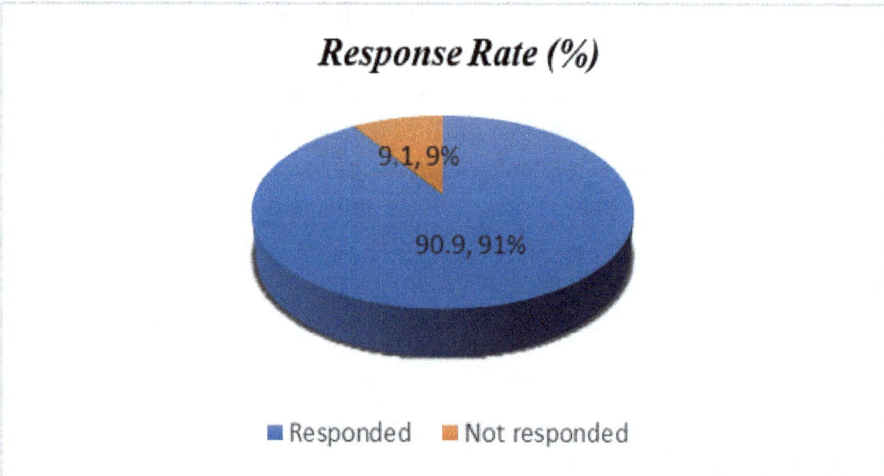

Gender Distribution of Respondents

In order to evaluate the degree of gender differences and validate the study's representativeness, the study figured out the respondents' gender distribution. When the genders of the participants were tallied, the results showed that there were 137 (47.1%) male students and 154 (52.9%) female students. This demonstrated that a higher proportion of female students than male students took part in the study. This suggests that there are more female students than male students in schools *(figure 2)*.

Figure 2: Gender distribution of respondents

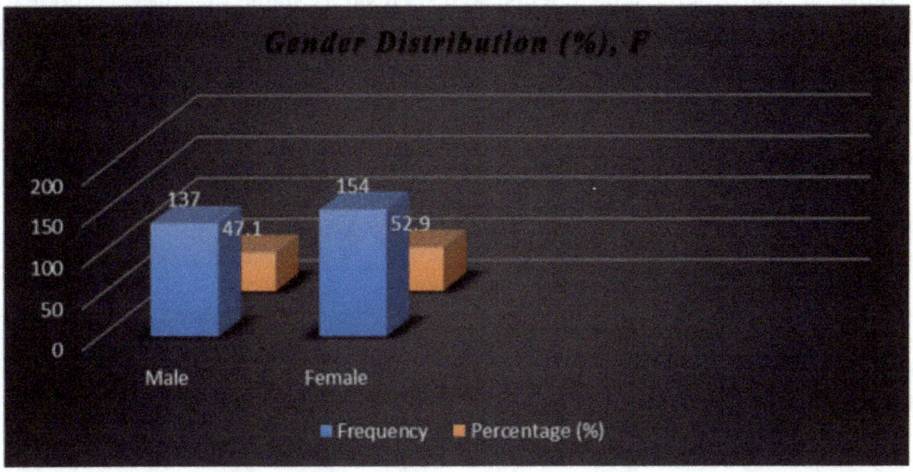

Age Distribution of Students Respondents

The age distribution of the student responses was also investigated in the study and further displayed in *table 1*.

Table 1: Age distribution of student respondents

Age (Years)	*Frequency*	*Percentage*
18-19	76	26.1
20-21	83	28.5
22-24	132	45.4
Total	291	100

The findings revealed that most students' respondents ranged from the age of 18 to 24 years, 132 (45.4%) 22-24 years, followed by 83 (28.5%) age between 20-21 years, while 76 (26.1%) ranged 18-19 years making the least respondent age bracket. This showed that most college students are at age between 20-21 years as captured by the study *(table 1)*.

Climate Change Knowledge

A large percentage of students 120 (41.3%) obtain their climate knowledge from class, independent research 47 (16.2%), online platforms 73 (25.1%), relatives and close friends 14 (4.8%), personal study 16 (5.5%), and other resources 21 (7.2%). This population is environmental students, so it seems reasonable that they would learn about climate change from their studies in class. It also explains why it is essential for this group that classroom education emphasize affective preparedness *(figure 3)*.

Figure 3. Climate change knowledge

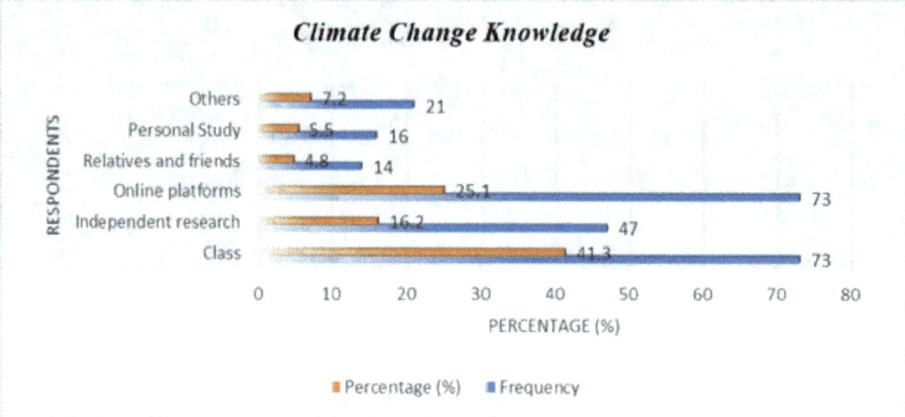

Psychological Effects

When asked to name the top four emotions they experience when thinking about climate change, the majority of students' responses were negatively responded, such as angry (29.7%), anxious (32.8%), alarmed (19.5%), and overwhelmed (18%). The focus group responses confirmed the survey's findings that students' feelings of anxiety and despair were high, while participants' feelings of overwhelm were low because climate change is not an actual problem to them, while 37.5% gave percentage summary of the psychological effects *(figure 4)*.

Figure 4. Psychological Effects

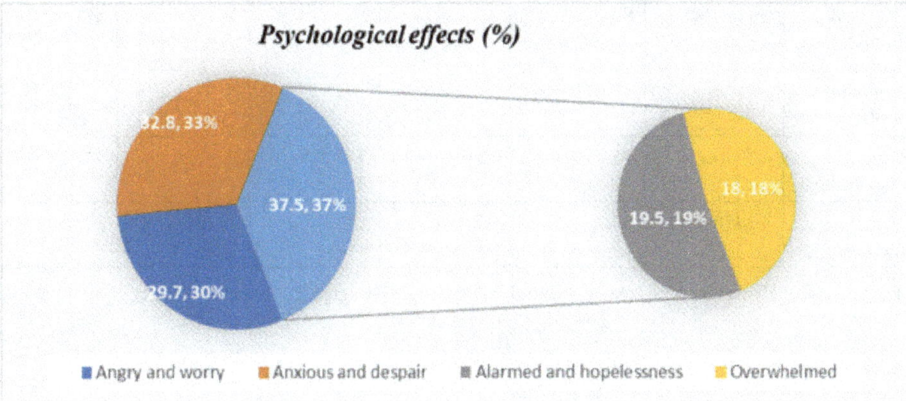

Using Likert scale questions to ask students to rate how much they agreed or disagreed that their climatic emotions hampered their academic performance, social lives, and mental health, the survey's findings also revealed that students felt that their psychological response to climate change impacted their mental health. These respondents agreed that their mental health was affected, but disagreed that their schoolwork and social interactions were affected. However, only 39.7% strongly agreed, 9.5% agreed that emotional response to climate change impacted their overall mental health, and 14.6% disagreed to varying degrees. Of all students, 64.3% fell on the strongly disagree-13.6% disagree side of the scale when questioned if climate emotions hindered their ability to do their schoolwork, 19.6% strongly disagreed that climate emotions negatively impacted their social life, and 15.6% strongly agreed *(table 2 and figure 5).*

Table 2. Effects of climate change on mental health, Frequency (%) N=291

Effects of climate change on mental health	**Mean (SD)**	**SA**	**A**	**N**	**D**	**SD**
School work	4.53 (0.716)	14.3	1.1	6.7	13.6	64.3
Social life	4.15 (0.719)	15.6	2.9	3.5	58.4	19.6
Discipline	4.31 (0.861)	25.2	31.2	16.5	15.2	21.9

continued on following page

Table 2. Continued

Effects of climate change on mental health	**Mean (SD)**	**SA**	**A**	**N**	**D**	**SD**
Peer relationship	4.29 (0.753)	41.8	32.6	9.1	2.3	14.2
Overall impacts	4.38 (0.851)	39.7	9.5	15.3	14.6	20.9

Key: SA; Strongly Agree, A; Agree, N; Neutral, SD; Strongly Disagree, D; Disagree

Figure 5: General effects of climate change on mental health

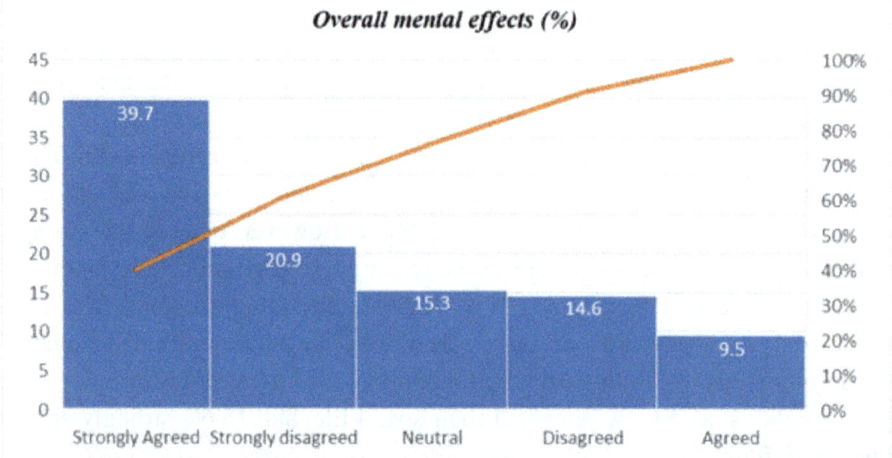

Coping Management and Strategies on Effects of Climate Change on Mental Health

Generational concerns compounded that it assists young adults to manage and cope with climatic change and it is necessary to empower them (Ojala & Bengtsson, 2019), 29% extremely agreed while 18% completely disagreed. Promoting positive emotions encourages the development of happy emotions in order to mitigate negative ones and eliminate them completely (Ojala & Bengtsson, 2019), 11% extremely agreed, 26% agreed while 40% strongly disagreed. Young people need to be educated good coping mechanisms for their climate emotions because of the debilitating consequences that negative climate emotions can have and because of their particular ability to impact change where 57% strongly agreed and 8% strongly disagreed. Students may have been hampered by the stigma associated with mental health since they may have felt ashamed or embarrassed about their own emotional state therefore its essential to support their mental state, 45% strongly agreed, 12% agreed and 17% strongly disagreed *(table 3 and figure 6)*.

Table 3. Coping strategies, Frequency (%) N=291

Strategies	Extremely	Very Much	A Little	Not at All
Generational concern	29	32	21	18
Promoting positive emotions	11	26	23	40
Preparedness environmental education	57	21	14	8
Reflective practices in environment	61	17	12	10
Supporting mental health stigma	45	12	26	17

Figure 6. Coping management and strategies

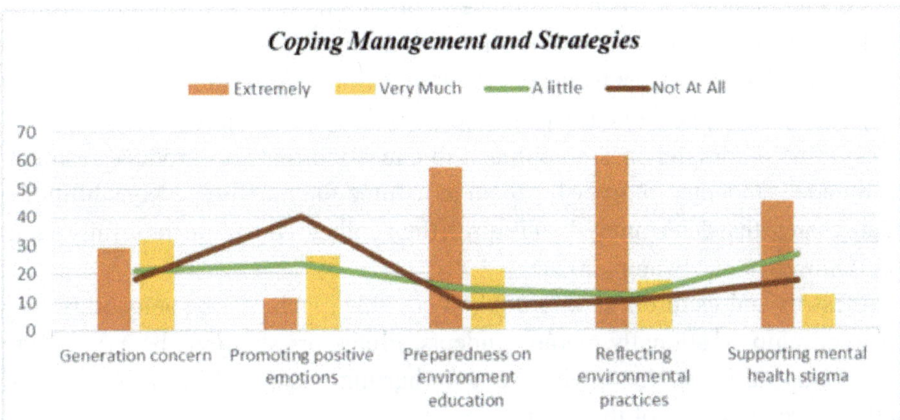

DISCUSSION

According to the study's data, climate change has a psychological impact on UVM environmental studies (ENVS) and environmental science (ENSC) majors, and this emotional reaction is frequently unfavorable. It's unclear, though, if this emotional reaction is only the result of growing climate awareness. This outcome is predictable and establishes the baseline that environmental education is succeeding in its long-term objective of raising student awareness. Given that the emotional response to climate change is identical in both years, climate information may contribute to negative climate feelings, but it does not present an accurate depiction.

Among the many emotions that students experience, motivation is another intriguing area of complexity. When students were asked to rank their top three feelings on climate change, “motivated” was the most frequently selected, positively connotated response. However, inspiration can originate from a number of sources, including negative feelings. Anger or fear may motivate a person just as much as inspiration and hope and inspiration can. Due to the term's ambiguity, it is difficult to determine with certainty whether pupils' motivation stems from a healthy source or from a point of desperation.

CONCLUSION

The development of environmental students into engaged, driven climate advocates depend on how they handle climate information. Students will have insufficient knowledge of climate change if affective education is not taught in addition to climate facts. Unbeknownst to the environmental movement, the inability to cope with loss, anxiety, and hopelessness hinders activity. The findings of this study suggest a more comprehensive approach to education that can produce resilient, well-rounded students. In order for them to assume the massive responsibility for climate change that determines their futures, their college education must fully equip them (Omwenga & Mwangi, 2024).

Professors and lecturers should think about how much of a responsibility they have to develop emotionally mature students. Professors shouldn't be expected to act as therapists for students who are truly hurting because they are not mental health experts. Because of the gender stereotypes that are imposed on women, female educators are particularly under pressure to assume this caregiving role (Ray, 2018). Professors can help students by de-stigmatizing mental health and advocating resources that are accessible to them, rather than taking entire responsibility for their mental health. People's resentment about having mental health issues can be lessened by simply bringing up the topic of mental health in the classroom and allowing for emotional reactions. Informing students about the community's and the university's counseling resources can assist ensure that they get in touch with the right individuals when they need to (Omwenga & Kayusi, 2024).

RECOMMENDATIONS

Although this study's sole focus was on student positionality, it raises issues that environmental lecturers should take into account even though it is unable to suggest any particular teaching strategies. Professors and lecturers should think about their level of accountability for developing emotionally mature students. Since professors are not trained in mental health, they shouldn't be expected to act as therapists for students who are experiencing severe difficulties. The gender norms that are enforced on women put particular pressure on female educators to assume this caregiving role (Ray, 2018). Rather than bearing sole responsibility for students' mental health, educators should help students by encouraging student-accessible services and de-stigmatizing mental health. People's guilt about having mental health issues can be lessened by simply bringing up the topic of mental health in the classroom and allowing for emotional reactions. Informing students about the community's and the university's counseling resources can assist ensure that they get in touch with the right individuals when they need to.

The findings of this study could be used in a variety of future investigations. Examining the ways in which race or gender influences a person's psychological reaction to climate change could connect issues of social justice and equality. Additionally, this study lays the foundation for future investigations into how climate awareness affects people's psychological reactions to climate change. Information regarding how climate anxiety impacts those who have not dedicated themselves to environmental work can be obtained by examining the differences in climate emotions between environmental students and students in other disciplines. The topic of whether climate fear is a barrier for those who are not active in environmental activities at all may be addressed by expanding the sample population in this way.

ETHICS APPROVAL AND PARTICIPATION CONSENT

Logistics refers to all of the procedures, actions, and activities a researcher must follow in order to successfully finish a research project. The Kisii University Ethics and Review Committee, which was approved by the Kisii county director of education and ministry of environment and the National Commission for Science, Technology, and Innovation (NACOSTI), provided the researcher with an ethical review certificate prior to the study. By not writing the respondents' names on the instruments, the researcher obtained their consent, which should be optional. All participants were kept anonymous and secret due to the sensitive nature of the material, and only those who indicated an interest in taking part were contacted. Particularly in data collection, participant selection, interpretation, and report writing, integrity and transparency were upheld throughout the study.

REFERENCES

Albrecht, G. (2005). Solastalgia: A new concept in health and identity. *PAN: Philosophy Activism Nature*, *3*, 41–55.

American Psychological Association. (2019). About EE and why it matters. Retrieved from https://naaee.org/about-us/about-ee-and-why-it-matters

Bethune, S. (2019). Gen Z more likely to report mental health concerns. *Monitor on Psychology*, *45*(3), 226–234.

. Chawla, D., & Atay, A. (2017). Introduction: Decolonizing autoethnography. *Cultural Studies Critical Methodologies, 18*(1), 3–8.

Clayton, S., Manning, C., Krygsman, K., & Speiser, M. (2017). Mental health and our changing climate: Impacts, implications, and guidance. Retrieved from American Psychological Association, 1-69.

Diprose, K., Liu, C., Valentine, G., Vanderbeck, R. M., & McQuaid, K. (2019). Caring for the future: Climate change and intergenerational responsibility in China and the UK. *Geoforum, 105*, 158–167. DOI: 10.1016/j.geoforum.2019.05.019

Eaton, M. (2017). Navigating anger, fear, grief, and despair. In *Contemplative approaches to sustainability in higher education: Theory and practice* (pp. 40–54). Taylor & Francis.

Fletcher, R. (2018). Beyond the end of the world: Breaking attachment to a dying planet. In Kapoor, I. (Ed.), *Psychoanalysis and the Global* (pp. 48–69). University of Nebraska Press. DOI: 10.2307/j.ctv47w9gt.9

Green Climate Fund. (2025). *Green Climate Fund.* Available online: https://www.greenclimate.fund/ (accessed on 30 March 2025).

Green Climate Fund. (2025). *Sectoral guide: Health & wellbeing*. Incheon City, Republic of Korea. Available online: https://www.greenclimate.fund/document/sectoral-guide-health-wellbeing (accessed on 30 March 2025).

Kayusi, F., Vallejo, R. G., Juma, L., Omwenga, M. K., & Chavula, P. (2025). A bibliometric analysis of Segment Anything (SA) research: Global trends, key contributors, and thematic insights. *Babylonian Journal of Machine Learning*, *2025*, 27–41. DOI: 10.58496/BJML/2025/003

Ma, X., & Huo, Y. (2023). Are users willing to embrace ChatGPT? Exploring the factors on the acceptance of chatbots from the perspective of the AIDUA framework. *Technology in Society*, *75*, 102362. DOI: 10.1016/j.techsoc.2023.102362

Maier, K. J., Whitehead, G. I., & Walter, M. I. (2018). Teaching psychology and climate change. *Teaching of Psychology, 45*(3), 226–234. DOI: 10.1177/0098628318779261

Maria, R. L., & Voss, J. (2019). American climate perspectives survey 2019. In In ecoAmerica and Lake Research Partners 2019. 1-4.

Naser, S. A., Zaqout, M. A., Ghosh, R., Atallah, E., & Alajrami, E. (2015). Predicting student performance using artificial neural network: In the faculty of engineering and information technology. *International Journal of Hybrid Information Technology, 8*(2), 221–228. DOI: 10.14257/ijhit.2015.8.2.20

NOAA. (2019). When was the first Earth Day? Retrieved from https://oceanservice.noaa.gov/facts/earth-day.html

Ojala, M., & Bengtsson, H. (2019). Young people's coping strategies concerning climate change: Relations to perceived communication with parents and friends. *Environment and Behavior, 51*(8), 907–935. DOI: 10.1177/0013916518763894

Omwenga, M., Anika, A., Mwangi, S., Omwenga, M., Anika, A., & Mwangi, S. (2024). Prevalence of conduct disorder indicators in young children from public primary schools in Masaba South, Kenya. *International Journal of Science and Research Archive, 13*(2), 2.

Omwenga, M., & Kayusi, F. (2024). Perception of parents of children with mental health in Kisii County, Kenya. *Asian Journal of Advanced Research and Reports, 18*(11), 264–278. DOI: 10.9734/ajarr/2024/v18i11794

Omwenga, M., & W, M. S. (2024). The influence of sociodemographic factors on anti-social behavior among children: A case of Bobaracho area, Nyaribari Chache, Kenya. *Archives of Current Research International, 24*(11), 375–385. DOI: 10.9734/acri/2024/v24i11979

Pihkala, P. (2018). Eco-anxiety, tragedy, and hope: Psychological and spiritual dimensions of climate change. *Zygon, 53*(2), 545–569. DOI: 10.1111/zygo.12407

Ray, S. J. (2018). Coming of age at the end of the world: The affective arc of undergraduate environmental studies curricula. In *Affective ecocriticism: Emotion, embodiment, environment*. University of Nebraska. DOI: 10.2307/j.ctv75d0g8.19

Ray, S. J. (2021). Climate anxiety is an overwhelmingly white phenomenon. *Scientific American*. Retrieved from https://www.scientificamerican.com/article/the-unbearable-whiteness-of-climate-anxiety/

Salas-Pilco, S. Z., & Yang, Y. (2022). Artificial intelligence applications in Latin American higher education: A systematic review. *International Journal of Educational Technology in Higher Education*, *19*(1), 1–20. DOI: 10.1186/s41239-022-00326-w PMID: 35013716

Stevenson, K. T., Peterson, M. N., & Bondell, H. D. (2018). Developing a model of climate change behavior among adolescents. *Climatic Change*, *151*(3/4), 589–603. DOI: 10.1007/s10584-018-2313-0

Tsai, M. L., Ong, C. W., & Chen, C. L. (2023). Exploring the use of large language models (LLMs) in chemical engineering education: Building core course problem models with ChatGPT. *Education for Chemical Engineers*, *44*, 71–95. DOI: 10.1016/j.ece.2023.05.001

UVM Office of Institutional Research. (2021). Enrollment. Retrieved from https://www.uvm.edu/oir/enrollment

Wallace, R. L., Greenburg, J., & Clark, S. G. (2020). Confronting anxiety and despair in environmental studies and sciences: An analysis and guide for students and faculty. *Journal of Environmental Studies and Sciences*, *10*(2), 148–155. DOI: 10.1007/s13412-020-00609-6

Zawacki-Richter, O., Marín, V. I., Bond, M., & Gouverneur, F. (2019). Systematic review of research on artificial intelligence applications in higher education – Where are the educators? *International Journal of Educational Technology in Higher Education*, *16*(1), 1–27. DOI: 10.1186/s41239-019-0171-0

Chapter 7
Reproductive Toxicities and Transgenerational Cancer Risk From Climate-Related Occupational Hazards

Rakshitha N. Ramesh
Sri Sathya Sai University for Human Excellence, India

Srushti Basavaraj Palled
Sri Sathya Sai University for Human Excellence, India

Amrutha Shivaraju
Sri Sathya Sai University for Human Excellence, India

Ramya Raghavan
http://orcid.org/0000-0002-9953-543X
Sri Sathya Sai University for Human Excellence, India

ABSTRACT

Climate change amplifies existing social and health inequities. Marginalized and low-income populations bear the greatest burden of heat-related illness and cancer risk. Climate stress is mediated by biological mechanisms (epigenetics), social structures (housing, occupation), and environmental exposures (heat, pollution). The convergence of exposome science, SDOH and DoHaD research reveals that global warming creates compounding risks. The molecular mechanisms, epigenetic modifications of environmental exposures have demonstrated that the consequences

DOI: 10.4018/979-8-3373-8470-2.ch007

of climate change extend to affect lifelong and transgenerational health trajectories. The extreme heat, air pollution, chemical exposures, and vector-borne diseases are linked to increased rates of congenital disorders, low birth weight, and reduced fertility. Outdoor manual labor, low-income workers are at intersecting vulnerabilities due to poverty, race, housing, and immigration status. The One Health framework, needs policy interventions as heat action plans, PPE standards, occupational exposure limits.

INTRODUCTION

The accelerating pace of climate change is an environmental crisis but also a public health emergency. Global warming affects all those whose livelihoods depend on outdoor labor and manual occupations. The intersection of climate change, occupational health, and social inequity is a critical domain for teratogen research and policy intervention.(H. Wang et al., 2026) Climate change is amplifying occupational exposures to a constellation of reproductive toxicants. Occupational hazards include heavy metals, pesticides, polycyclic aromatic hydrocarbons (PAHs), heat stress, and endocrine-disrupting chemicals (EDCs) across agricultural, construction, emergency response, and outdoor worker populations. Concurrently, mechanistic research in epigenetic toxicology has established that germline epigenomic alterations induced by such exposures can transmit reproductive dysfunction and cancer susceptibility across multiple generations. Despite the scale and urgency of occupational safety literature, no narrative review has examined reproductive toxicities and the intersection of transgenerational cancer risk.

The existing literature treats reproductive toxicity and cancer risk as largely parallel, non-intersecting endpoints. This review positions reproductive toxicity, germline epigenetic disruption induced by climate-amplified occupational exposures, as the mechanistic bridge between first-generation occupational injury and F2/F3 transgenerational cancer incidence. Teratogens from occupational exposure reframe reproductive harm not as a terminal endpoint but as a propagating transmission mechanism. (Korolenko et al., 2026) Teratogens are physical, chemical, biological, and psychosocial agents capable of disrupting normal fetal development. One health perspective integrates planetary health dynamics with occupational epidemiology, emphasizing proactive prevention across agriculture, veterinary services, and wildlife management. (Bao et al., 2026) Heavy metals (As, Cd, Pb) produce germline-level epigenetic disruption. Lineage fidelity is altered by histone deacetylase 1 (HDAC1) upregulation and DNA repair gene silencing. Developmental robustness modifications have transgenerational implications, as documented in F2 and F3 generations of animals. Environmental pesticide exposure induces global DNA hypomethylation and promoter hypermethylation of oncosuppressor genes (CDKN2A, CDH1) in exposed workers. (Korolenko et al., 2025) Wildfire-exposed female firefighters face elevated preterm birth risk; firefighter cohorts show accelerated epigenetic aging. Occupational heat stress impairs spermatogenesis via testicular hyperthermia and oxidative stress, with meta-analysis confirming reduction across all sperm quality parameters. (Gagnon et al., 2026) Foundational transgenerational studies (Skinner et al.) demonstrate that F0 exposures to vinclozolin, DDT, and jet fuel hydrocarbons generate unique sperm epimutation profiles in unexposed F3 descendants, including copy number variation increases. (De Santos & Skinner, 2026) The combined evidence supports the model: climate-amplified occupational exposure → germline epigenetic reprogramming → transgenerational cancer and reproductive dysfunction. Occupational safety frameworks (including OECD TG 443) do not currently account for climate-dynamic exposure intensification or multigenerational heritable risk. Immediate research priorities include climate-stratified occupational cohort epigenome-wide association studies (EWAS) and regulatory revision to incorporate F2/F3 risk endpoints. (Korolenko et al., 2026)

The social determinant of health SDOH framework and the developmental origins of health and disease DOHaD layers risks that climate change poses to reproductive health and transgenerational health. The magnitude of transgenerational cancer risk is dose-dependent and modified by exposure timing preconception vs. gestational, sex of the exposed parent, and genetic susceptibility. (Gretchev et al., 2026) The present work explores how complex, climate-amplified occupational exposures affect reproductive and developmental endpoints. It advances understanding of how climate-driven occupational exposures translate into heritable biological changes. Long-term disease risk effects propagate cancer incidence in parents and offspring. Evidence-based human research links climate-related occupational exposures to measurable germline or epigenomic changes to warn about cancer risk across generations. Mutagen and carcinogen exposures are shaped by social structures, housing, and occupation, and are further intensified by the occupational environment, heat, air pollution, and chemical hazards. (Nath et al., 2026) These disrupt reproductive physiology, impair gametogenesis, and induce heritable epigenetic DNA methylation, histone modifications, and non-coding RNAs that elevate cancer susceptibility in future generations. (Cheng et al., 2026) This chapter is timely to advocate for prevention strategies, workplace standards, and intergenerational risk communication. The objectives are to map the landscape of climate-amplified occupational hazards with documented reproductive toxicity and germline epigenetic effects; to synthesize evidence for transgenerational cancer risk transmission via sperm and oocyte epimutations; and to identify regulatory and research gaps in occupational health frameworks. (Figure 1) Protecting pregnant women, children, and vulnerable workers from climate-related hazards is critical to preventing lifelong and transgenerational health consequences. This chapter synthesizes current evidence on reproductive toxicities and cancer risk from climate-related occupational exposures, elucidating the underlying epigenetic mechanisms. The case studies examine the One Health framework as a holistic approach to surveillance, policy, and intervention. (Izabayo et al., 2026) The discussion concludes with a call for integrated, equity-focused strategies encompassing heat action plans, urban greening, and climate-resilient health systems. (Iyer et al., 2026) The present work advocates advanced surveillance systems to address the urgent occupational health challenges posed by a warming world.

METHODS

A systematic search of PubMed/MEDLINE, Scopus, Web of Science, and EMBASE (January 2020 – March 2026), and also from ILO, WHO, OECD, and Lancet Countdown (2020–2024). Search terms combined exposure classes (heavy metals, pesticides, PAH, heat, EDC) with outcome terms (reproductive toxicity, DNA methylation, sperm epimutation, transgenerational, cancer risk). Final inclusion: 123 studies after PRISMA screening of 4,817 records. Search terms were organized into concept blocks combined with Boolean operators.

- Exposure: "heavy metals" OR arsenic OR cadmium OR lead OR mercury OR chromium OR organophosphate OR organochlorine OR pesticide OR polycyclic aromatic hydrocarbon OR PAH OR wildfire smoke OR heat stress OR endocrine-disrupting chemical OR PFAS OR phthalate
- Reproductive/Germline Outcome: reproductive toxicity OR spermatogenesis OR sperm DNA OR DNA methylation OR histone modification OR epimutation OR germline epigenome OR transgenerational OR intergenerational OR microRNA OR ncRNA
- Cancer / Health Endpoint: cancer risk OR oncogenesis OR tumor suppressor OR carcinogenesis OR oxidative stress OR copy number variation

Two independent reviewers screened titles and abstracts. Conflicts were resolved by consensus; a third reviewer adjudicated irreconcilable disagreements. Full-text assessment was performed for all records that passed the abstract screen. PRISMA flow is presented. (Figure 1)

Figure 1. The present work shows how climate-driven amplification of occupational hazard intensity modifies the germline epigenetic burden of today's outdoor workers, and that cancer and reproductive risks may be transmitted to their children and grandchildren.

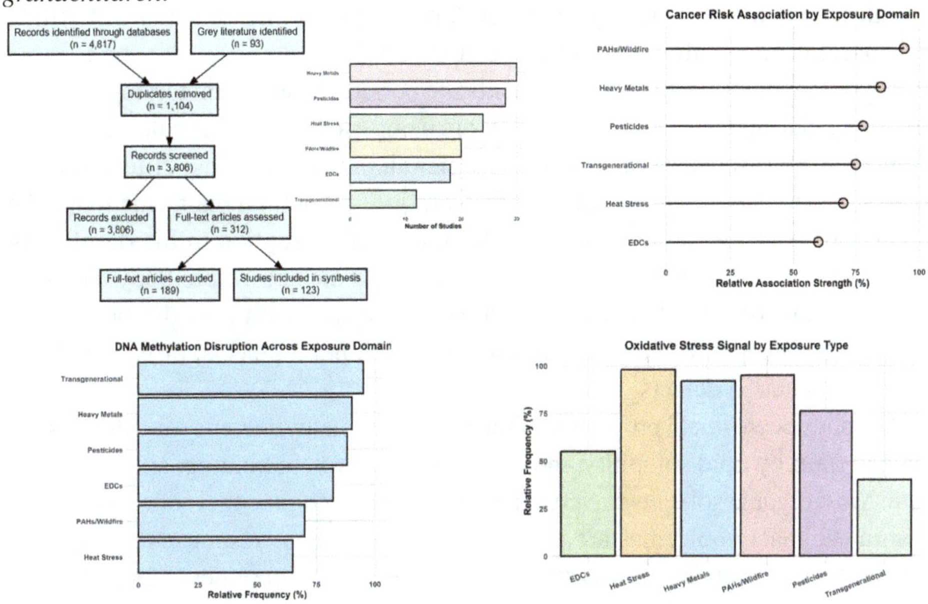

PRINCIPLES OF TERATOGENESIS: TRANSGENERATIONAL CANCER RISK

The term 'teratogen' is defined by Wilson's six principles of teratology (1959) as any agent that crosses placental barriers during developmental windows and is associated with structural or functional abnormalities in the developing organism. (Izabayo et al., 2026) Teratogens in the exposome are beyond thalidomide and rubella to include environmental pollutants, occupational chemicals, thermal, and psychoneuroendocrine perturbations. Molecular biomarkers and mechanistic pathways link environment, reproduction, and cancer risk from maternal trauma susceptibility to teratogenic insult governed by reproductive toxicology. (Souza et al., 2026) The genotype of the embryo determines differential vulnerability to specific agents. Single-nucleotide polymorphisms in CYP1A1 and GSTM1 modulate the embryo's ability to detoxify during hepatobiliary system development. Human embryo preimplantation insults cause resorption, while mild insults may allow repair via totipotent cell replacement. Phthalates, organochlorines, and reactive oxygen species generated by thermal or chemical stressors cause inhibition of nucleic acid synthesis and energy metabolism, disruption of cell membrane integrity, receptor-mediated endocrine mimicry, and induction of epigenetic alterations. These hazards interact with social and biological vulnerabilities to produce a spectrum of adverse reproductive outcomes, preterm birth, low birth weight, stillbirth, impaired fertility, and congenital anomalies. (Andonotopo et al., 2025)

Organogenesis (weeks 3–8) represents the period of maximum teratogenic susceptibility, during which organ primordia are being established. Climate change and toxins challenge timed cascades of cell proliferation, migration, and differentiation. Exposure to teratogens during this window produces structural malformations. The fetal period remains highly vulnerable to functional disruption in the developing central nervous system, immune system, and endocrine axes. Core body temperature elevations of ≥1.5°C above normal are associated with neural tube defects, including anencephaly and spina bifida, congenital heart defects, orofacial clefts, and limb reduction defects.

The physicochemical properties of the molecules methylmercury and dioxins are characterized by lipid solubility, molecular weight, protein binding, and ionization state. Most organic solvents traverse the placenta freely due to their lipophilicity, or biomimetic macromolecules are transported to create dose-response relationships in teratology. The endocrine disruptors lead to pediatric tumor development, atrial or ventricular structural malformation, heart functional impairment, neuroendocrine anomalies, or growth restriction. (Raghavan & Ahamad, 2025) The evidence across epidemiology, toxicology, and epigenetics validating climate-exacerbated occupational reproductive toxicities drives intergenerational cancer burdens. The findings underscore an urgent need for sex-stratified occupational health standards, climate-adaptive prenatal care, and biomonitoring of vulnerable gestational populations. Case studies examine etiological mechanisms and equity implications for future work on vulnerability-informed metrics and policy reforms to mitigate occupational toxicant exposure.

Extreme climate events or anthropogenic climate change are leading to unprecedented cumulative lifetime exposure. Existing studies typically examine single exposures to heat, pollutants, or chemicals in isolation, rarely modeling co-occurring climate-driven exposures that workers experience in real settings. The exposome in the context of reproductive toxicity seeks to quantify the totality of environmental exposures that affect gametogenesis and fetal growth. Climate change leads to increased heat, ground-level ozone, air pollutants, infectious agents, pollen, and psychosocial stress. (Souza et al., 2026) Temperature changes in the exposome contribute to reproductive toxicities and transgenerational health effects, mediated by gene expression changes that affect physiological components.(Barouki, 2024) The IPCC Sixth Assessment Report (2021) identified human reproductive health, pregnancy outcomes, congenital anomalies, and infant mortality as indicators of climate-attributable health effects. Heat stress, pesticides, volatile organic compounds (VOCs), and mycotoxins, including aflatoxins, also affect mental health.(Xiao et al., 2026) Biotic and abiotic mechanisms mediate occupational risks via epigenetic modifications. There is a critical gap at the intersection of occupational epidemiology, reproductive toxicology, epigenetics, and cancer risk modeling.(Andonotopo et al., 2025) Each of these exposure classes carries documented reproductive toxicity; several carry carcinogenic classifications (IARC Groups 1 and 2A).

A separate but maturing scientific literature establishes that occupational and environmental chemical exposures can reprogramme the germline. (Korolenko et al., 2025) The sperm and oocyte epigenomes transmit disease phenotypes to subsequent generations without altering the DNA sequence itself. This phenomenon, epigenetic transgenerational inheritance, has been demonstrated across multiple toxicant classes in rodent models (vinclozolin, DDT, atrazine, jet fuel hydrocarbons, permethrin, DEET) and is increasingly correlated in human epidemiological cohorts examining pesticide and heavy metal exposures. (De Santos & Skinner, 2026)

Figure 2. Climate Change May Intensify Occupational Exposures And Contribute To Reproductive Toxicity, Developmental Risk, Epigenetic Disruption, And Intergenerational Or Transgenerational Disease Burdens

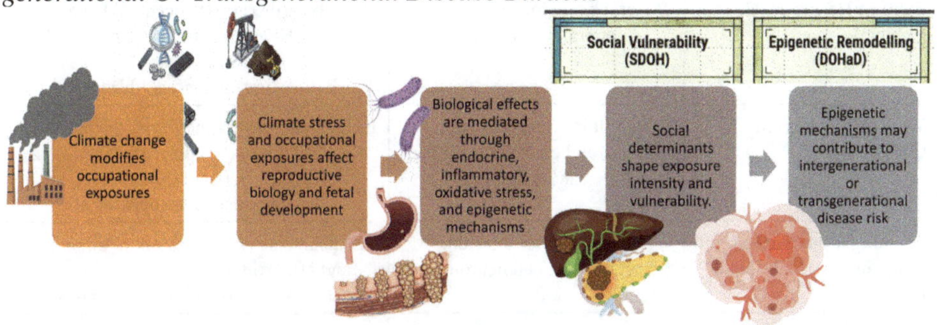

ONE HEALTH: REPRODUCTIVE TOXICITIES AND CANCER RISK

Carcinogenesis is a multifaceted process influenced by a myriad of environmental factors. The identification of biomarkers that indicate environmental influences on cancer risk is crucial for early detection, prevention, and treatment strategies. Table 1 aims to synthesize existing research on biomarkers associated with environmental carcinogenesis. Several chemical, physical, and biological agents are the etiologies, with underlying mechanisms and epidemiological evidence. Occupational hazards reveal how environmental degradation directly influences the workplace, ultimately driving reproductive toxicities and intergenerational cancer risks.

Climate change in the One Health reveals occupational health risks that affect human, animal, plant, and environmental interactions. One Health shows that zoonotic shifts and mycotoxin surges are accelerating threats. Large-scale epidemiologic datasets need high-resolution occupational exposure histories tied to reproductive outcomes for long-term cancer surveillance. Stressors include extreme heat, wildfire smoke, flooding, vector-borne infections, and psychosocial disaster trauma. Common occupational hazards, such as heavy metals, ionizing radiation, pesticides, solvents, polycyclic aromatic hydrocarbons, and endocrine-disrupting chemicals, exert reproductive toxicity. Climate change is reshaping the landscape of occupational health among marginalized and low-income populations. Workers engaged in outdoor and manual labor bear a disproportionate burden of heat-related illness, developmental risk, and exposure to hazardous pollutants.

Table 1. Reproductive Toxicities from Climate-Related Stress

Teratogen	Reproductive Outcome	Mechanistic Gestational Effects
Extreme Heat (>35°C ambient)	Preterm birth, low birth weight, neural tube defects	Hyperthermia-induced oxidative stress, HSP disruption, and placental insufficiency Weeks 3–8 (organogenesis) third trimester
Wildfire Smoke (PM2.5, CO, PAHs)	Stillbirth, reduced fetal growth, and congenital cardiac anomalies	Inflammatory cytokines cross the placenta, and PM2.5 induces epigenetic alterations in fetal DNA All trimesters dose-dependent
Flooding Contaminated Water heavy metals, mold toxins	Spontaneous abortion, preeclampsia, fetal neurological deficits	Lead/arsenic chelate, essential minerals, and aflatoxin disrupt steroidogenesis First trimester critical ongoing exposure
Drought-Related Malnutrition (food insecurity)	IUGR, skeletal malformations, and immune dysregulation	Deficiencies in folate, zinc, and iodine impair neural and skeletal morphogenesis Periconceptional first trimester
Climate-Driven Vector-Borne Infections (Zika, dengue)	Microcephaly, cortical malformations, fetal loss	Viral neurotropism, immune evasion, and transplacental viral transmission First trimester (highest risk)
Psychosocial Stress / Disaster Trauma (cortisol elevation)	Preterm birth, low birth weight, behavioral teratogenicity	Glucocorticoid excess suppresses the fetal HPA axis and alters placental CRH signaling All trimesters cumulative effect
Rising Ozone & Air Pollution (O_3, NOx, CO)	Reduced birth weight, respiratory malformations, and cardiac defects	Oxidative stress impairs trophoblast invasion and mitochondrial dysfunction First and second trimester

CLIMATE-EXACERBATED OCCUPATIONAL HAZARDS WITH SOCIAL DETERMINANTS OF HEALTH

Climate change is increasing the frequency and severity of extreme heat events, leading to a significant rise in heat-related disease burden. (Figure 1) The growing public health challenge of heat-related illness (HRI) due to global warming emphasizes the need for prompt diagnosis and management to reduce morbidity and mortality. Climate change and heat in a toxicological framework describes the biological pathways linking climate change stressors to adverse health outcomes. (P. et al., 2025) Chemical and biological agents intensify risk in agriculture, construction, manufacturing, and informal labor sectors in low- and middle-income countries (LMICs). Susceptibility to heat exposure affects tropical regions throughout their life course.

Socioeconomic disparities intensify efforts to mitigate health impacts and integrate heat-exposure considerations into clinical practice guidelines. (Xu et al., 2025) Social determinants of health, such as poverty, race, housing, and immigration status, influence vulnerability to climate-related occupational hazards. Workers in low-income, minority, and migrant populations are disproportionately employed in high-risk sectors. (Do et al., 2024) The intensification of climate-related occupational hazards has implications for reproductive health. Agriculture, construction, and waste management are more exposed to heat, pollutants, and hazardous chemicals. These groups often lack access to adequate housing, healthcare, and social protections, further compounding their risk. Health outcomes follow a social gradient, with the most deprived groups experiencing the highest burden of disease and the lowest life expectancy. (Andonotopo et al., 2025) Climate change alters the distribution and bioavailability of toxic chemicals while increasing the frequency of extreme weather, heat stress, and wildfires. For the workforce, this translates directly into extreme heat exposure, micro- and nanoplastic pollution, endocrine disruptors and industrial chemical exposures, and vector-borne diseases. (Simonato et al., 2026) (Figure 2) Chronic exposure is increasing economic and productivity losses, as well as healthcare costs, among industrial and physical labor workers. (Priya et al., 2026)

Epidemiological data on heat-related deaths prove disparities in heat illness and reproductive toxicity between 2000 and 2019. Mortality among people over 65 is increased by 70%, with the most severe impacts observed among outdoor and manual laborers in low-income and marginalized communities. Migrant and minority workers are overrepresented in high heat exposure, and workplaces often lack access to cooling, hydration, and rest breaks. The burden of toxic exposures is not evenly distributed; low-income, minority, and migrant workers are more likely to be employed in jobs with high chemical exposures and less likely to have access to protective equipment or health monitoring. (Patel et al., 2026)

GEOGRAPHIC AND DEMOGRAPHIC FACTORS IN OCCUPATIONAL HAZARDS

Geographic and demographic factors influence workers in agriculture, construction, mining, manufacturing, and waste management. Rising global temperatures, extreme weather events, altered ecosystems, and intensified industrial activity are transforming gene-environment interactions (GxE).(Gagnon et al., 2026) Understanding these variations can aid in developing targeted prevention strategies and identifying at-risk populations. The biomarkers reflect environmental exposure to heat stress, airborne pollutants, pesticides, heavy metals, endocrine-disrupting chemicals, and combustion byproducts. Heat exhaustion, respiratory illness, and injuries have been shown as chemical, physical, and biological agents. Heritable molecular changes persist across generations through the interplay between these factors. Further research is needed to elucidate the complex relationships between climate-driven occupational hazards, reproductive health, and transgenerational disease risk in diverse populations.(Nath et al., 2026)

Urban heat islands exacerbate heat exposure in densely populated, poorly vegetated urban areas, disproportionately affecting low-income and minority communities. These populations are more likely to live and work in areas with limited green space, higher ambient temperatures, and greater air pollution, compounding their risk of heat-related illness and other health impacts. Translating the theory of transgenerational toxicology into measurable clinical data requires massive, highly integrated multi-omics cohorts. By layering genomics, epigenomics, transcriptomics, and metabolomics, researchers can trace the precise molecular pathways from an occupational or environmental exposure in a parent to an altered phenotype in their offspring.

Figure 3. The cancer risk and adverse reproductive outcomes affected by Carcinogenic factors found as occupational hazards. (Andonotopo et al., 2025). (Souza et al., 2026)

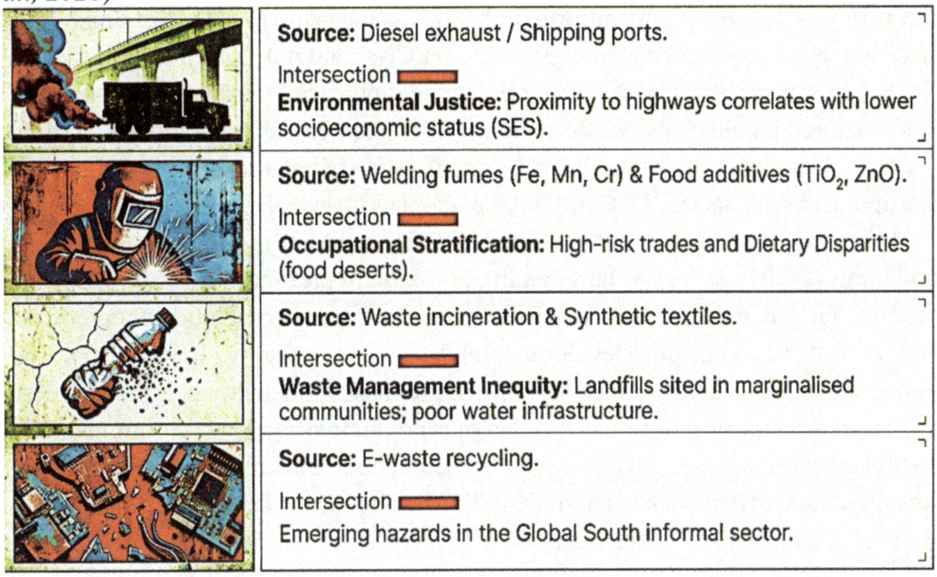

OVERVIEW OF CARCINOGENIC FACTORS

Climate-driven environmental changes and occupational health are critical in molecular oncology and environmental toxicology. The One Health lens recognizes the inextricable links among human, animal, and ecosystem health. Dioxins, from food sources or municipal waste incineration, pose a significant threat to human, animal, and environmental health due to their persistence, bioaccumulation, and toxicity. Polychlorinated compounds readily cross biological barriers, directly threatening fetal development and maternal health. The immunosuppression, hepatotoxicity, neurotoxicity, endocrine disruption, and reproductive impairment are consistent across vertebrate species, supporting cross-species risk assessment. Climate-related ecological shifts also impact agricultural and laboratory environments. (Figure 2) The pregnant women, infants, and indigenous communities are at heightened risk from mercury exposure. Biomagnification and cumulative neurotoxic effects of methylmercury occur during mercury's global cycling and transport create exposure pathways. (Souza et al., 2026)

Health impacts of polychlorinated naphthalene (PCNs) in densely populated urban areas of Asia, identifying metallurgy industries as the primary source of emissions. The sources of polychlorinated naphthalene (PCNs) and polychlorinated biphenyls (PCBs) in urban industrial surface soils include industrial thermal processes, fuel combustion, and electrical waste handling. Pregnant researchers and agricultural workers face unique, acute risks from the mobilization of organic solvents, endocrine-disrupting chemicals (EDCs), and engineered nanomaterials.

The polybrominated Diphenyl Ethers (PBDEs) are identified as persistent organic pollutants (POPs), serve as flame retardants, and are present in dietary intake and through dust inhalation. POPs via maternal transfer led to toxicological outcomes, endocrine disruption, neurodevelopmental deficits, reproductive impairments, and immunotoxicity, indicating that children and pregnant women face heightened risks. For example, exposure of Italian women of reproductive age to polybrominated diphenyl ethers (PBDEs), a class of environmental contaminants, has been reported. Results indicated that the most abundant PBDE congeners were BDE-153, BDE-47, BDE-100, and BDE-99. (Abballe et al., 2026) Environmental distribution of lower-chlorinated PCN congeners (mono to tetra) varies, contributing to contamination in air, soil, water, sediments, and biota near industrial areas.

Inhaled Pollutants

Air pollution and wildfire smoke represent additional, compounding hazards for reproductive health. Outdoor workers are disproportionately exposed to fine particulate matter (PM2.5), nitrogen dioxide (NO2), sulfur dioxide (SO2), and polycyclic aromatic hydrocarbons (PAHs), all of which are linked to reduced fertility, lower ovarian reserve, decreased sperm motility, and impaired in vitro fertilization outcomes. Globally, PM2.5 exposure accounts for 18% of preterm births, while each day of parental tobacco smoke exposure increases the risk of preterm birth by 0.49%. (Ianosi et al., 2025) Wildfire smoke, which contains high concentrations of PM2.5, is up to ten times more harmful than ambient PM2.5, with firefighters and agricultural workers facing the greatest risks during wildfire events. Air pollutants induce oxidative stress, inflammation, and endocrine disruption, impairing gametogenesis and embryonic development, and increasing the risk of congenital disabilities. Increased ambient ozone, greenhouse gases, and particulate matter (PM2.5, PM10) from worsening air quality and wildfires act as systemic environmental toxicants. Airborne pollutants have been shown to alter reproductive hormones and reduce sperm count and quality (Akhigbe et al., 2025). Furthermore, occupational and residential exposure to inhaled pollutants is heavily correlated with adverse reproductive outcomes, disproportionately impacting marginalized communities.

Table 2. Reproductive Toxicities from Occupational Hazards

Industry	Reproductive Toxicity	Mechanism	Regulatory Limit (OSHA/NIOSH)
Lead (Pb) Battery smelting, painting	Spontaneous abortion, premature delivery, impaired neurogenesis, reduced sperm motility	Inhibits δ-aminolevulinic acid dehydratase crosses placenta freely competes with Ca^{2+} in neurons	BLL <10 μg/dL (NIOSH) <5 μg/dL pregnant
Organophosphate Pesticides (OPs) Agriculture, pest control	Neural tube defects, cleft palate, IUGR, male reproductive toxicity	AChE inhibition oxidative stress endocrine disruption of thyroid/sex hormones	Chlorpyrifos is banned (EPA 2021) for food uses
Ionizing Radiation (X-rays, γ-rays) Healthcare, nuclear, aerospace	Microcephaly, growth retardation, childhood leukemia, fetal death	DNA double-strand breaks in rapidly dividing fetal cells apoptosis of neural progenitors	<1 mSv/month fetal dose (ICRP)
Toluene Organic Solvents Printing, shoe or leather laboratory	Fetal alcohol syndrome (FAS-like): craniofacial, limb, renal anomalies preterm birth	CNS lipophilicity disruption of neurotrophic signaling teratogenic epigenomic remodeling	50 ppm TWA (OSHA)
Polycyclic Aromatic Hydrocarbons (PAHs) Foundry, coke oven, asphalt, firefighting	Low birth weight, placental DNA adducts, impaired immune programming	AhR activation epigenetic silencing of tumor suppressor genes transplacental genotoxicity	No federal PEL NIOSH minimal feasible level
Endocrine Disrupting Chemicals (BPA, phthalates, PFAS) Plastics, coatings, food packaging, firefighting foam	Hypospadias, cryptorchidism, reduced anogenital distance, polycystic ovary induction	ER/AR/TR ligand mimicry epigenetic reprogramming of gonadotropin axis altered imprinting	Occupational PEL for most EDCs varies
Mercury (Hg) Methylmercury Artisanal mining, fish processing, dentistry	Minamata disease in offspring: cerebral palsy, intellectual disability, blindness, deafness	Selectively concentrates in fetal CNS Inhibits microtubule polymerization in neuronal migration	25 μg/m³ ceiling (OSHA)

Chemical Agents

Gastrulation is a crucial developmental stage in humans, involving germ layer specification and axis formation. Gastrulation failure is linked to pregnancy loss and congenital malformations from chemical agents as environmental teratogens. Pollutants such as polycyclic aromatic hydrocarbons (PAHs), heavy metals, and pesticides have been linked to transgenerational effects. The neurotoxicity and teratogenicity of the pesticides ametryn and carbendazim, occupational pollutants, affect fertility and embryonic development. (Honles et al., 2026) The pregnancy outcomes, embryonic malformations, from transient hypoxia and reactive oxygen species (ROS) are key mechanisms in organ failures. FATE-MAP is an integrated platform that utilizes high-throughput perturbations of human 2D gastruloids, quantitative phenotypic mapping, predictive deep learning, and mechanistic morphogen modeling to analyze failure modes in human gastrulation. (Souza et al., 2026) FATE-MAP combines a transformer that links chemical structures to phenotypes with clinical molecules identified as potential teratogens. Key parameters influence gastruloid patterning, while chronic exposure to heavy metals can cause oxidative stress and DNA damage. Cell density and SOX2 stability factors lead to dysfunction of vital organs, including the nervous system, kidneys, liver, and cardiovascular system. (Rufo et al., 2026) Biomonitoring strategies, testing blood, urine, hair, and nails, are essential for assessing internal exposure to heavy metals and for detecting subclinical toxicity. The combinations of acetaminophen with other medications, as well as hydrocodone, were linked to increased risks for more severe defects. Toxicants can disrupt hormonal regulation, impair gamete quality, and damage reproductive tissues. Teratogens interfere with critical developmental processes during gestation. Excessive ROS production disrupts cellular homeostasis by damaging DNA, proteins, and lipid membranes, leading to genomic instability. (Domingo, 2026) For instance, studies have shown that exposure to benzene is associated with leukemia. Elevated odds ratios (aORs) were observed for certain combinations of analgesics, such as acetaminophen with aspirin and caffeine, hydrocodone, codeine, or propoxyphene, for congenital disabilities with survival rates below 70%. (Ekenze et al., 2026) Dietary components, aflatoxins, and nitrosamines lead to oxidative stress and molecular damage. Aflatoxins, produced by molds in certain crops, are known carcinogens linked to liver cancer. (Nanjundeshwara et al., 2025)

Physical Agents

Physical agents, ionizing radiation and ultraviolet (UV) light, are well-documented carcinogens. UV radiation is a significant risk factor for skin cancers, with biomarkers such as p53 mutations frequently observed in melanoma. Ionizing radiation's role in carcinogenesis has been extensively studied in populations exposed to nuclear accidents, with biomarkers like chromosome aberrations serving as indicators of exposure. Plastics are widely used in healthcare applications, such as single-use syringes, coatings, and implantable devices. Microplastics (MP) and nanoplastics (NP) are resistant to degradation. Microplastics (diameter smaller than five millimeters) and nanoplastics (diameter smaller than one micrometer) can accumulate as endocrine disruptors, affecting multiple organ systems. (Mandlazi et al., 2026)

EDCs linked to gynecomastia, bisphenols, phthalates, and polycyclic aromatic hydrocarbons.

Gynecomastia (GYN) is a common benign breast condition in men caused by an imbalance between estrogen and androgens, leading to glandular proliferation and adipose hypertrophy.

Endocrine-disrupting chemicals (EDCs) can disrupt hormonal balance by mimicking estrogen, antagonizing androgens, or interfering with hormone metabolism, posing a significant environmental risk for male breast development. The coexposure to the ultraviolet filter homosalate and polystyrene nanospheres amplifies estrogenic responses. Endogenous 15(S)-hydroxyeicosatetraenoic acid (15(S)-HETE) is identified in estrogenic responses observed in mammary glands due to HMS-PNS coexposure. It shows the association between endocrine-disrupting chemicals (EDCs) and gynecomastia, a common benign breast condition in men caused by an estrogen-androgen imbalance. (H. Wang et al., 2026) The coexposure of HMS and PNS elevates 15(S)-HETE levels in MCF-7 cells, promoting cell proliferation through the estrogen receptor alpha-arachidonate 15-lipoxygenase (ERα-ALOX15) signaling pathway. Pharmacological inhibition of the PI3K/AKT pathway and knockdown of ALOX15 suppresses HMS-PNS-induced cell proliferation, indicating that the estrogenic amplification occurs via the ERα-ALOX15-dependent production of 15(S)-HETE and subsequent PI3K/AKT/SGK1 signaling. (Li et al., 2026)

Table 3. Environmental Toxicants and Primary Epigenetic or Genetic Targets

Biomarkers	Source	Role in Toxicity & Carcinogenesis
Reactive Oxygen Species (ROS) markers (8-OHdG)	Tobacco smoke, air pollution, heavy metals.	Induces oxidative stress, leading to base modifications and subsequent epigenetic dysregulation.
DNA damage markers (γ-H2AX)	UV radiation, air pollutants, and endocrine disruptors.	Flags double-strand DNA breaks, heavily influencing chromatin remodeling and epigenetic repair mechanisms.
Hormonal receptors (Estrogen, Progesterone)	Endocrine disruptors.	Aberrant receptor binding alters downstream transcription and epigenetic marking of target genes.
Exosomal cargo (miRNAs and proteins)	Chronic inflammation, occupational toxins.	Facilitates cell-to-cell communication, propagates epigenetic changes, and establishes pre-metastatic niches.

Biological Agents

Biological agents, including viruses and bacteria, contribute to cancer risk through various mechanisms. Human papillomavirus (HPV) is a well-established risk factor for cervical cancer, with specific viral biomarkers (E6 and E7 oncoproteins) indicating oncogenic potential. Similarly, *Helicobacter pylori* infection is linked to gastric cancer, with biomarkers such as cagA gene status providing insights into risk stratification. Glyphosate and Atrazine are endocrine disruptors widely used in agriculture that interfere with hormonal balance, impacting male reproductive function beyond just the testes. Atrazine disrupts the hypothalamic-pituitary-gonadal (HPG) axis, affects steroidogenesis enzymes, induces oxidative stress, and disrupts spermatogenesis, leading to reproductive abnormalities. (Martins-Santos & Oliveira, 2026)

Intergenerational and transgenerational effects of atrazine or other herbicide exposure have been observed, with potential links to increased risks of prostate cancer, cryptorchidism, and hypospadias in offspring. We also need a framework to understand the totality of environmental, chemical, and psychosocial exposures resulting from armed conflict.

Chronic inflammation has been recognized as a key player in cancer development. Environmental factors such as pollution can induce inflammatory responses, leading to the release of cytokines and reactive oxygen species that promote tumorigenesis. Biomarkers like C-reactive protein (CRP) and interleukin-6 (IL-6) have been implicated in various cancers, serving as indicators of inflammation-related cancer risk. The use of targeted synthetic disease-modifying antirheumatic drugs (tsDMARDs) during pregnancy among women with autoimmune diseases has an impact on maternal and neonatal outcomes. It finds a significant association between tsDMARD exposure and preterm birth. (Cheng et al., 2026)

Heavy Metals

Climate-driven processes (permafrost thaw, coastal erosion, drought-intensified mining dust, and flooding of legacy industrial sites) are actively remobilizing historically sequestered deposits of arsenic, lead, cadmium, and mercury into occupational air and water pathways. Agricultural workers, miners, quarry workers, and construction workers represent the primary exposed cohorts. Heavy metals such as cadmium and arsenic, pesticides, organophosphates and carbamates, and endocrine disruptors like bisphenols and phthalates have been linked to infertility, hormonal disruption, developmental abnormalities, and increased cancer risk. These environmental changes can disrupt immune regulation and may lead to autoimmune rheumatic diseases through mechanisms such as oxidative stress, dysbiosis, and trained immunity. (Borba et al., 2026) Both increased temperature and longer exposure times lead to progressive dysfunction of the electron transport chain (ETC) and the mitochondrial phosphorylation system. Significant impairments in mitochondrial function occur even with moderate thermal stress. Thermal damage, effects on cell death and tissue integrity, becomes irreversible when the temperature exceeds 43 °C.

(Mezzacappo et al., 2025) Elevated temperatures increase arterial stiffness and blood pressure. Multi-omics analyses reveal heat-induced changes in pathways related to systemic inflammation, oxidative stress, coagulation, and lipid metabolism.

The principal mechanisms of heavy metal-induced reproductive toxicity include DNA strand damage, oxidative stress via reactive oxygen species (ROS) generation, ionic mimicry of calcium and zinc signalling, hormonal disruption of the hypothalamic–pituitary–gonadal (HPG) axis. The transgenerational risk occurs via epigenetic and intergenerational alterations. Epigenetic dysregulation altered DNA methylation at promoters of steroidogenic enzymes, changes in histone acetylation at tight-junction genes, and misregulation of non-coding RNAs targeting spermiogenesis and endometrial receptivity provides a mechanistic basis for long-term and potentially transgenerational reproductive effects.(De Santos & Skinner, 2026)

Cadmium (Cd) is particularly well-characterized at the preimplantation embryo level. Maternal Cd exposure before pregnancy induces significant loss of the active histone marks H4K8ac and H4K12ac, upregulates the deacetylase HDAC1 in preimplantation embryos, and silences the DNA repair genes XRCC6, RAD51, and PARP1 in 8-cell embryos, creating conditions for germline epigenetic instability that may propagate to subsequent generations. (Ianosi et al., 2025)

Table 4. Intracellular and Extracellular Molecular Mediators of Environmental Toxicity

Signaling Pathway	Genes / Proteins	Environmental Source	Oncological Systemic Outcome
Hypoxia Response Pathway	HIF-1a expression	Hypoxic tumor microenvironments: response to occupational toxins.	Drives angiogenesis and metabolic reprogramming; heavily influences therapeutic resistance.
Growth Factor / MAPK Signaling	KRAS mutations	Chronic inflammation, smoking, and occupational hazards.	Uncontrolled cellular proliferation and survival are heavily implicated in pancreatic cancer.
Tumor Suppression / Apoptosis	TP53, p16INK4a	Aflatoxins, heavy metals, and industrial solvents.	Loss of cell cycle checkpoints; failure to trigger apoptosis in response to severe environmental DNA damage.
TGF-β Signaling	SMAD4 loss	Toxins associated with pancreatic and gastrointestinal cancers.	Disruption of cellular differentiation and promotion of epithelial-to-mesenchymal transition (EMT).

Industrial Toxicity and Dysbiosis

Many industrial chemicals mimic or block hormone signaling pathways, affecting the regulation of estrogen, androgen, and thyroid hormones. Hormonal imbalance can impair gamete formation, disrupt menstrual cycles, and alter spermatogenesis. Intestinal damage is a critical dysbiosis factor in heatstroke, resulting from heat stress and hypoxia, which compromise the integrity of the intestinal epithelial barrier and lead to colorectal cancer and systemic inflammation. (X. Wang & Qin, 2026) The mechanisms of intestinal damage following heatstroke include the body's stress response, oxidative imbalance, disruption of tight junction proteins, extensive cell death, and microbial imbalance. A total of 113 biomarkers were studied for the diagnosis of heat-related illnesses, with the most common being troponin I, IL-6, platelets, and ALT. (Palasz et al., 2025) The liver sustains acute injury during heat stress through mechanisms such as direct hepatocyte damage from hyperthermia, the role of heat shock proteins (HSPs), gut barrier damage leading to endotoxemia, ischemia-reperfusion injury, and various modes of cell death, including apoptosis, pyroptosis, and ferroptosis. Clinical manifestations of liver injury in heat stroke include elevated transaminases, abnormal bilirubin metabolism, and coagulation dysfunction, often accompanied by damage to other organs like the heart, brain, and kidneys. Current treatment strategies focus on rapid cooling and organ support, with potential future therapies including cell death inhibitors, antioxidants, gut barrier protection, and, in severe cases, artificial liver support systems (ALSS) or liver transplantation. (Dou et al., 2026)

FLOODING AND CONTAMINATED WATER

Flooding events liberate sequestered soil contaminants, lead, arsenic, cadmium, and mercury. Pregnant communities use the persistent organic pollutants released into surface and groundwater systems for drinking and food cultivation. Aflatoxins produced by Aspergillus species that thrive in post-flood agricultural conditions further compound exposure to reproductive teratogens. (Nanjundeshwara et al., 2025) Lead and arsenic both cross the placenta efficiently and accumulate in fetal hepatic and neural tissues. Arsenic specifically inhibits the activity of enzymes critical for folate-dependent one-carbon metabolism, predisposing to neural tube defects and epigenetic hypomethylation of fetal DNA.

Transgenerational Mechanisms of Environmental Carcinogenesis

The most unrecognized threat of climate-exacerbated occupational hazards lies in their molecular latency. Toxicological effects often span generations through germline epigenetic reprogramming. Single-cell RNA sequencing helps analyze the transcriptomic responses of human embryonic stem cells to retinoic acid treatment, revealing immediate and delayed endocrine-disruptor effects. Environmental toxicants perturb cellular homeostasis by altering specific intra- and extracellular molecules that serve as measurable biomarkers of epigenetic toxicity and cancer susceptibility.(Acharya et al., 2025)

The association between exposure to oil and gas wells during pregnancy and the risk of structural congenital disabilities among residents of the San Joaquin Valley, California, focused on both active and inactive wells. A case-control study was conducted using data from the California Center of the National Birth Defects Prevention Study, analyzing births from 1997 to 2011 and considering 16 different structural congenital disability phenotypes. Results indicated that exposure to active wells was associated with higher odds of certain congenital heart defects (CHDs). Altered effects are reported for atrial septal defect secundum and gastroschisis, while exposure to inactive wells was associated with higher odds of cleft palate. (Stone et al., 2026) The risk of congenital disabilities associated with living near both active and inactive oil and gas wells may have causal mechanisms. Epigenetic alterations induced by trace elements acting as environmental carcinogens cause severe oxidative stress and directly alter the epigenome. For example, cadmium exposure has been linked to significant DNA methylation changes of imprint control regions in mother-child pairs. Nanoparticles such as titanium dioxide, silver, and cerium oxide have been shown to penetrate the placental barrier. (Andonotopo et al., 2025) Embryo-fetal exposure to environmental pollutants affects developmental signaling pathways, including Wnt and Hedgehog. Parental exposures affect embryonic growth and tissue differentiation. leaving developing fetal tissues highly susceptible to mutagens. Global DNA hypomethylation and hypermethylation of tumor suppressor genes before structural genetic mutations impair fetal development and increase the risk of congenital abnormalities or developmental disorders. (Zhang et al., 2025)

Table 5. Perturbed Signaling Pathways and Downstream Effects

Environmental Toxicant /Hazard	Cancer Association	Primary Epigenetic Alteration / Genetic Target	Biological Impact
Heavy Metals & Aflatoxin B1	Liver Cancer, Colon Cancer	DNA methylation patterns; p16INK4a hypermethylation; TP53 and CYP1A1 mutations.	Silencing of tumor suppressor genes; dysregulation of cell cycle control.
Air Pollution & PAHs	Lung Cancer	p16INK4a methylation; DNA adduct formation (BPDE-DNA).	Early-onset epigenetic silencing; direct structural DNA damage.
Endocrine Disruptors	Breast Cancer	Altered miRNA signatures (miR-21, miR-155).	Disruption of post-transcriptional gene regulation; abnormal cellular proliferation.
Pesticides & Solvents	Colon Cancer, Liver Cancer	miRNA expression changes (miR-21 overexpression); p53 mutations.	Evasion of apoptosis; sustained proliferative signaling.

Environmental carcinogens often cause DNA damage, leading to mutations that drive cancer development. The role of biomarkers such as 8-oxoguanine (a marker of oxidative DNA damage) has been highlighted in studies linking air pollution exposure to lung cancer. (Nath et al., 2026) Understanding the interplay between DNA repair mechanisms and environmental exposures shows deficiencies in repair pathways that exacerbate cancer risk. Environmental nanoparticles have been shown to alter placental steroidogenesis, thereby affecting fetal growth and maintenance of pregnancy. Heat stress disruption of mitochondrial function and cellular metabolism in placental tissues may further compromise nutrient transport and energy production. (Mezzacappo et al., 2025)

EPIGENETIC MODIFICATIONS

Epigenetic changes, including DNA methylation and histone modifications, are significant mechanisms within the framework of the developmental origins of health and disease (DOHaD). Many organochlorines and organophosphates act as mutagens, causing DNA strand breaks, chromosomal abnormalities, or oxidative lesions. Global warming accelerates histone modifications, and microRNA signaling is damaged in germ or early embryonic cells. (Bao et al., 2026) For example, exposure to heavy metals has been associated with altered methylation patterns in genes involved in cell cycle regulation, which can serve as biomarkers for cancer risk. Chronic exposure to toxicants can promote inflammatory responses and immune dysregulation, creating a cellular microenvironment that favors tumor initiation and progression. These modifications can provide insights into the long-term effects of environmental exposures. Epigenetic disruptions in genes are involved in neuronal survival and detoxification pathways. Evidence linking pesticide exposure to neurodegenerative diseases such as Parkinson's disease further illustrates the long-term implications of chronic exposure. (Urasa et al., 2024) There is an association between residing near oil and gas wells during early pregnancy and the risk of structural congenital disabilities. Residents of the San Joaquin Valley, California, find that exposure to active wells is linked to increased odds of certain congenital heart defects. (Stone et al., 2026)

Table 6. Epigenetic Mechanisms Associated with Health Outcomes

Exposure	Epigenetic Mechanism(s)	Outcome(s)	Key Findings
Heat Stress	DNA methylation, histone modification	Reduced sperm quality, low birth weight, and preterm birth	Heat exposure accelerates epigenetic aging and impairs gametogenesis
Air Pollution (PAHs, PM2.5)	Global DNA hypermethylation, miRNA changes	Childhood leukemia, low birth weight, and infertility	Cord blood hypermethylation linked to PAH exposure
EDCs (BPA, DES, Phthalates)	Promoter hypomethylation, histone acetylation	Uterine/prostate cancer, infertility, transgenerational effects	BPA/Phthalate exposure alters methylation, increases cancer risk
Heavy Metals (Lead, Arsenic)	DNA methylation, histone modification	Infertility, congenital disabilities, and increased cancer risk	Prenatal arsenic exposure alters methylation in cord blood
Pesticides	DNA methylation, ncRNA dysregulation	Testicular cancer, congenital disabilities	Pesticide exposure linked to epigenetic changes and cancer

CASE STUDIES AND EPIDEMIOLOGICAL EVIDENCE

The biological mechanisms by which climate-related occupational exposures translate into reproductive toxicity and cancer risk are increasingly understood through the lens of epigenetics. Epigenetic modifications, such as DNA methylation, histone modification, and non-coding RNA expression, mediate the effects of environmental exposures on gene expression, with consequences that can persist across generations.

Occupational exposures to heat, PAHs, PM2.5, and various chemicals induce epigenetic modifications in germ cells. DNA methylation changes, such as global hypomethylation and locus-specific hypermethylation, disrupt gene expression patterns critical for tumor suppression and cell cycle regulation. For example, exposure to PAHs and PM2.5 in industrial and urban occupational settings leads to global hypomethylation and promoter hypermethylation of genes such as GSTP1 and RASSF1A, both of which are implicated in cancer development. Heat stress alters histone acetylation and methylation patterns in germ cells, thereby affecting chromatin structure and gene accessibility and increasing the risk of oncogenic transformation in progeny. Non-coding RNAs, including microRNAs and piRNAs, are dysregulated by chemical exposures, thereby further modulating gene expression and contributing to heritable cancer risk. (Lakshmaiah & Raghavan, 2026)

Agricultural Carcinogens in African Communities

Epidemiological studies have established associations between environmental carcinogens and epigenetic modifiers. For instance, the link between asbestos exposure and mesothelioma is well-documented, with biomarkers such as pleural effusion cytology providing diagnostic utility. Similarly, the correlation between tobacco smoke and lung cancer has led to the identification of biomarkers. (Ianosi et al., 2025) Agricultural communities in many African regions are frequently exposed to pesticides through farming activities, contaminated food, and environmental residues. (Orikpete et al., 2025) Organochlorines and neonicotinoids are highly stable and can remain in soils, water bodies, and food systems for extended periods. In Africa, residues of neonicotinoid pesticides such as imidacloprid, acetamiprid, and thiamethoxam have been detected in water, agricultural produce, honey, and human biological samples, including urine. (Mengistu et al., 2024) Long-term exposure to neurotoxic pesticides has been associated with degeneration of dopaminergic neurons in the brain, a hallmark of Parkinson's disease. Maternal exposure to organophosphates and DDT causes neurodevelopmental disorders and microcephaly. (Izabayo et al., 2026)

Case Study: Agricultural Workers and Pesticides (India)

In India, heatwaves (>35°C) coincide with peak pesticide spraying, exposing laborers to organochlorines. (Dou et al., 2026) A cohort of 5,000 farmworkers showed that paternal glyphosate exposure was linked to a 2.5-fold increase in ALL risk in offspring (F1), mediated by sperm hypomethylation of Runx1. (Patel et al., 2026) Transgenerational effects show granddaughters exhibited 1.8-fold breast cancer incidence, echoing DES legacies. Climate link drought-heat expands glyphosate-resistant weeds, prolonging exposure. (Gagnon et al., 2026) Climate change is also intensifying chemical exposures in occupational settings through increased pesticide use and drift. Agricultural workers are at the highest risk of chronic pesticide exposure, which is consistently associated with reproductive disorders, reduced fertility, congenital disabilities, and increased cancer risk. (Priya et al., 2026) Many pesticides act as endocrine disruptors, interfering with hormone signaling even at low doses. Chronic exposure leads to bioaccumulation and biomagnification, increasing long-term health risks. Studies have documented a 68% increase in sperm malformation rates among agricultural workers exposed to pesticides, with female workers experiencing significant reductions in fertility and increased rates of congenital disabilities. (Ananthakumar & Raghavan, 2026)

Heat-Exposed Pregnancies

Yale analysis of US birth cohorts reports that first-trimester heat effects are >90th percentile associated with childhood cancers and leukemia, with a HR 1.48. Heat shock proteins indicate heat damage; HSP overload is likely due to disrupted hematopoiesis. Heat impairs detoxification, thereby potentiating pesticide genotoxicity and DNA adduct formation in farmworkers. (Gagnon et al., 2026) Wildlife interface in Heat-stressed bats or rodents sheds Nipah and Henipah viruses to date palm sap harvesters in Bangladesh clusters, leading to 75% mortality. Indian garment workers suffer from heat and phthalate stress to exhibit 25% preterm birth rise, with F1 metabolic reprogramming heightening adult leukemia. These cases illustrate that equity gradients need updating among LMIC informal workers, where 80% of female agricultural workers bear a 70–90% burden. Heat and occupational hazards are perpetuating cancer disparities call for workplace reforms (H. Wang et al., 2026)

Zoonotic and Vector Risks in Occupational Settings

Vector-borne diseases are expanding in range and seasonality due to climate change, exposing outdoor workers to new and intensified risks.(Bhuvana et al., 2026) In the United States alone, there are approximately 476,000 annual cases of Lyme disease, with outdoor workers in agriculture, forestry, landscaping, and firefighting at increased risk. Pregnant workers are vulnerable, as vector-borne infections such as malaria and Zika can cause pre-eclampsia, preterm birth, low birth weight, and congenital malformations. One Health views heat as a force multiplier for chemical pesticides and insecticides, which pose zoonotic risks. These diseases can directly infect reproductive tissues or induce systemic inflammation, leading to adverse pregnancy outcomes and impaired fertility. Warming expands pathogen ranges, exposing workers to emerging infections. (Anderer, 2025)

- Lyme disease/ tick vectors: Forestry workers in expanding temperate zones face 2–5x seroprevalence; climate models predict 30–50% northern shift by 2050, heightening chronic neurotoxicity.

- Avian influenza (H5N1): Poultry farmers endure heat-stressed birds shedding virus; occupational clusters (2025 US outbreaks) link to 20% case fatality.
- Schistosomiasis resurgence: Irrigation workers in drought-flood cycles contact Schistosoma-infested snails; heat prolongs snail survival, elevating fibrosis/cancer risks.

War and Conflicts as Amplifiers of Occupational Toxicants

Armed conflict generates a diverse range of stressors, including chemical, physical, biological, and psychosocial factors. The risk factors, long-term light exposure, viral infections, and chemical pollutants, are significant causes of circadian rhythm disorders. Xenobiotics, Steroid hormones interact dynamically over time and can influence biological systems and disease risk in complex ways. The exposures associated with war, such as environmental degradation and psychosocial stress, can lead to immune dysregulation and epigenetic modification. Rapid urbanization, extensive wastewater reuse, and inadequate treatment infrastructure affect vulnerable populations during critical developmental periods. Here are landmark cohort studies and consortia that have successfully utilized multi-omics to identify transgenerational and intergenerational epigenetic biomarkers in high-risk populations.

The FROM Study: Intergenerational Biomarkers of VOC Exposure

The Forensic Research via Omics Markers (FROM) group specifically targets communities and occupational groups living in environmentally vulnerable areas. Housing near refineries, abandoned metal mines, and waste incinerators shows global patterns of reproductive toxicities. A mother-infant cohort study assesses the intergenerational impact of chronic exposure to volatile organic compounds (VOCs), including toluene and xylene. It reports that VOCs are ubiquitous in industrial and painting occupations. The results integrated blood DNA methylation data (epigenomics) with global gene expression profiling (transcriptomics) to perform a molecular network analysis. The study mapped how maternal occupational-level toluene exposure led to epigenetic alterations and subsequent upregulation of AQP10, which is linked to keratinocyte dysfunction. The multi-omics network analysis revealed that the infants of these mothers possessed mirrored epigenetic changes, specifically in the IL31RA and CCL20 genes. Data provided concrete evidence that VOC exposure disrupts immune response and skin barrier function networks across generations before clinical symptoms of atopic dermatitis manifest. (Nagashri et al., 2024)

Figure 4. The Developmental Origins of Health and Disease shows that climate stress, or global warming, reprograms cancer risk via altered gene expression. (Do et al., 2024) (Xu et al., 2025)

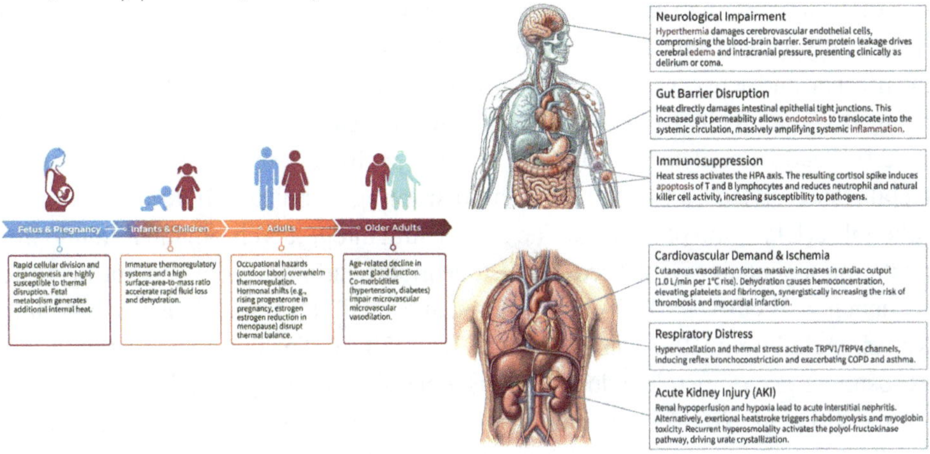

The HELIX Project: The Human Early-Life Exposome

The HELIX (Human Early Life Exposome) project is one of the most comprehensive multi-center cohort studies designed to map how early-life and maternal environmental exposures impact health across the life course. (Figure 3) Over 1,300 mother-child pairs underwent longitudinal tracking of more than 100 chemical, physical, and lifestyle exposures, including heavy metals, endocrine-disrupting chemicals, and air pollution. HELIX uniquely utilizes a four-layer omics pipeline. The methylome (DNA methylation arrays), transcriptome (RNA-seq), metabolome, and proteome were studied in the early-life exposome. (Borba et al., 2026) By deploying advanced bioinformatic dimension-reduction techniques to integrate these layers, HELIX identified over 1,100 specific molecular associations with environmental hazards. They discovered that the methylome acts as the primary molecular memory bank. For instance, maternal exposure to inhaled particulate matter and smoking created persistent DNA methylation signatures in the cord blood of neonates. These specific epigenetic markers were then bioinformatically linked to altered metabolomic profiles in the children, predisposing them to metabolic and neurodevelopmental disorders years later. (P. et al., 2025)

The TaRGET II Consortium: Mapping the Toxicoepigenome

The Toxicant Exposures and Responses by Genomic and Epigenomic Regulators of Transcription (TaRGET II) program, funded by the NIEHS, was established specifically to determine how environmental toxicants perturb the epigenome across the lifespan and across generations. Human epidemiological data in TaRGET utilize controlled mammalian models, such as murine maternal-fetal cohorts, to isolate specific occupational hazards. In vitro studies specifically show the effects of heavy metals (such as lead, Pb) and endocrine-disrupting chemicals (EDCs), such as bisphenol A (BPA) and phthalates. The consortium integrates chromatin immunoprecipitation sequencing (ChIP-seq) for histone modifications, whole-genome bisulfite sequencing (WGBS) for DNA methylation, and small RNA-seq to track non-coding RNAs, including miRNAs and piRNAs. TaRGET studies have demonstrated that exposure to heavy metals induces a transgenerational reprogramming of the epigenome. (M et al., 2026) For example, parental exposure to lead or phthalates alters the expression of specific microRNAs and DNA methylation patterns in germ cells. Through multi-omics tracking, researchers demonstrated that these exact epigenetic signatures bypass the normal demethylation clearance process during embryogenesis, persisting into the F1 and F2 generations and serving as early-warning biomarkers for neurotoxicity and increased oncological risk. (Biradar & Raghavan, 2026)

Table 7. Landmark Case Studies Across Domains on Heat Stress in workplace

Case Study	Population Setting	Key Findings	Mechanisms	Policy Implications
EXPOsOMICS (Europe)	Urban adults/ children	↑ Asthma, dermatitis, hay fever; oxidative stress biomarkers	Air pollution + heat	Integrated mitigation
Ahmedabad, India	Urban poor, outdoor workers	1,344 excess deaths (2010); deaths avoided post-HAP	Poverty, lack of AC, and outdoor work	Heat Action Plan
Urban US Cities	Minorities, low-income, elderly	↑ Heat-related morbidity/ mortality in poor/minority areas	Segregation, poor housing	Cooling centers, urban greening
Global DOHaD Review	Pregnant women, infants	↑ Preterm birth, stillbirth, low birth weight	Fetal programming, epigenetics	Early-life interventions
Kenya Drought Epigenetics	Sibling pairs, Samburu, Kenya	DNA methylation changes, ↓ child weight, ↑ and adiposity	Epigenetic mediation	Nutrition, maternal support
Lancet Countdown Framework	Global, vulnerable populations	↑ Heatwave days, heat deaths; infants/elderly most at risk	Exposome-SDOH-DOHaD integration	Climate-resilient health systems

FUTURE DIRECTIONS: IMPLICATIONS FOR HEALTH EQUITY

Global frameworks (WHO, ILO) should prioritize transgenerational endpoints in chemical risk assessment and extend REACH/REACH-like climate-modulated vulnerability metrics across the life course. Heat action plans, which provide early warning systems, hydration protocols, and work-rest cycles, have been shown to reduce heat-related morbidity and adverse pregnancy outcomes among outdoor workers by 20–30%. PPE standards and stricter exposure limits for chemicals such as anesthetic gases, antineoplastic agents, and heavy metals have reduced phthalate-associated reproductive risks in healthcare and industrial settings, and reproductive toxicity. (Biradar & Raghavan, 2026) Urban greening initiatives, such as increasing tree canopy and green infrastructure, reduce urban heat island effects and improve air quality, with pilot studies showing improved birth outcomes. Climate-resilient health systems are being developed to ensure that health services remain accessible and effective during extreme weather events, with a focus on protecting the most vulnerable.

Surveillance systems that integrate exposome and SDOH data are essential for targeting interventions and monitoring progress. (Ananthakumar & Raghavan, 2026) The One Health approach calls for integrated surveillance systems that combine environmental, occupational, and health data, enabling the identification of high-risk populations and the evaluation of policy effectiveness.

- Climate-stratified occupational cohort EWAS: prospective epigenome-wide association studies in agricultural, construction, and emergency response workers with climate-adjusted exposure modeling
- Oocyte epimutation characterization: current transgenerational epigenetic inheritance evidence is heavily sperm-centric (due to sample quantity constraints); single-cell epigenomic technologies now enable oocyte-level analysis
- F2/F3 cancer registry linkage: occupational cohort registries should be linked to multi-generational cancer incidence data to enable epidemiological detection of transgenerational signals
- Co-exposure modeling: computational and in vitro work to characterize epigenotoxic interaction effects of climate-amplified exposure cocktails (heat + pesticide + PM2.5)
- Regulatory framework revision: OECD TG 443 should be updated to require F3 generation endpoints and to allow climate-dynamic exposure modeling as an input parameter

Genomic and synthetic biology: CRISPR/Cas9 and Cas13 employ gene editing to correct environmental mutations. Bioengineered T-cells (CAR-T) enable personalized immune therapies. (Pallavi et al., 2026) AI and computational models on cross-sector platforms link occupational health records, veterinary diagnostics, crop mycotoxin monitoring, climate forecasts, WBGT, and aflatoxin risk maps. Vulnerable indices combine occupational heat, chemical biomonitoring, and spatial multi-omics to enable proactive epimutation screening. Prospective cohorts tracking F0–F3 epigenomes in high-risk occupations need spatial multi-omics to map tissue-specific risks. Employing AI-driven drug design permit model tumor behavior, and utilizing predictive algorithms for liquid biopsies to enable early detection and real-time monitoring. Spatial multi-omics profiling of worker-plant-animal interfaces adds aflatoxin-heat germ cell epimutations. (Korrapati et al., 2026)

Table 8. Emerging Technological Innovations for Health Monitoring

Technology	Application	Innovation Potential	Future Integration
Wearable biosensors	Continuous monitoring of body temperature, hydration, and heart rate	Early detection of physiological strain	Integration with occupational safety and public health databases
Satellite remote sensing	High-resolution urban industrial or pollutant mapping	Identification of micro-scale exposure hotspots	Integration with epidemiological datasets
Artificial intelligence models	Predictive modeling of disease risk	Real-time forecasting of population vulnerability	Smart warning systems and adaptive urban management
Internet of Things (IoT) climate sensors	Distributed environmental monitoring networks	Localized risk, exposure assessment	Smart city climate resilience systems
Multi-omics biomarkers	Identification of molecular responses	Understanding biochemical pathways of vulnerability	Precision public health approaches

THE ONE HEALTH IMPERATIVE: NEED FOR OCCUPATIONAL SAFETY

Traditional, isolated models of occupational safety are insufficient for addressing these deeply systemic issues. The California Center of BD-STEPS aims to enhance understanding of the causes and prevention of congenital disabilities, which are a leading cause of infant mortality in the US. Through participation in the Birth Defects Study to Evaluate Pregnancy Exposures (BD-STEPS) and collaboration with other research centers, the Center seeks to identify modifiable risk factors associated with congenital disabilities to reduce their public health burden. The One Health approach correctly asserts that human health relies entirely on well-functioning ecosystems to limit the spread of disease and stabilize the climate. (Yuan et al., 2026) Agricultural workers may simultaneously encounter heat stress, ultraviolet radiation, pesticides, and dust. Firefighters and emergency responders are exposed to combustion products, polycyclic aromatic hydrocarbons (PAHs), and volatile organic compounds released during wildfires intensified by climate change. Waste and recycling workers may face exposure to endocrine-disrupting compounds and persistent organic pollutants that accumulate in environmental systems affected by temperature changes and flooding. By integrating occupational health with environmental history via the One Health framework, we can better address persistent global health disparities and environmental justice. (Nath et al., 2026) The California Center for Birth Defects Research aims to enhance understanding of the causes and prevention of congenital disabilities, which are a leading cause of infant mortality in the U.S. Through participation in the (Grant number: U01DD001302-e) BD-STEPS III initiative, the Center will conduct collaborative studies to identify modifiable risk factors associated with congenital disabilities, leveraging its expertise and diverse data sources to reduce the public health burden. Occupational standards integrate heat-WBGT into pesticide/chemical regs to mandate shaded breaks, hydration for pregnant workers. ILO/WHO One Health occupational standards can subsidize climate-resilient farming. Climate-smart agriculture should give subsidies for protective gear in heat-prone sectors. Biocontrol, heat-adapted PPE with zoonotic barriers, and vaccination drives for high-risk occupations are essential. Moving forward, developing standardized early-risk screening frameworks to monitor epigenetic shifts in high-risk occupational groups will be vital for predicting and mitigating metabolic and oncological risks before they manifest clinically. (Bao et al., 2026)

Table 9. Evidence Summary by Exposure Domain

Exposure Domain	Reproductive Effects	Epigenetic Inheritance
Heavy Metals (As, Cd, Pb, Hg, Cr^{6+})	DNA methylation disruption, histone mod., oxidative stress, germline reprogramming; Cd alters H4K8ac/ H4K12ac in preimplantation embryos	Germline epigenomic alterations from toxicant exposure transmit disease phenotypes (obesity, PCOS, germ cell apoptosis) to F2/F3 generations in animal models.
Organophosphate & Organochlorine Pesticides	Global DNA hypomethylation (dominant signal); CDKN2A, CDH1 hypermethylation; miRNA-223, miR-133b dysregulation	Transgenerational POPs transmission documented via breast milk
PAHs / Wildfire Smoke (Firefighters, cleanup crews)	PAH–DNA adducts; preterm birth in wildland firefighters; epigenetic aging acceleration; carcinogenicity classified IARC Group 1	Miscarriage, preterm birth, low birth weight, and gestational complications
Occupational Heat Stress	Testicular hyperthermia, sperm DNA fragmentation, ROS accumulation, germ cell apoptosis	miscarriage risk in pregnant outdoor workers
Endocrine-Disrupting Chemicals (EDCs: BPA, phthalates, PFAS)	HPG axis disruption; androgen/ estrogen receptor interference;	Heritable sperm epimutations; non-monotonic dose–response
Transgenerational Inheritance (F2/F3 Cancer Risk)	F3 generation CNV increases in sperm after F0 vinclozolin/DDT/jet fuel exposure; ncRNA and piRNA transmission confirmed	Linked to leukemia, lymphoma, and organ-specific cancers

LIMITATIONS

The transgenerational cancer risk model is mechanistically supported primarily by animal model evidence; direct human epidemiological evidence for F2/F3 cancer phenotypes traceable to occupational exposures does not yet exist. Second, occupational cohort studies are heterogeneous in exposure assessment methods, biological sample types, and epigenetic platforms (450K vs. EPIC BeadChip methylation arrays), limiting quantitative meta-analysis across domains. The climate-amplification of exposure dose was inferred from mechanistic and physiological evidence rather than measured in prospective occupational studies. Climate-stratified exposure data are largely absent from the primary literature. (Korolenko et al., 2026)

CONCLUSION

The integration of exposome science, social determinants of health (SDOH) frameworks, and the Developmental Origins of Health and Disease (DOHaD) paradigm reveals that global warming creates compounding risks that are mediated by epigenetic mechanisms, social structures, and environmental exposures. Climate change-related exposures during global temperature rise can make occupational hazards fatal. Climate events include extreme heat, tropical cyclones, air pollution, and food and water supply issues. Vulnerable worker groups facing significant socioeconomic challenges are at higher risk from climate extremes. Climate change amplifies effects of reproductive toxicants, seeding transgenerational cancer risks through epigenetic reprogramming of germ cells and fetal development. Chronic exposure to climate-exacerbated hazards and environmental toxicants has been reported to cause occupational effects. The increasing occupational incidence of heat-related illnesses due to climate change emphasizes the need for reliable diagnostic tools for heat stroke and heat exhaustion. Heat stress affects next generations via hormonal signaling, placental aging, and microbial ecology in maternal, placental, and fetal tissues. Heavy metals, endocrine disruptors, and industrial pollutants induce distinct molecular and epigenetic biomarker signatures, thereby proving their role as etiological agents in malignancies. Environmental exposures are linked to oxidative stress, inflammation, and immune regulation. One Health works towards mitigating environmentally driven cancer risks. The literature suggests integrating oncological biomarkers into advanced bioengineering frameworks, leveraging precision nanomedicine, CRISPR-based gene editing, and AI-driven diagnostics to develop precision therapies. The standardized exposure measurement needs to improve predictive performance and support precision-guided interventions. The authors propose future research directions that integrate the exposome with the One Health framework to understand better extreme environmental changes and their implications for human health.

Addressing the reproductive toxicities and transgenerational cancer risks arising from climate-related occupational hazards requires integrated, equity-focused policy interventions and robust surveillance systems. A range of policy interventions have been implemented or proposed to mitigate these risks, including heat action plans, personal protective equipment (PPE) standards, occupational exposure limits, urban greening, and climate-resilient health systems.

The convergence of exposome science and the DOHaD framework has transformed our understanding of how climate-related occupational exposures impact reproductive and transgenerational health. By elucidating the molecular mechanisms and epigenetic modifications induced by environmental exposures, researchers have demonstrated that the consequences of climate change include lifelong and transgenerational risks. (Iyer et al., 2026) The climate-related occupational hazards, extreme heat, air pollution, chemical exposures, and vector-borne diseases are linked to increased rates of preterm birth. Further, low birth weight, stillbirth, reduced fertility, and sperm damage prove the transgenerational cancer risk. Marginalized and engaged in outdoor and manual labor, they bear a disproportionate high-risk burden. Low-income workers are at intersecting vulnerabilities due to poverty, race, housing, and immigration status. The One Health framework needs policy interventions, such as heat action plans, PPE standards, and occupational exposure limits. (Korolenko et al., 2025) Urban greening and climate-resilient health systems have demonstrated effectiveness in reducing reproductive and cancer risks, but broader implementation and enforcement are needed. Low-resource settings can implement global or satellite surveillance systems that incorporate exposome and SDOH data, which are critical for identifying high-risk populations. (De Santos & Skinner, 2026) The integration of exposome science and DOHaD research supports the development of targeted, equity-focused surveillance and intervention systems that protect the most vulnerable populations in a warming world.

In conclusion, addressing the reproductive toxicities and transgenerational cancer risks arising from climate-related occupational hazards requires a holistic, integrated approach. Occupational risk arises from environmental exposures and social determinants that affect critical windows in developmental timing. Equity-focused policies that prioritize workers' needs can mitigate the health impacts of occupational hazards and build a resilient future.

REFERENCES

Abballe, A., Felip, E. D., Dellatte, E., Iacovella, N., Marra, V., Miniero, R., Valentini, S., & Ingelido, A. M. (2026). Polybrominated Diphenyl Ether (PBDE) Serum Concentrations in Italian Women of Reproductive Age. *Toxics*, *14*(1), 72. Advance online publication. DOI: 10.3390/toxics14010072 PMID: 41600621

Acharya, B., Behera, A., Moharana, S., Prajapati, B. G., & Behera, S. (2025). Nanoparticle-Mediated Embryotoxicity: Mechanisms of Chemical Toxicity and Implications for Biological Development. *Chemical Research in Toxicology*, *38*(4), 521–541. DOI: 10.1021/acs.chemrestox.4c00472 PMID: 40105412

Ananthakumar, B., & Raghavan, R. (2026). Social and rural determinants of glycemic control in type 2 diabetes mellitus. *The Indian Journal of Medical Research*, *163*(2), 231–236. DOI: 10.25259/IJMR_1884_2025 PMID: 41949130

Anderer, S. (2025). Beyond Genes: Human Exposome Project to Tackle External Drivers of Disease. *Journal of the American Medical Association*, *334*(1), 7–8. DOI: 10.1001/jama.2025.8666 PMID: 40445605

Andonotopo, W., Bachnas, M. A., Dewantiningrum, J., Pramono, M. B. A., Bernolian, N., Yeni, C. M., Wiradnyana, A. A. G. P., Sanjaya, I. N. H., Akbar, M. I. A., Darmawan, E., Sulistyowati, S., Stanojevic, M., & Kurjak, A. (2025). The fetal exposome and Preterm Birth: A systematic synthesis of environmental exposures and multi-omics evidence. *Journal of Perinatal Medicine*. Advance online publication. DOI: 10.1515/jpm-2025-0231 PMID: 41242981

Bao, M., Nobile, F., Koureas, M., Bottini, I., Fomenko, E., Brimicombe, C., Tirado, V., Mouchtouri, V., Hadjichristodoulou, C., Ljungman, P., Chersich, M., Jackson, D., Luchters, S., Stephansson, O., Stafoggia, M., & de Bont, J. (2026). Trimester-specific exposure to multiple heat indicators and adverse birth outcomes across four European countries. *Environmental Research*, *296*, 124092. DOI: 10.1016/j.envres.2026.124092 PMID: 41740699

Barouki, R. (2024). A toxicological perspective on climate change and the exposome. *Frontiers in Public Health*, *12*, 1361274. Advance online publication. DOI: 10.3389/fpubh.2024.1361274 PMID: 38651121

Bhuvana, G. N., Umesh, S. M., & Raghavan, R. (2026, January). (2025). Expanding the frontiers of Schistosomiasis Treatment: The promise of Nanoparticle-Based Therapies. *Journal of Vector Borne Diseases*, *63*(1), 16–26. Advance online publication. DOI: 10.4103/jvbd.jvbd_180_23 PMID: 40662683

Biradar, S. P., & Raghavan, R. (2026). The Paternal Link to Congenital Heart Disease: Phthalate Exposure Alters Sperm Gene Expression and Increases Offspring Cardiac Risk. *The Multifactorial Etiology of Congenital Heart Disease: Insights from Teratology, Genetics, and Exposomics*, 99–130. DOI: 10.4018/979-8-3373-3735-7.ch004

Borba, V., Carrera-Bastos, P., Hernández, A. L., & Shoenfeld, Y. (2026). Climate Change and the Exposome in Autoimmune Rheumatic Diseases. *Rheumatic Diseases Clinics of North America*, *52*(1), 13–31. DOI: 10.1016/j.rdc.2025.09.004 PMID: 41265939

Cheng, V., Sayre, E. C., Amiri, N., Cragg, J. J., Harrison, M., Proulx, L., & De Vera, M. A. (2026). Perinatal use of targeted synthetic DMARDs and maternal-neonatal outcomes in autoimmune disease: A population-based cohort study. *Journal of Autoimmunity*, *159*, 103538. DOI: 10.1016/j.jaut.2026.103538 PMID: 41780201

De Santos, S., & Skinner, M. K. (2026). Environmental Transgenerational Epigenetics: Effects on Fertility and Offspring Health. *Advances in Experimental Medicine and Biology*, *1506*, 153–167. DOI: 10.1007/978-3-032-18376-7_8 PMID: 42036612

Do, V., Parks, R. M., Casey, J. A., Goin, D. E., & Kioumourtzoglou, M.-A. (2024). Use, limitations, and future directions of mixtures approaches to understand the health impacts of weather- and climate change-related exposures, an under-studied aspect of the exposome. *Exposome*, *4*(1), osae007. Advance online publication. DOI: 10.1093/exposome/osae007 PMID: 39444644

Domingo, J. L. (2026). Developmental toxicity of metals: Three decades of insights into mechanisms, epigenetics, and transgenerational inheritance. *Environmental Research*, *292*, 123703. DOI: 10.1016/j.envres.2026.123703 PMID: 41506432

Dou, J., Cui, Q., Li, W., Jia, K., & Wang, F. (2026). Heat stroke and the liver: Mechanisms of injury and therapeutic strategies. *European Journal of Medical Research*, *31*(1), 263. Advance online publication. DOI: 10.1186/s40001-025-03775-x PMID: 41530884

Ekenze, S. O., Huezo-Garcia, M., Kerr, S. M., Fisher, S. C., Nestoridi, E., Werler, M. M., & the National Birth Defects Prevention Study. (2026). Periconceptional Use of Analgesics and Severity of Structural Birth Defects. *Clinical Pharmacology & Therapeutics*, *n/a*(n/a). DOI: 10.1002/cpt.70259

Gagnon, D., Schlader, Z. J., & Jay, O. (2026). The Physiology behind the Epidemiology of Heat-Related Health Impacts. *Physiology (Bethesda, MD)*, *41*(1), 30–42. DOI: 10.1152/physiol.00012.2025 PMID: 40602791

Gretchev, A., Temples, H. S., England-Mason, G., Dewey, D., Sarasua, S. M., Farrell, C., & Haley-Zitlin, V. (2026). Unpacking developmental programming: A conceptual and historical analysis in the context of DOHaD. *Journal of Developmental Origins of Health and Disease, 17*, e17. DOI: 10.1017/S2040174426100518 PMID: 41919437

Honles, J., Cerapio, J. P., Monge, C., Marchio, A., Ruiz, E., Fernández, R., Casavilca-Zambrano, S., Contreras-Mancilla, J., Vidaurre, T., Condom, T., Zerathe, S., Dangles, O., Deharo, É., Herrera-Zuñiga, J., Pineau, P., & Bertani, S. (2026). Mapping pesticide mixtures to cancer risk at the country scale with spatial exposomics. *Nature Health, 1*(5), 520–531. DOI: 10.1038/s44360-026-00087-0 PMID: 42111112

Ianosi, E. S., Tomoroga, D. M., Văsie iu, A. M., Grigorescu, B. L., Vultur, M., & Ianosi, M. B. (2025). Epigenetic Alterations Induced by Smoking and Their Intersection with Artificial Intelligence: A Narrative Review. *International Journal of Environmental Research and Public Health, 22*(11), 1622. DOI: 10.3390/ijerph22111622 PMID: 41302568

Iyer, K. A., Koynova-Tenchov, R., Sasso, J. M., Thite, T., Deng, Y., & Zhou, Q. A. (2026). The Future of Epigenetics: Emerging Technologies and Clinical Applications. *ACS Pharmacology & Translational Science, 9*(3), 506–544. DOI: 10.1021/acsptsci.5c00729 PMID: 41852637

Izabayo, P., Hakizimana, J. C., Uwineza, A., & Alagbonsi, A. I. (2026). Environmental factors associated with microcephaly in Africa: A systematic review. *Neurotoxicology and Teratology, 113*, 107577. DOI: 10.1016/j.ntt.2025.107577 PMID: 41419017

Korolenko, A. A., Nilsson, E. E., De Santos, S., & Skinner, M. K. (2025). Generational stability of environmentally induced epigenetic transgenerational inheritance of adult-onset disease over ten mammalian generations. *Environmental Epigenetics, 11*(1), dvaf033. Advance online publication. DOI: 10.1093/eep/dvaf033 PMID: 41438350

Korolenko, A. A., Nilsson, E. E., De Santos, S., & Skinner, M. K. (2026). Stability of epigenetic transgenerational inheritance of adult-onset disease and parturition abnormalities. *Proceedings of the National Academy of Sciences of the United States of America, 123*(8), e2523071123. DOI: 10.1073/pnas.2523071123 PMID: 41701841

Korrapati, C., Chowdareddy, S., Suresh, C., & Raghavan, R. (2026). Cancer Cell Membrane-Coated Nanoparticles: Therapeutic Advancements and Hybrid Systems. *Applications of Nanomaterials for Oncology*, 131–160. DOI: 10.4018/979-8-3373-3765-4.ch005

Lakshmaiah, A., & Raghavan, R. (2026). Deep Learning-Enhanced Hyperspectral Imaging: Applications in Oncology Cancer Diagnosis and Tumor Margin Identification. *Biomedical Applications in Deep Learning-Enhanced Hyperspectral Imaging*, 335–368. DOI: 10.4018/979-8-3373-7183-2.ch011

Li, Z., Chen, X., Yang, X., Huang, X., Ye, R., Du, J., Tang, S., Qiu, A., Huang, Y., Deng, Y., Zhong, Y., Liang, B., Hui, C.-Y., Chen, D., & Huang, Z. (2026). Endogenous 15(S)-hydroxyeicosatetraenoic acid mediates amplified estrogenic responses under nanoplastic–homosalate coexposure. *Environment International, 208*, 110083. DOI: 10.1016/j.envint.2026.110083 PMID: 41592362

M, M. K., P, D. A., J, S. P., Moger, S. S., & Raghavan, R. (2026). Spatial Multi-Omics in Cancers: Integrating Molecular Landscapes for Precision Oncology. *Spatially Resolved Single-Cell and Multiple Omics in Human Diseases*, 217–248. DOI: 10.4018/979-8-3373-4849-0.ch007

Mandlazi, N. B., Vilakati, G. D., Madikizela, L., & Kgomo, H. (2026). Non-targeted screening of emerging pollutants across wastewater and receiving waters: A case study in Eswatini. *The Science of the Total Environment, 1014*, 181353. DOI: 10.1016/j.scitotenv.2026.181353 PMID: 41512340

Martins-Santos, E., & Oliveira, C. A. (2026). A Review on the Effects of Atrazine on Male Rat Reproductive System Cytoarchitecture, Steroidogenesis and Oxidative Pathway. *Journal of Molecular Histology, 57*(1), 45. DOI: 10.1007/s10735-025-10671-5 PMID: 41528512

Mengistu, D. A., Geremew, A., & Tessema, R. A. (2024). Pesticide safety practice and its public health risk in African regions: Systematic review and meta-analysis. *BMC Public Health, 24*(1), 2295. DOI: 10.1186/s12889-024-19764-4 PMID: 39180029

Mezzacappo, N. F., Inada, N. M., Vollet-Filho, J. D., Denton, M. L., Yakovlev, V. V., & Bagnato, V. S. (2025). Uncovering the time-temperature thresholds of in vitro mitochondrial bioenergetics dysfunction under hyperthermic stress. *International Journal of Hyperthermia, 42*(1), 2560017. DOI: 10.1080/02656736.2025.2560017 PMID: 40977260

Nagashri, N., Archana, L., & Raghavan, R. (2024). Microbial Volatile Compounds (MVOCs) in Food Industries and Food Safety Applications. *Innovations, Securities, and Case Studies Across Healthcare, Business, and Technology*, 75–96. DOI: 10.4018/979-8-3693-1906-2.ch005

Nanjundeshwara, N., Lakshmaiah, A., & Raghavan, R. (2025). Aflatoxigenic Fungi. *Aflatoxigenic Fungi*, 225–239. DOI: 10.1201/9781003487043-29

Nath, A., Sahu, S., & Lee, J. K. W. (2026). An Umbrella Review of Systematic Reviews and Meta-Analyses on Occupational Heat Exposure, Health Risks, and Productivity Losses Globally. *Current Environmental Health Reports*, *13*(1), 2. DOI: 10.1007/s40572-025-00520-8 PMID: 41499074

Orikpete, O. F., Kikanme, K. N., Falade, T. D. O., Dennis, N. M., Ejike Ewim, D. R., & Fadare, O. O. (2025). Neonicotinoid pesticides in African agriculture: What do we know and what should be the focus for future research? *Chemosphere*, *372*, 144057. DOI: 10.1016/j.chemosphere.2024.144057 PMID: 39746486

P., A., D., K. M., M., J., P., S., & Raghavan, R. (2025). The Cancer Exposome. *Cancer Exposomics and Environmental Influences on Carcinogenesis*, 161–192. DOI: 10.4018/979-8-3373-2165-3.ch004

Palasz, J., Farooqi, W., Musharraf, M. B., Rippon, B., Jedlicka, C., & Razzak, J. (2025). Diagnostic Biomarkers for Heat Stroke and Heat Exhaustion: A Scoping Review. *Disaster Medicine and Public Health Preparedness*, *19*, e153. DOI: 10.1017/dmp.2025.10069 PMID: 40524268

Pallavi, V. N., Chowdareddy, S., Byalahunasishetter, B. I., & Raghavan, R. (2026). CRISPR-Cas9 Technology for Precision Cancer Medicine: Target Discovery, Disease Modeling, and Therapeutics. *CRISPR-Cas9 Technology for Precision Medicine*, 35–72. DOI: 10.4018/979-8-3373-4857-5.ch002

Patel, C. J., Ioannidis, J. P. A., & Manrai, A. K. (2026). An atlas of exposome–phenome associations in health and disease risk. *Nature Medicine*, *32*(4), 1501–1510. DOI: 10.1038/s41591-026-04266-0 PMID: 41851348

Priya, G., Kotwal, N., Madhu, S. V., Patra, S., Pandit, K., Anne, B., Shenoy, M. T., Swain, J., Das, S., Kopalle, L. N., Nagendra, L., Bhattacharya, S., Shaikh, S. S., Mondal, S., Bhat, G. K., Joshi, A., Kalra, S., Biswas, K., Kumar, H. K. V. S., & Mittal, S. (2026). Endocrine Society of India's National White Paper on Endocrine Disrupting Chemicals – A Call for Action. *Indian Journal of Endocrinology and Metabolism*, *30*(1), 4–19. DOI: 10.4103/ijem.ijem_976_25 PMID: 41918600

Raghavan, R., & Ahamad, S. (2025). Developmental Origins of Health and Disease. *Cancer Exposomics and Environmental Influences on Carcinogenesis*, 1–34. DOI: 10.4018/979-8-3373-2165-3.ch001

Rufo, J., Qiu, C., Han, D., Baxter, N., Daley, G., Dhillon, J., Wong, F., Collins, J. J., & Wilson, M. Z. (2026). FATE-MAP predicts teratogenicity and human gastrulation failure modes by integrating deep learning and mechanistic modeling. *Nature Communications*, *17*(1), 3327. Advance online publication. DOI: 10.1038/s41467-026-69596-6 PMID: 41714622

Simonato, M., Brener, M. I., Ferrari, G., Graves, C., Yan, B., & George, I. (2026). Impact of microplastics and nanoplastics on cardiovascular health. *Trends in Cardiovascular Medicine*. Advance online publication. DOI: 10.1016/j.tcm.2026.02.008 PMID: 41720227

Souza, M. C. O., Barbosa, F.Jr, & Domingo, J. L. (2026). Mercury contamination in a changing climate: One health approaches to human, wildlife, and ecosystem protection. *Chemosphere*, *394*, 144821. DOI: 10.1016/j.chemosphere.2025.144821 PMID: 41478202

Stone, J., Mayo, J. A., Shaw, G. M., & Gonzalez, D. J. X. (2026). Structural birth defects and exposure to oil and gas wells during pregnancy. *Pediatric Research*, •••, 1–7. DOI: 10.1038/s41390-025-04719-1 PMID: 41540200

Urasa, S. J., Dekker, M. C. J., Howlett, W. P., Mwezi, R. J., Dorsey, E. R., & Bloem, B. R. (2024). Parkinson's Disease in Sub-Saharan Africa: Pesticides as a Double-Edged Sword. *Journal of Parkinson's Disease*, *14*(3), 437–449. DOI: 10.3233/JPD-230409 PMID: 38517806

Wang, H., Wang, M., Fan, C., Sun, Y., Feng, J., He, M., Li, N., Shang, F., & Liu, B. (2026). Potential association between endocrine disrupting chemicals (EDCs) and gynecomastia: A systematic review based on partial experimental evidenceendocrine-disrupting chemicals. *Frontiers in Endocrinology*, *17*, 1734644. Advance online publication. DOI: 10.3389/fendo.2026.1734644 PMID: 41710411

Wang, X., & Qin, W. (2026). Effect of gut microbiome and blood metabolites on colorectal cancer: A bidirectional Mendelian randomization and mediation analysis based on STROBE-MR guidelines. *Medicine*, *105*(1), e46936. DOI: 10.1097/MD.0000000000046936 PMID: 41496083

Xiao, W., Cao, Y., Ma, F., Zhang, H., & Xia, D. (2026). Volatile organic compound exposure and depression risk: Insights from individual, mixture, and sex-stratified analyses. *BMC Public Health*, *26*(1), 1748. Advance online publication. DOI: 10.1186/s12889-026-27444-8 PMID: 41998567

Xu, J., Wang, Q., Anikeeva, O., Zhu, P., Bi, P., & Huang, C. (2025). Effects of extreme heat on physiology, morbidity, and mortality under climate change: Mechanisms and clinical implications. *BMJ (Clinical Research Ed.)*, *391*, e084675. DOI: 10.1136/bmj-2025-084675 PMID: 41161807

Yuan, X., Wang, H., Li, X., Lyu, J., Zhou, S., Ji, Y., Wang, H., Liu, J., Li, Q., & Wang, H.-J. (2026). The association of heat stress with fetal umbilical artery blood flow: A birth cohort study in Beijing, China. *Environmental Research*, *288*, 123167. DOI: 10.1016/j.envres.2025.123167

Zhang, Y., Zhang, Y., Que, H., Lu, C., & Zhou, S. (2025). Occupational nanoparticles: Major sources, physicochemical properties, multi-organ toxic effects, and associated mechanisms. *Toxicology Mechanisms and Methods*, *35*(8), 847–864. DOI: 10.1080/15376516.2025.2505629 PMID: 40394907

Chapter 8
Workplace Impacts of Climate Change on Mental Health, Burnout, and Occupational Stress

Sanjeev Kumar
http://orcid.org/0000-0002-7375-7341
Lovely Professional University, India

Pardeep Kumar
http://orcid.org/0000-0003-0308-7634
Lovely Professional University, India

ABSTRACT

Climate change is not only a proven environmental issue but also health and well-being priority in the work place environment of the employees. The occupational stress levels, burnout, and other adverse effects on the psychological state of different industries could increase due to extreme weather, hot weather, and other environmental disturbances. The hazards in the environment are known to create physical stress and environment that bring about psychological stress that is felt to a certain degree in climate sensitive sectors like the agricultural sector, emergency services and outdoor work. Moreover, anxiety and uncertainty related to climate, which is one of the results of the ecological crisis awareness, may change mental functioning, decision-making and general job performance even in less climate-related professions. This is a dual challenge as companies have to defeat operational hazards and offer workforce durability through flexible approaches and psychological assets, climate-conscious workplace guidelines.

DOI: 10.4018/979-8-3373-8470-2.ch008

INTRODUCTION

The problem of climate change has already become one of the most topical world issues of the twenty-first century, which does not only influence the environmental system, but the social, economic, and organizational structure as well. Global warming, altered climatic conditions, prolonged heat waves, floods, droughts and extreme weather are not yesterday, but are presently determining the mode of conducting the work places, and the perception of the workers regarding the working conditions. The ecological degradation, biodiversity declines and economic expenses have been the traditional topics of the discussion of climate change. Nonetheless, a new understanding of climate change is that it is one of the factors that contribute to human health significantly with mental and psychosocial wellbeing at work (Brooks and Greenberg 2023). The employees are faced with new forms of stresses as the organizations cope with the shocks that are brought about by the instability in the environment and these are not necessarily the pressure of work only. The dynamics of work place is altering with these changes in the environment. Examples include the direct effect of increase in temperatures to productivity especially in the industries that involve outside activities like agriculture, building, mining and transportation. Physical effects of heat stress include physical exhaustion, dehydration, irritability and poor cognitive functioning that ultimately cause an individual to be accident prone and subject to work-related errors. Floods and Cyclones are extreme weather condition that tend to cripple supply chains, infrastructures and occupational insecurity and is one of the factors that lead to the rise of occupation stress. Workers in emergency services and health sector are usually traumatized in the event of the disaster related to climate and, thus, they are more vulnerable to burnout, compassion fatigue, and psychological trauma. The even the laborers of the climate-controlled indoor spaces are not secure as the economic uncertainty, the renewal of the organization and the remote working technologies are introducing a novel trend in work and job security. The psychological indirect effects of climate changes include other physical effects. This has resulted in what scholars call climate anxiety or eco-anxiety a steady fear of an environmental doomsday (Adamopoulos et al. 2025). Employees can also build the sense of helplessness, be ambivalent in future and moral distress, especially in a sector that is viewed to bring about environmental pollution. This type of emotion may affect concentration, decision-making and motivation hence affecting the performance in general. Moreover, the climate change has not been fairly distributed to the vulnerable groups such as low-income workers and people in the developing world further increasing disparities in the workplace and stress factors. The climate change in this case is not like environmental issue but an issue of health and human resources in the work place (Levi, Kjellstrom, and Baldasseroni 2018). The rationale behind the overlap of climate change and mental health in the work place reasonableness of the study is that there is an increased body of research demonstrating that the effects of environmental stressors are extremely critical on the occupational health. Organizations are also being called upon to ensure that health of the employees is compromised along with ensuring that business is carried on as usual in case of climatic disturbances. High turnover, absenteeism, burnouts, and depression are mental problems that are linked to loss of productivity, high costs in form of healthcare and depression. Disregarding the psychological aspect of climate change can hence compromise organizational resiliency and sustainability. Besides, those regulatory bodies and occupational health systems concerned are only starting to perceive the dangers of climate as hazardous conditions of the working environment, which requires the assumption of the holistic approach to adaptation. This association is quite important especially in the present times when environmental insecurity is on the rise (Tang and Ho 2025). The latter can increase in number as the rate and severity of the climate phenomena increase, therefore having the potential to overload the psychological resources of the employees. Emotional exhaustion, depersonalization, and low personal accomplishment are the signs of burnout that can be otherwise reduced once the employees are exposed to the long-term environmental threats and conventional work-related pressures. The occupational stress models postulated claim that there is occupational stress in case the demands surpass the resources in place; climate change literally raises the demands since it creates unpredictability, physical and emotional stress. The employees would not be able to survive without the appropriate support systems and this will lead to low well-being and performance in the organization. The research papers concerning the effect of climate change in the mental health of the workplaces should be applied to a particular number of mutually related goals. The first one is aimed at establishing and researching the role of environmental stressors in determining and defining psychological consequences of anxiety, stress and burnout (Nowrouzi-Kia et al. 2026). This is through the analysis of the instantaneous exposures, heat and disaster response, and the marginal, the inaccuracy that climate causes, and the economic turmoil. Second, it is to determine the vulnerabilities within the industries in the specific sectors as some industries and occupational groups might be at a greater risk because of their jobs. Third, more general organizational consequences of the climate-based stress, such as decreased productivity and both participation of the employees and their retention will also be pursued in the paper. The second crucial goal is to emphasize the effectiveness in the organizational interventions that will ensure that the effects are alleviated. The paper has also specified the value of proactive programs by taking into account adaptive programs like flexibility in the workplace, support programs, mental health support and climate-specific regulations in Occupational Safety (Syahrir and Falah 2021). Climate adaptation in the workforce planning can be adopted to increase the individual and organizational resiliency. In the end, this finding is supposed to add to the better perception of climate change as a multidimensional phenomenon that slices the environmental sustainability, the well-being of the employees, and organizational performance. These self-reinforcing issues must be resolved so that individuals create healthy work environments where the successes can be optimized in order to provide in an age that has been characterized by an environmental change. Conceptual Framework The theoretical models that underpin this study are the environmental science theory, the occupational health psychology and organizational behavior theories in describing the impacts that climate change has on the mental health in the workplace, occupational stress and burnout.

The term climate change is used to describe the change in temperature, precipitation and the occurrences and occurrence of extreme weather over long period of time mainly due to human emission of greenhouse gasses. All these environmental changes have many stressors that not only impact the workers but it affects the entire sectors as well. The environmental stress sources are the increased temperature, air pollution, water deficit, unusual weather calamities, disturbance of ecosystem and resource instability. Work insecurity, mental stress, and physical risks are the translations of the occupational stressors (Tsakonas et al. 2025). A good example is when prolonged exposure to heat can lead to the impairment of concentration, fatigability, and repeated exposure to disaster-related climate changes can lead to traumas, anxiety, and the chronic stress theory. The second group of stressors may be a form of economic turmoil or employment security or the environmental sustainability even to those employees who do not experience the physical risks directly. Mental health within the work place is founded on the emotional, psychological and social condition of the workers at the workplace as well as within the organization. It dictates thoughts and feelings, behaviors and earnings of people at the work place. When mind is in proper working cause better productivity, engagement, creativity and collaboration and when it is improper causes absenteeism, presenteeism, quitting and poor performance. Climate change is known to bring new psychosocial stressors to disrupt such a balance. Eco-anxiety or climate anxiety refers to the constant feeling of anxiety with regard to environmental degradation and lack of knowledge of the future. When the employees are directly related to the environmental sensitive sectors, they might experience weak feelings, ambivalence or even fear of the long term effects on the environment particularly when the employees have direct relationship with the environmental issue that affects them (Kjellstrom, Holmer, and Lemke 2009). In addition, when exposed to climatic disasters more than once (e.g., floods, wildfires, heatwave), people can get acute stress reactions and, in the most extreme cases, post-traumatic stress symptoms. Mental health at working places should then not merely be understood as it is merely seen through the organizational strains of workload and role conflict but go understanding that it is seen through ecological imbalances as an entirety that constitutes the lived experience by employees. The main constructs that have been used to elaborate the impact of stressors in the environment on the well-being of employees are occupational stress and burnouts (Lan et al. 2020). Occupational stress is caused by job demands that surpass the resources of a particular individual whereby it brings about psychological and physiological pressure. The theoretical background is rather beneficial, and that is why, under the condition of the high demands (e.g., workload, time pressure, environmental hazards) and the lack of resources (e.g., support, autonomy, and recovery time), the exhaustion and disengagement are most likely to become the outcomes. Climate change may not only add physical but also unpredictable and emotional working pressure but drain resources by depriving of either the job security or social stability. The symptoms of burn out are the emotional exhaustion, depersonalization or cynicism and decrease in personal achievement and develop as a result of failure to manoeuvre the chronic stress over time. Working on industries that are climate sensitive like agriculture, emergency services, health and construction might end up accumulating stresses in the long run because of the frequent exposure to the environment hence increasing the number of burnouts. The other theory that supports this framework is the Conservation of Resources (COR) theory, which contends that people aim at gaining, maintaining and accumulating valuable resources like energy, time, health and social support. Climate change poses a threat to the resources as it poses physical danger, economic imbalance and environmental devastation. Where the resources lost are more than those that are gained it is likely that the individuals will be stressed and prone to mental illnesses. The component of cognitive appraisal to the measures of stress among the employees is also enshrined in the transactional models of stress; fear towards the environment and having faith in coping strategies. Therefore, perceived organizational support and resilience could be the cause of different psychological reactions of two employees who are exposed to similar weather (Schulte and Chun 2009). The connection between the change of the surrounding and the well-being in the occupations is made both directly and indirectly. Direct exposures are first hand exposure to environmental hazards, which entails losing physical stamina due to extreme heat or responding to disasters leading to traumas. These indirect paths are the reorganization of the organization as a result of the economic variation caused by the change in climate, and the hindrance of the supply chain and long-term uncertainty that cause career insecurities. Individual variables that moderate such pathways include resilience, coping skills and past mental health condition and organizational variables include the leadership, safety culture and access to mental health services. Stressors on the environment may be progressive and lead to poor health, low productivity and turnover in the long run. This conceptual framework downgrades climate change as a cross dimensional occupational threat element, which cuts across environmental contexts, psychological action and company designs. Mediating Climate Change to Work Mental Health (Trombley, Chalupka, and Anderko 2017). Climate change and psychological health at the workspace are symbiotic and complex and take place in a physical, psychological, and economical level. The processes are familiar with decoding environmental impact into work pressure, fatigue and general health issues in the industries. The short-term and long-term stressors that are modifying the working experience of the employees are the hike in temperatures, weather extremes, environmental insecurity, and economic disruptions. These routes will be required to formulate some organizational interventions and resilience plans. The heating process and heating stress are some of the most immediate processes. The rising rate of frequency and intensity of the heat waves is also due to the rising temperatures of the world and it directly affects the workers, especially the workers working in the field of agriculture, construction, transport and mining among other non-climate-controlled sectors (Lundgren et al. 2013). Auditory noise exposure exaggerates heat exposure and also reduces physical endurance, cognitive functioning and enhances fatigue and aggravates the irritability. Mechanisms Linking Climate Change and Workplace Mental Health are shown in table 1

Table 1. Mechanisms Linking Climate Change and Workplace Mental Health; Source: Author Contributions.

Mechanism	Description	Psychological Impact	Workplace Outcomes	Vulnerable Sectors
Rising Temperatures and Heat Stress	Increased exposure to high temperatures and prolonged heatwaves affecting physical and cognitive functioning	Irritability, fatigue, emotional exhaustion, reduced concentration	Lower productivity, higher accident rates, increased absenteeism	Agriculture, construction, mining, transportation
Extreme Weather Events and Disaster Exposure	Exposure to floods, cyclones, wildfires, and climate-related disasters	Acute stress, anxiety, trauma, post-traumatic stress symptoms	Burnout, job disruption, turnover, reduced engagement	Emergency services, healthcare, disaster management
Climate Anxiety and Eco-Distress	Chronic worry and fear about environmental degradation and future uncertainty	Persistent anxiety, helplessness, moral distress	Reduced job satisfaction, impaired decision-making, disengagement	Cross-sectoral (especially youth and sustainability-focused roles)
Job Insecurity and Economic Disruptions	Organizational restructuring, layoffs, and economic instability due to climate impacts	Psychological strain, fear, uncertainty, reduced morale	Increased occupational stress, decreased performance, higher turnover	Energy, manufacturing, informal sector workers

The issue of climate change is not necessarily this problematic in all industries, so, psychological and occupational effects of climate change are different in relation to the line of work, exposure to the environmental hazard, and the ability of the organization to adapt to the changes. The industry has an influence on the work places implying that various work places are vulnerable and resilient to various types of stress, and both of them are potent and weak. The reality that must be realized is that there is a necessity to understand such variations so as to develop some mental health and work safety policies.

There are other climate sensitive industries like agriculture, among others which has been a blow to some of the sectors. It boasts a predictable weather, no radical changes on the temperature which remains constant, water is also provided to the farming industry, fisheries and forestry industries among others. The alteration in climatic conditions has posed a threat to the crop yield and stability of the revenue since there is no rainfall, prolonged drought, floods, and pests' infestations. The climatic change imposes both physical and economic cost to the agricultural farmers. Physiologically, excessive body heat raises the degree of fatigue, risk of water and heat loss. In the economic dimension, it undergoes economic pressures and debts and future of livelihood insecurities because crops fail or animals are lost. The pain in the body and the loss of money add such worry, depressive moods and constant tension. Most of the rural settlements are not reachable by mental health facilities and it is also used to complement the vulnerability. The provided routine climate catastrophes are able to annihilate hope and power and the future of burnout and emotional burnout among the employees, the lives, and personalities of which are inevitably interconnected with the condition of the environment.

The other yet not less significant impact of the climate change on emergency and healthcare services is its impact on the emergency and healthcare services. The workers in such industries are typically the downstream side when a disaster is the issue in regard to the climate like floods, cyclones, wild fires and heat waves among others. The paramedics, the firefighters, the emergency responders and the health care professionals may be exposed to the traumatic events of witnessing a massive victimization and great stress with decision making. Crisis enhances the working conditions, as climate, duration of increased working hours and rest period enhances the incidence of burnouts. The effects of the heat diseases, respiratory illnesses due to air pollution or facial injury during severe weather conditions could also be the cause of the influx of patients in the healthcare systems. Re-experiencing pain, life threatening events might lead to compassion fatigue, secondary traumatic stress and emotional burn out. Moreover, as the damages are matched with losses and property destruction in case of a disaster, such professionals may also suffer losses, and it is one of the factors contributing to the psychological stress. The cumulative impact of the multiple disaster response that cannot be accommodated by the respective organizational resources may cause a colossal harm to the mental well-being and job satisfaction.

Outdoor industries and construction in general are very sensitive to the weather conditions and the rise in temperatures. The workers in the outside, construction workers, road repair workers, sanitation workers, transportation workers and other workers are usually physically demanding in the sun of the sun or in the bad weather. There is heightened heat stress, thirstiness and cardiac overstret due to increased heat waves. Heat may lead to lack of concentration, coordination and response hence is likely to be the cause of accidents at the place of work. Regarding the psychological side, a person can become petulant, devoid of morale, and worn out due to the living conditions which are unsafe and uncomfortable all the time (Kjellstrom et al. 2016). The employers can establish productivity goals when the situation demands the employees to be strained by the adversity of the different conditions. Other than this, other unpleasant weather conditions like heavy rains and storms, may also stall the work and create financial instability particularly among the contractual workers or the daily wage workers. Unemployment would also contribute to the burnout and anxiety levels of most of the working places in the outdoor set-ups. It does not simply inhibit that only such physical climate stressors that happen in corporate and indoor work places are less susceptible to it. Nonetheless, some indirect effects can also be observed in the office employees e.g., fear of the weather, their company is reorganizing, their chain system is not organized and they work at home due to the severe weather. The climatic change may also affect the corporate-level strategy of the company in terms of regulation compliance and sustainability expectations that impose an extra, prudential, and emotional responsibility on the employees (Tsakonas et al. 2024). It might be connected with fossil fuel, or it might be connected with the fact that the companies of the industry are environmentally-sensitive, the transition to greener features will expose the workers to moral distress or fear of a long-term job. Moreover, such climatic tragedies like the disconnection of power supply, destruction of infrastructure, resettlement, etc. also could disrupt the labor process and introduce the influx of work in excess in the case of the re-appearance of the work. The extreme weather precautionary telecommuting to save the population brings about work-life imbalance and stress in addition to the merge of work and personal life.

In general, the impact of sector-specific can be revealed to possess a very heterogeneous effect on the workplace mental health because of climate change. The economy is vulnerable and uncertain about the environment, which takes over the agricultural sector and the climatic sensitive industries. The intensities of the stresses are the exposure to the trauma and severity of the workload within the emergency and healthcare services. The company has some of the characteristics of these industries, which include physical stress and physical safety in the construction and outdoor labor sector, indirect psychological and organizational strain in corporations. Such inequalities in the sectors will be established and will help the policymakers and the organizations to make sectoral responses in addressing the environmental exposure and the psychosocial wellbeing. This treatment is necessary to preserve the psychological well-being of the employees and productivity within the flexible framework of the conditions of the work in the era of the rapidly growing climate change.

INFLUENCE ON THE EMPLOYEE RESULTS

The effects of climate change stressors are extensive in terms of its implication on the outcomes of the workers, particularly the psychologic, emotive, cognitive, and organizational commitments. These impacts are not restricted to the physical safety aspect but other more psychosocial impacts of the work which directly affect the work performance as the environmental pressure mounts. They complement each other and are more apt to cumulative i.e. when exposed to the stressors of a weather condition, chances are high that either would become long term fatigued of the wellbeing and the overall organizational performance. Psychological distress and anxiety are one of the short-term effects. Stressful responses are also likely to be developed by the exposed employees who are subjected to weather extremes, whether heat or environment insecurity (Pawirosumarto, Sarjana, and Muchtar 2017). Fear, helplessness and hypervigilance of people can be a result of short-term exposure to disaster e.g. floods, or heat wave but permanent anxiety can be the result of long-term exposure to climatic variability. The climatic issues are the personal safety issues, family, and economic prosperity, and they result in an emotional stress. Moreover, the data regarding the deterioration of the environment in the long-term can also create the eco-anxiety within the interminable orientation on the future. The psychological discomfort is likely to be expressed as irritability, sleeping disorder, the lack of concentration and mood change. The disease is chronic, and the issue of anxiety may introduce other serious mental illness, in the event of the lack of treatment, such as the depression or stress disorder. This lowers emotional stability and capacity of an individual to cope in a working place where the individual gets more agitated and overestimates the daily job demands of the employees (Pousa and Mathieu 2014).

Climate strain has a longer-term organizational impact that is more long term like absenteeism and turnover of work force. The employees are quite stressed, burnt out or mentally health dysfunctional whenever the workers fall ill or have to take a leave to deal with the physical and psychological burnouts. They can as well be climate catastrophe ships which result in individuals being displaced, destruction or transportation damage of property hence short term absences. The constantly shifting distractions destroy a sense of permanence and integration amongst the workers in the company. Such knowledge will render the workforce less committed because they will be aware that no organizational support will be given in the event that there will be disastrous weather conditions and they will be willing to resign under their own volition. The second factor which can lead to a workforce instability is that the jobs are not stable due to the economic changes that are posed by the climate in which the workers are subjected to less risky jobs. It is financially implicative to the organization since the rate of turnover is high and it is time consuming to train and lose organization knowledge. One must indicate that the result of these employees is not left in isolation but rather connected to each other (Prasetya 2018). The stress may end up becoming a burn out provided that the stress is not pressing. Besides augmenting the absenteeism, burnout may reduce the productivity. The poor performance and cycle of stress are likely to increase the burden on the management. Climate change is therefore a secondary issue surrounding the existing imminent challenges of the workplace that makes it more vulnerable and a challenge to the organizational strength. These effects are in most cases alleviated by the forces such as leadership and mental health services, accommodation at the workplace and social support systems, which provide the stipulation of the degree of outcome.

The outcomes of the performance of the workers need a solution that is dynamic and systematic. The other psychosocial risk, which the organization must evaluate in addition to the environmental risk, is the climate change, which has factual implications on the sustainability of the work force. The employers can avoid the psychological pressure, burnout, protective mental functioning and labor force sustainability, which are becoming increasingly necessary, especially in the context of the increasing environmental uncertainty.

RISKS AND ISSUES WITHIN THE COMPANY

Climate change is a complicated issue within the company that cannot only be found within the environmental responsibility and the sustainability promise. The severity of the climatic risks is increasing and the reason why the organizations are expected to react to the turbulence in their operations, its exposure to the employees and law and policy changes. These are not only the problems where business continuity implication applies but also the wellbeing of the workers, equity and long term strategic planning. They are risks, which are absolute to be learnt and make strong and accommodating organizations to learn how to behave in the period of the environment ambiguity. One of the closest organizational issues that are associated with the climate change is the operation disruption. One can speak about flooding, cyclones, heatwave, and wildfires as the elements of extreme weather and it might also result in the destruction of infrastructure, the disruption of the supply chain, and the end of production (Heikkilä et al. 2010). The electrical problems, the failure of the transportation and communication system are a complication of the usual activities and cause the economy losses. Even in the industries, which are exposed to the natural conditions or the surrounding environment, such as in agriculture, in construction and logistics the slightest change in the climate conditions will diminish the working process and the output of the work. Organizational Challenges and Risks Associated with Climate Change are shown in table 2

Table 2. Organizational Challenges and Risks Associated with Climate Change; Source: Author Contributions

Challenge Area	Key Issues	Impact on Organization	Impact on Employees	Risk Level
Operational Disruptions	Extreme weather events, supply chain breakdowns, infrastructure damage, power outages	Financial losses, delayed production, reduced efficiency, business interruption	Increased workload, job stress, uncertainty, fatigue	High
Workforce Vulnerability and Inequality	Heat exposure, unsafe conditions, income instability, limited healthcare access	Reduced workforce stability, higher turnover, decreased morale	Physical strain, anxiety, burnout, unequal risk exposure	High
Legal Compliance Risks	Occupational safety regulations, environmental standards, liability for unsafe conditions	Legal penalties, reputational damage, increased compliance costs	Concern about job security, reduced trust in management	Medium–High
Policy and Regulatory Changes	Climate reporting requirements, sustainability mandates, labor transition policies	Strategic restructuring, additional operational costs, adaptation investments	Job restructuring stress, skill gaps, fear of displacement	Medium–High
Insurance and Financial Risks	Rising insurance premiums, climate litigation, investment risks	Increased operational expenses, reduced profitability	Indirect stress due to financial instability	Medium

INTERVENTIONS AND STRATEGIES OF ADOPTION

Climate change is a new area of work, and the organizations must wish to tap into the oriented and participatory intervention to not only tame the mind of its employees, but also ensure that the business is not stalled. Adaptive strategies should be applied to the psychosocial stressors and physical climatic threats. All the levels of policy, programmatic and operational levels have good policies and programs, which ensure resiliency and reduced vulnerability among the workers. Climate change is now an official occupational health and safety and wellbeing issue with such policies (Alfaro-Serrano et al. 2021). Adaptation of the working schedules during the excessive weather, implementation of rest breaks, hydration, and installation of cooling systems are the changes that the organizations can implement as heat action plans. Disaster preparedness, emergency communications plans and business continuity plans are also to be incorporated in the policy structures. Additionally, regarding the climate-sensitive policy, the provisions can be made within the area of mental health such as post-disaster mental health care, paid emergency leaves and flexible working schedules in case of the existence of climate-related crisis. The advantages of the inclusion of climate risk assessment in the procedure of the development of strategic plans are that the organizations in question can anticipate future difficulties and not respond to particular incidences (Badham, Kee, and Hunter 2018). The uncertainty is mitigated through strategic policies, which are aimed at preparing and committing the organization to the security of the employees thus contributing to the reduction of stress and increasing trust of those who are under fear or traumatized by climatic conditions. Employee Assistance Programs (EAPS) play a key role in this context to help the workers who are under stress because of climatic conditions, traumatic experiences or fears. Certain services that such programs provide are confidential counseling services, crisis intervention and referral to the specialized mental health workers. The EAPs could be able at locations where the weather is atrocious most of the seasons to involve themselves in offering traumatic counseling and grief counseling services and financial advisory services to the displaced or the victims who lost their loved ones. Early help-seeking behavior and mental stigmatization may become the results of active mental health awareness programs as the EAP programs. The other channel through which the organizations can incorporate the digital counseling platforms is by ensuring the platforms can be accessed in case of emergencies or remote work situations. EAPs can offer the management of stress and resilience training hence that the employees are equipped with the working skills in managing the vagaries of the surrounding and the professional pressures owed to the possibility of diminishing the development of chronic stress into burnout, depression or long-term mental illness (Sheeran et al. 2023). The programs include certain amount of mindfulness, relaxation, cognitive-behavioral, emotional control and time management programs. Resilience programs focus on the objective of being flexible, optimistic and problem solvers in an event of disruption. The eco-anxiety can be further overcome with the help of the climate literacy workshops to present the factual information and create an accent concerning the actions that might be undertaken to enhance the sustainability (Read and Wainger 2023). The resilience training that is team based facilitates peer support, co-operative coping, and co-operative coping.

These work arrangement flexibilities are highly crucial as adaptive resources in the event of climatic variability due to the fact that employees can be busy, productive and psychologically fit even in the event of the climatic disturbances. Working in case of extreme weather reduces the number of physical contacts with the risk factor and does not halt the running of the business. The flexibility of the schedule will also allow employees to work during down seasons or even address personal problems when they are interrupted by the weather conditions, yet will have to be arranged in a manner that it would not lead to isolation, digital burnout and working lines. To gain a sense of balance, one must be straight forward within the realms of communication, work assignment and performance review. The flexible working arrangement must be taken to control well because it will bring in more flexibility, reduce stress, and allow the employees to cope with any arising climate risks.The occupational health and safety systems must be adaptable to the arising climate risks. This includes the modernization of the ventilation system to regulate the extent of air pollution, introduction of the respective cooling systems in the areas where there is a high temperature and provision of the corresponding safety gears to the workers operating in open field. The key prevention measures are regular health screening of the employees during the heatwave and educating the employees on heat stress or dehydration symptoms (Balducci et al. 2017).

In the future, risk assessment of climatic forecasts should also be done as opposed to only using previous data. The safety measures used are to be updated with the disaster responses, evacuation, and mental health first aid training. The extra physical safety systems will assist not only in keeping their employees safe of the dangers of the environment, but also in reducing the level of psychological strain as they will be safe and prepared. Overall, the adaptive measures and interventions can bring a comprehensive management of how to tackle the issues of climate related in the workplace. The shift of policies, mental health support, training on resilience, flexible working policies, and safety concerns allow firms to invest in workforce well-being and invest in the long-term business sustainability during the period of changes in the environment. Leadership role and Organization Culture. The most defining aspects that are influencing the interpretation, management and reduction of the impacts of the climate issues in the workplace include organizational culture and leadership (Mu et al. 2025). Though the structural support is based on policies and formal measures, the efficiency of the implementation and maintenance of adaptive strategies is established by the behavior of the leader and cultural norms. The necessity of the organizations to possess a leader that is receptive, sensitive, open, and committed to the climate and wellbeing of the employees is the purpose of the climate change. The resilience also occurs with the assistance of an enabling organizational culture which is the establishment of trust, collaboration and psychological safety. The leaders must go out of their way to assess the dangers associated with climate, be open of the potential disturbances and the sustainability should become a component of vision and decision making in the organization (McKay, Margaret Dolcini, and Hoffer 2017). Communication, especially where there is a climatic disaster, is free, thus confusion and misinformation which otherwise would have compounded the level of anxiety among the employees is reduced. The reactive leaders are empathetic and will be concerned about the well-being of the employees in case of an extreme weather or any interruption of their job. They are able to accommodate working flexibility, provide additional mental health services or invest on safety. The leaders show the examples of adaptive behavior, a calm attitude towards the situation when the pressure is experienced, thereby, creating the climate of organizational resilience. As well, a leader who takes personal interest in the workers throughout the sustainability efforts like green workplace or climate action plan can help in the transformation of the fear or green-anxiety into the beneficial engagement (Martin et al. 2018). This participatory may be applied in an attempt to elevate morale and make the employees feel that the group is responsible. Strategy foresight is another significant factor in climate-responsive leadership. The leaders must be in a position to anticipate the dangers of the climate change long term, invest in the resilience infrastructure and resilience in cases where reskilling of the work force is needed. The balanced and progressive shape that the leaders have modeled promotes continuity of the operations and psychological stability by balancing the environmental sustainability in the wellbeing of the employees and the growth of the organization. Resilience cannot be said to be the ability to stabilize the situation, once destabilized by the climate and instead, learn to anticipate, absorb and adapt to change (Schoemaker et al. 2020). Open communication, collaboration, and innovation that will make sure the aspect of flexibility is injected into the working process, promote cross-functional cooperation, and allow the unceasing study of the issues of climate hazards and sustainability are the organizational policies that could be followed in the development of the resilience.

This will motivate and encourage the employees and raise their morale, as they will be rewarded in the process of giving the response to the crisis and recovery, as well as giving them a sense of purpose. The culture of wellbeing and culture of productivity also help the resiliency of the companies and ascertaining that the sense of belonging and identity is propagated and hence limiting the instances of stress and burnout being chronic. Employees are not scared since, on the contrary, they are motivated by instilling common values with strength rather than discussing them in the presence of the personal need. The term psychological safety is used to communicate the nature of work relationship in which the employees do not fear to share their thought, challenges and vulnerability with no fear of being coerced, reprimanded and punished (Estrela et al. 2025). The psychological safety has specifically been applicable in the climate change situation where the employees can be requested to report on any environmental risks, request accommodation in case of severe weather or discuss mental issues openly. Other ways through which leaders can make sure that they improve psychological safety are active listening, emotional response and openness of the discussion. The workers should be empowered to develop sustainability plans, air safety concerns concerning the heat, or air quality, and contribute to climate change management. The employees believe that they are listened to thus bringing in increased trust and participation. The reduction in the mental health-related stigma is another attribute of the proposed change that will be encouraged through the facilitation of psychological safety. Normalizing stress, burnout, and climate anxiety discussion process would be facilitated through incorporating the well-being issues discussion into the regular team discussion and leadership messages. The interpersonal relationships of emotional intelligence and mental health awareness are also built in the process of training of the managers. The final but not the least is the concept of the leadership and the organizational culture which can be perceived as the pillars where all the adaptive strategies are pegged on. All these assist in managing stressors in the environment among the employees: climate-sensitive leadership, influential psychological safety, and powerful organizational culture. Their collaboration creates operationally prepared workplaces to manage climatic upheavals and psychologically advantageous to provide long term well as organizational stability to climate change that just continues to grow.

POLICY IMPLICATIONS AND FUTURE DIRECTIONS

The problem of global warming is structural and, both, long-term, as well as the one that is demonstrated with interagency response regarding the policies and organizational strategies. The leaders of the organization and the policymakers should offer integrated orientations, provided that the environmental risks are even more closely tied to the stability of the workforce and their occupational health, and it is the climate resilience that would have to be adjusted in accordance with the health of the employees. The sustainability of the work places will be pegged on the necessity of ensuring that the adaptation strategies are developed in the workforce systems, cooperation that will be witnessed in the different sectors and research that will be carried out to address the arising issue. The idea of Climate Adaptation in the Workforce Planning was added to the 9.1 that was to be adopted. Climate adjustment integrity in the workforce approach implies that the organizations must not simply focus on the immediate crisis management to plan towards the strategy. The climate risk assessment should be included in the workforce planning to predict the ways the alterations in the environment can affect the demand of the labor and skills requirements, health of the employee and distribution of work. The examples would be the high-temperature industries, the floods, or the scarcity of resources that will lead to the implementation of the staffing models, the safety and continuity of operations approaches. The human resource strategies must address the reskilling and upskilling programmes which would involve moving employees to sustainable and green economies. As the regulatory frameworks would have changed, and the requirements of the market shifted to the low-carbon functioning, the former jobs would lose their jobs, and the new ones would appear in the field of the renewable energy, the environment management and the climate risks assessment. The training plans that are basic reduce the employment insecurity and alleviate the employment transition. Climate adaptation should also be provided in the occupational health policies. The entire workforce planning manuscript must incorporate the heat stress management plans, contingency plans of the staffing requirement in the event of the disaster occurrence and how the psychological well-being of the employees would be addressed. The governments can implement this integration by setting country level labor regulations that respond to the occupational hazards which are associated with the climatic changes and the creation of sustainable business. The adaptation to the workforce planning enhances the security and resiliency of the employees in the long-term.

The issues of mental health at the workplace that climate causes require intervention on the realms of the civil societies, the private, and the public. Risks of climate are systemic and interrelated enough to ensure that the solutions are coordinated and multidisciplinary. The governments are able to consult the businesses in formulating the level of occupational safety in the severe weather conditions. The infrastructure development program, the retraining of the working forces program, the disaster preparedness programs can be provided as the public-privatization cooperation. The health care systems and occupation health workers are very instrumental in aiding the climate-prone workers, by performing trauma counseling, preventive services and mental health services to the workers. This may be through a research team and an academic institution developing evidence-based models of intervention and training the future professionals on how to handle the resistance of workforce to climate change. With the incorporation of the climate risk assessment in the lending and coverage designs, financial institutions, and insurers can promote resilience investments. Knowledge and best practices are also disseminated and distributed with ease by the cross sector working. By the employers, local authority and the community organizations withdrawing the case of climate-related disasters, the affected employees will receive the required support in time. Such relationships reduce the risks of overlapping and also making a community stronger. It has been seen that the cross-sector collaboration offers complete solutions beyond the organizations through the centralization of resources and expertise. 9.3 Research Gaps and New Trends. Although the problem of climate-related mental health challenges at the workplace is increasingly gaining popularity, the research is still lacking in multiple aspects. They need long-term research that will assist in exploring the effects of the chronic exposure to climatic stressors to the long-term psychological effects. The recent research is inclined to focus on short term response to disasters, but it is the long-term pressure that should be discussed more thoroughly due to the long-term lack of security of the environment. The sector analyses are also weak in their own way particularly the informal and low-income employment that may offer the utmost level of vulnerability. Further research is needed to record the overlap between the demographic variable age, gender, socioeconomic status, and geographic location with climate-related occupational risks. This will render the policy making accommodative and fair. The desire to become a more resilient entity and a usage of digital technologies is one of the recent tendencies. This is in the form of remote working systems and digital mental health platforms, predictive climate model platforms, and all these come alongside coping with the workplace. In addition, the eco-anxiety/climate-related moral distress is becoming more popular, and it is suggested that the flexible enough psychological models related to the environmental issues should be presented. The future research should compare the usefulness of resilience training and flexible work policies and leadership intervention in lessening mental health risks.

CONCLUSION

The problem of climate change has turned out to be the feature of determining the working relations, psychological condition of the workers and the sustainability of the organization. The environmental conditions that increase the psychological disturbance, work burnout, low productivity and labour turnover include the increasing temperatures, weather extremity, economic changeover and environmental uncertainty. Such dilemmas demand composite policy interventions that will provide and protect physical and psychosocial health. The greatest feature of resilient organizations is that climate adaptation should also be included into the workforce planning, the cross-sectoral collaboration should be improved and new trends should be researched. Inclusive policies and preemptive mental health response; also encourage organizational competencies to respond to the uncertainty in the environment by means of leadership. Lastly, the notion that the workplaces are climate-resilient might be defined as where individuals are concerned not only about their staff and sustainability of their companies. The fact that climate change is an environmental and human resource issue will assist policymakers and organizational administrators in adopting the adjustive mechanisms that will protect the mental health, establish equity, and sustainable development in the age of the fast-evolving environmental change.

REFERENCES

Adamopoulos, I., Valamontes, A., Tsirkas, P., & Dounias, G. (2025). Predicting Workplace Hazard, Stress and Burnout Among Public Health Inspectors: An AI-Driven Analysis in the Context of Climate Change. *European Journal of Investigation in Health, Psychology and Education*, *15*(5), 65. DOI: 10.3390/ejihpe15050065 PMID: 40422294

Alfaro-Serrano, D., Balantrapu, T., Chaurey, R., Goicoechea, A., & Verhoogen, E. (2021). Interventions to Promote Technology Adoption in Firms: A Systematic Review. *Campbell Systematic Reviews*, *17*(4), e1181. DOI: 10.1002/cl2.1181 PMID: 36950339

Badham, J., Kee, F., & Hunter, R. F. (2018). Simulating Network Intervention Strategies: Implications for Adoption of Behaviour. *Network Science*, *6*(2), 265–280. DOI: 10.1017/nws.2018.4

Balducci, S., D'Errico, V., Haxhi, J., Sacchetti, M., Orlando, G., Cardelli, P., Vitale, M., Bollanti, L., Conti, F., Zanuso, S., Nicolucci, A., & Pugliese, G.Balducci, Stefano, Valeria D'Errico, Jonida Haxhi, Massimo Sacchetti, Giorgio Orlando, Patrizia Cardelli, Martina Vitale, Lucilla Bollanti, Francesco Conti, Silvano Zanuso, Antonio Nicolucci, Giuseppe Pugliese, and for the Italian Diabetes and Exercise Study 2. (2017). Effect of a Behavioral Intervention Strategy for Adoption and Maintenance of a Physically Active Lifestyle: The Italian Diabetes and Exercise Study 2 (IDES_2). *Diabetes Care*, *40*(11), 1444–1452. DOI: 10.2337/dc17-0594 PMID: 28821576

Brooks, S. K., & Greenberg, N. (2023). Climate Change Effects on Mental Health: Are There Workplace Implications? *Occupational Medicine (Oxford, England)*, *73*(3), 133–137. DOI: 10.1093/occmed/kqac100 PMID: 36170162

Estrela, M., Leitão, C., Neto, V., Martins, B., Santos, J., Branquinho, A., Figueiras, A., Roque, F., & Herdeiro, M. T. (2025). Educational Interventions for the Adoption of Healthy Lifestyles and Improvement of Health Literacy: A Systematic Review. *Public Health*, *245*, 105788. DOI: 10.1016/j.puhe.2025.105788 PMID: 40466525

Heikkilä, A.-M., Malmén, Y., Nissilä, M., & Kortelainen, H. (2010). Challenges in Risk Management in Multi-Company Industrial Parks. *Safety Science*, *48*(4), 430–435. DOI: 10.1016/j.ssci.2009.12.008

Kjellstrom, T., Briggs, D., Freyberg, C., Lemke, B., Otto, M., & Hyatt, O. (2016). Heat, Human Performance, and Occupational Health: A Key Issue for the Assessment of Global Climate Change Impacts. *Annual Review of Public Health*, *37*(1), 97–112. DOI: 10.1146/annurev-publhealth-032315-021740 PMID: 26989826

Kjellstrom, T., Holmer, I., & Lemke, B. (2009). Workplace Heat Stress, Health and Productivity – an Increasing Challenge for Low and Middle-Income Countries during Climate Change. *Global Health Action*, *2*(1), 2047. DOI: 10.3402/gha.v2i0.2047 PMID: 20052422

Lan, Y.-L., Huang, W.-T., Kao, C.-L., & Wang, H.-J. (2020). The Relationship between Organizational Climate, Job Stress, Workplace Burnout, and Retention of Pharmacists. *Journal of Occupational Health*, *62*(1), e12079. DOI: 10.1002/1348-9585.12079 PMID: 31407446

Levi, M., Kjellstrom, T., & Baldasseroni, A. (2018). Impact of Climate Change on Occupational Health and Productivity: A Systematic Literature Review Focusing on Workplace Heat. *La Medicina del Lavoro*, *109*(3), 163–179. DOI: 10.23749/mdl.v109i3.6851 PMID: 29943748

Lundgren, Karin, Kuklane, Kalev, & Gao, Chuansi, and Ingvar HOLM^|^Eacute;R. (. (2013). Effects of Heat Stress on Working Populations When Facing Climate Change. *Industrial Health*, *51*(1), 3–15. DOI: 10.2486/indhealth.2012-0089 PMID: 23411752

Martin, N. A., Hulland, K. R. S., Dreibelbis, R., Sultana, F., & Winch, P. J. (2018). Sustained Adoption of Water, Sanitation and Hygiene Interventions: Systematic Review. *Tropical Medicine & International Health : TM & IH*, *23*(2), 122–135. DOI: 10.1111/tmi.13011 PMID: 29160921

McKay, V. R., Dolcini, M. M., & Hoffer, L. D. (2017). The Dynamics of De-Adoption: A Case Study of Policy Change, de-Adoption, and Replacement of an Evidence-Based HIV Intervention. *Translational Behavioral Medicine*, *7*(4), 821–831. DOI: 10.1007/s13142-017-0493-1 PMID: 28397157

Mu, L., Li, Y., Liu, H., & Wang, Q. (2025). Implementation or Inaction: How Governmental Multi-Interventions Catalyze Farmers' Adoption for Climate Adaption Technology? *Journal of Cleaner Production*, *521*, 146193. DOI: 10.1016/j.jclepro.2025.146193

Nowrouzi-Kia, B., Choi, C., Premji, R., Sathananthan, A., Balakrishnar, K., Haritos, A., Long, B.-Z. S., Mazur, M., & Fatemi, A.-B. 2026. "The Impacts of Climate Change on Occupational Health and Work among Outdoor Workers: A Scoping Review" edited by J. Robinson. *PLOS Global Public Health* 6(2):e0005888. DOI: 10.1371/journal.pgph.0005888

Pawirosumarto, S., Sarjana, P. K., & Muchtar, M. (2017). Factors Affecting Employee Performance of PT.Kiyokuni Indonesia. *International Journal of Law and Management*, *59*(4), 602–614. DOI: 10.1108/IJLMA-03-2016-0031

Pousa, C., & Mathieu, A. (2014). The Influence of Coaching on Employee Performance: Results From Two International Quantitative Studies. *Performance Improvement Quarterly*, *27*(3), 75–92. DOI: 10.1002/piq.21175

Prasetya, A. (2018). ANALYSIS OF FACTORS THAT INFLUENCE EMPLOYEE PERFORMANCE (Study on Permanent Employees in Operational Section of PT WIMCycle Indonesia - Surabaya). *Profit*, *12*(1), 1–12. DOI: 10.21776/ub.profit.2018.012.01.1

Read, D. J., & Wainger, L. (2023). Assessing Intervention Effectiveness at Promoting Voluntary Conservation Practice Adoption in Agrienvironments. *Conservation Biology : The Journal of the Society for Conservation Biology*, *37*(1), e14009. DOI: 10.1111/cobi.14009 PMID: 36178035

Schoemaker, N. K., Wentholt, W. G. M., Goemans, A., Vermeer, H. J., Juffer, F., & Alink, L. R. A. (2020). A Meta-Analytic Review of Parenting Interventions in Foster Care and Adoption. *Development and Psychopathology*, *32*(3), 1149–1172. DOI: 10.1017/S0954579419000798 PMID: 31366418

Schulte, P. A., & Chun, H. K. (2009). Climate Change and Occupational Safety and Health: Establishing a Preliminary Framework. *Journal of Occupational and Environmental Hygiene*, *6*(9), 542–554. DOI: 10.1080/15459620903066008 PMID: 19551548

Sheeran, P., Wright, C. E., Listrom, O., Klein, W. M. P., & Rothman, A. J. (2023). Which Intervention Strategies Promote the Adoption and Maintenance of Physical Activity? Evidence From Behavioral Trials With Cancer Survivors. *Annals of Behavioral Medicine : A Publication of the Society of Behavioral Medicine*, *57*(9), 708–721. DOI: 10.1093/abm/kaad002 PMID: 37078961

Syahrir, M. S., & Falah, F. (2021). Occupational Burnout and Workload: Mental Health Challenges in the Workplace and Solutions. *Journal of Health Literacy and Qualitative Research*, *1*(2), 69–83. DOI: 10.61194/jhlqr.v1i2.528

Tang, W. S. W., & Cyrus, S. H. H. (2025). A Systematic Review on the Impact of Climate Change on Occupational Mental Health: A Focus on Vulnerable Industries. *Social Psychiatry and Psychiatric Epidemiology*, *60*(10), 2275–2287. DOI: 10.1007/s00127-025-02936-x PMID: 40473944

Trombley, J., Chalupka, S., & Anderko, L. (2017). Climate Change and Mental Health. *The American Journal of Nursing*, *117*(4), 44–52. DOI: 10.1097/01.NAJ.0000515232.51795.fa PMID: 28333743

Tsakonas, K., Badyal, S., Takaro, T., & Buse, C. (2025). Occupational Health Impacts of Climate-Related Emergency Events on Health System Workers in British Columbia. *International Journal of Disaster Risk Reduction*, *118*, 105227. DOI: 10.1016/j.ijdrr.2025.105227

Tsakonas, K., Badyal, S., Takaro, T., & Buse, C. G. (2024). Rapid Review of the Impacts of Climate Change on the Health System Workforce and Implications for Action. *The Journal of Climate Change and Health*, *19*, 100337. DOI: 10.1016/j.joclim.2024.100337 PMID: 41647318

Chapter 9
Vector-Borne and Zoonotic Diseases at Work:
A One Health Perspective in the Face of Climate Change

Armand Namekong Fokeng
http://orcid.org/0000-0002-1247-1331
Università di Corsica Pasquale Paoli, France

Archile Paguem
University of Buea, Cameroon

ABSTRACT

Climate change is reshaping ecological systems and altering the risks of zoonoses and vector-borne diseases in the workplace. By altering temperatures, precipitation and seasonal dynamics, it affects the distribution and activity of vectors (mosquitoes, ticks, sandflies, Culicoides, tsetse flies, triatomines) as well as that of animal reservoirs. Combined with urbanisation and habitat fragmentation, these changes open up new routes of transmission. The most exposed occupations include agriculture, forestry, wildlife management, slaughterhouses, and the human and animal health sectors. Through a One Health approach, this chapter analyses these risks by integrating human, animal and environmental dimensions, examines the most relevant diseases globally, and proposes strategies for integrated surveillance and targeted prevention for exposed workers (Figure 1).

DOI: 10.4018/979-8-3373-8470-2.ch009

INTRODUCTION

Figure 1. Graphical abstract: interrelationship between climate change, environmental changes, vectors or animal host dynamics and occupational exposure of workers to vector-borne zoonoses in a One Health perspective. (±) = positive or negative influence

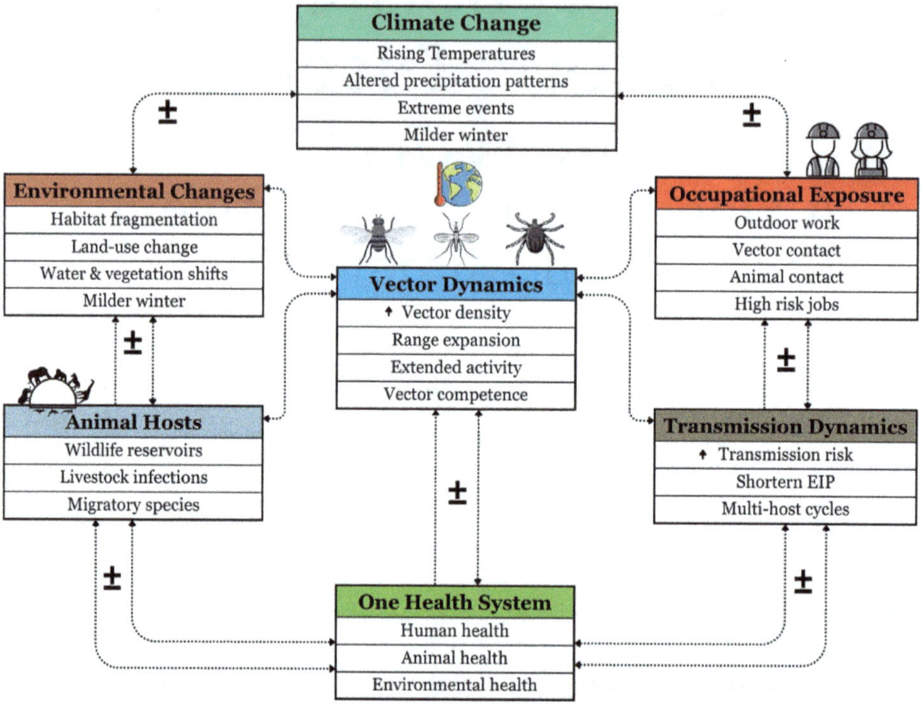

Zoonoses and vector-borne diseases (VBDs) account for a substantial proportion of global infectious morbidity and mortality (Bidaisee & Macpherson, 2014). A significant share of these infections is associated with occupational exposure (OE) to animals, vectors, or altered ecological environments (Durand-Billaud et al., 2024). Approximately 60% of human infectious diseases are zoonotic in origin, and up to 75% of emerging infectious diseases are associated with zoonoses (Lesne, 2021). Collectively, these diseases represent over 17% of all infectious diseases worldwide and are responsible for nearly one million deaths annually (Durand-Billaud et al., 2024). Many zoonoses involve arthropod vectors or animal reservoirs (e.g. rodents, bats, birds) with which outdoor workers particularly those in agriculture, forestry, and landscape management, are frequently in contact (Bidaisee & Macpherson, 2014; Richard & Oppliger, 2015). The dynamics of infectious agents, vectors, reservoir hosts, and pathogen replication are strongly influenced by climatic conditions (IPCC, 2023). Rising temperatures, for instance, promote vector proliferation, extend transmission seasons, and alter ecosystems, leading to shifts in both vector and human population distributions (IPCC, 2023).

Climate change refers to long-term changes in temperatures and weather patterns occurring over several decades (de Souza & Weaver, 2024). It is increasingly recognised as a major determinant of global health dynamics (de Souza & Weaver, 2024). This phenomenon is largely driven by anthropogenic activities including fossil fuel combustion, intensification of the greenhouse effect and deforestation, resulting in a gradual increase in global surface temperatures (Paz, 2020). However, climate change is not limited to warming ; it also involves disruptions to precipitation patterns and wind regimes, and an increased frequency and intensity of extreme weather events (Lee et al., 2013). Acting as a risk multiplier, it modifies temperatures, precipitation, humidity, and seasonality thereby influencing the survival, reproduction and geographical distribution of vectors such as mosquitoes, ticks and sandflies, as well as the pathogens they carry (Semenza & Suk, 2018). In addition broader environmental consequences including rising sea levels, ecosystem degradation and biodiversity loss can profoundly alter the interactions between hosts, vectors and pathogens (Lee et al., 2013; Nnko et al., 2021). However, vulnerability to VBDs is not solely determined by environmental factors; it also depends on socio-demographic and socio-economic factors, including age, housing quality, access to healthcare and occupational activities (IPCC, 2023). These combined changes may increase the risk of the emergence or re-emergence of VBDs, and facilitate their geographical spread, particularly in occupational settings (Lee et al., 2013; Magnavita et al., 2022).

Recent studies in Europe already report measurable effects, such as the northward and altitudinal spread of *Ixodes ricinus* and the emergence of local outbreaks of dengue, chikungunya and West Nile virus in previously unaffected areas (Magnavita et al., 2022; Semenza & Suk, 2018; Stufano et al., 2025).

Crimean-Congo haemorrhagic fever (CCHF) illustrates the importance of occupational exposures (OEs) in this context (Ahmed et al., 2022). This zoonotic disease, transmitted by *Hyalomma* ticks, primarily affects livestock farmers, veterinarians, abattoir workers, and healthcare professionals exposed to infected animals or humans (Fanelli & Buonavoglia, 2021). Climatic conditions favourable to tick survival and activity, combined with agricultural practices and work organisation, may influence the risk of human transmission (Gale et al., 2012; Kuehnert et al., 2021).

The aim of this chapter is to propose a One Health framework tailored to occupational settings, in order to identify, monitor and prevent risks associated with climate-sensitive vector-borne zoonoses.

In this context, we conducted a narrative review with two main objectives. First, to examine how VBDs and zoonoses constitute an emerging risk to occupational health and safety in the context of climate change. Second, to highlight that the One Health approach is essential for understanding, monitoring, and preventing these risks. This chapter first defines key concepts (zoonoses, VBDs, OEs, climate change and One Health) before analysing the ecological mechanisms through which climate change affects vector dynamics. It then examines OEs patterns, disease incidence and distribution trends, surveillance and prevention tools, and methodological challenges and future challenges.

BACKGROUND

Climate change is one of the most complex threats to global public health in the 21st century (IPCC, 2023). Beyond its direct effects on human health, it acts as an amplifier of infectious risks by profoundly reshaping ecosystems and interactions between species (Mora et al., 2022). Among the most well-documented consequences is the geographical and temporal spread of vector-borne diseases and zoonoses, the global prevalence of which is constantly increasing (IPCC, 2023).

Zoonoses account for approximately 60% of emerging infectious diseases (Taylor et al., 2001). They are a growing concern in occupational health, as many professions involve direct or indirect contact with animals, biological vectors or natural environments conducive to their transmission (Haagsma et al., 2012). The International Labour Organisation (ILO) also recognises diseases of biological origin, including zoonoses, as priority occupational risks requiring enhanced surveillance (Scott et al., 2024).

The role of arthropod vectors ticks, mosquitoes, sandflies, Culicoides, triatomines and tsetse flies in the transmission of pathogens is well established (Kilpatrick & Randolph, 2012). However, climate change significantly alters their biology, phenology and range. Rising average temperatures accelerate the vectors' life cycles, expanding their ecological niches into latitudes and altitudes previously unaffected (Semenza & Suk, 2018). For example, *Ixodes ricinus*, the main vector of Lyme disease and tick-borne encephalitis in Europe, is gradually expanding its range northwards and to higher altitudes (Boulanger et al., 2024). Similarly, *Aedes albopictus*, a competent vector for dengue and chikungunya, is colonising new temperate regions, including the Mediterranean basin (Oliveira et al., 2021).

Changes in rainfall patterns also play an important role. Periods of drought followed by heavy rainfall create conditions conducive to the proliferation of vectors such as mosquitoes of the *Culex* genus, which transmit West Nile virus (Paz, 2020). At the same time, the fragmentation of natural habitats due to human activity disrupt ecological balances, forcing reservoir animals to move closer to inhabited areas and workplaces ; this may increase the risk of contact with human populations (Gottdenker et al., 2014).

The One Health approach which recognises the interdependence of human, animal and environmental health offers a conceptual framework particularly suited to analysing these risks at the interface between work, wildlife and the environment (Zinsstag et al., 2023). Several studies have demonstrated that integrating this approach into occupational health policies improves the early detection of emerging pathogens and optimises prevention strategies (Rüegg et al., 2017).

In this context, certain diseases provide a striking illustration of the interplay between climate change, occupational activities and zoonotic risk. Crimean-Congo haemorrhagic fever (CCHF), transmitted mainly by ticks of the genus *Hyalomma*, is emblematic of this dynamic: its endemic range is gradually expanding towards southern and western Europe, in association with global warming and the spread of its vectors (Estrada-Peña, 2025; Spengler et al., 2016). Agricultural workers, livestock farmers, veterinary staff and healthcare workers constitute the occupational groups most vulnerable to this infection (Ergonul & Whitehouse, 2007; Vonesch et al., 2019).

Building on this evidence, integrated strategies for epidemiological surveillance and prevention are essential. These must combine ecological habitat management, the strengthening of laboratory detection capabilities, and the adoption of individual and collective protective measures tailored to the exposed occupational contexts.

MAIN FOCUS OF THE CHAPTER

Issues, Controversies, Problems

Despite the growing body of research on climate change and infectious diseases, several scientific and operational tensions remain and warrant discussion.

The first controversy concerns the respective roles of climatic and non-climatic factors in the emergence of vector-borne and zoonotic diseases. Whilst the role of global warming in the spread of vectors such as *I. ricinus* or *Ae. albopictus* is well documented, some authors emphasise that the globalisation of trade, changes in land use and habitat fragmentation play a role that is at least as decisive (de Souza & Weaver, 2024; Gottdenker et al., 2014). For TBE in particular, the relative contribution of climate, animal trade and migratory birds to the spread of *Hyalomma* ticks in Europe remains uncertain (Desvars-Larrive et al., 2024; Estrada-Peña et al., 2021). Attributing epidemic trends to a single factor risks leading to overly simplistic preventive policies.

A second challenge concerns the production and comparability of occupational health data. The majority of available studies focus on Europe and North America, leaving regions that are nevertheless highly exposed, such as sub-Saharan Africa or South-East Asia, in the shadows (Diarra et al., 2023; Msellemu et al., 2024).

Methodological heterogeneity across studies, systematic under-diagnosis and the lack of occupational stratification in surveillance systems make international comparisons difficult and likely underestimate the true burden of these diseases among workers (Richard & Oppliger, 2015).

Finally, the One Health approach is unanimously presented as the conceptual framework of reference. However, its operational implementation faces persistent structural obstacles: the compartmentalisation of human, animal and environmental surveillance systems, insufficient multisectoral governance, and a lack of resources in low-income countries (FAO-UNEP-WHO-OIE, 2022; Gashema et al., 2025). One Health integration remains more of an institutional ideal than a reality on the ground, particularly in contexts where occupational exposures are most severe.

1. FUNDAMENTAL CONCEPTS: ZOONOSES, VECTOR-BORNE DISEASES, "ONE HEALTH", AND OCCUPATIONAL EXPOSURE

1.1. Definition of Zoonoses

A zoonosis is defined as a disease or infection that can be naturally transmitted between vertebrates and humans through various routes (Bidaisee & Macpherson, 2014). Zoonotic diseases encompass a wide spectrum of pathogens including viruses, bacteria, parasites, and fungi that may be transmitted through multiple pathways such as food consumption, direct contact, aerosol inhalation, or via blood-feeding arthropod vectors (Bidaisee & Macpherson, 2014).

Among these, VBDs represent a major subset of zoonoses in which a vector such as mosquito, tick, or phlebotomine sandfly transmits the pathogen from one host to another (Paz, 2020).

1.2. Vector-borne zoonoses

Vector-borne zoonoses are caused by pathogens transmitted to humans by blood-sucking arthropods including ticks, phlebotomines and *Culicoides species* (Paz, 2020). These vectors play an active biological role, often enabling pathogen development or replication before transmission to a susceptible host (Paz, 2020). Transmission typically involves a complex cycle between a reservoir host (animal or, in some cases, human), the vector, and a susceptible human host **(Table 1)** (Paz, 2020).

Environmental changes such as land-use modification, deforestation, agricultural expansion, irrigation, and habitat fragmentation can intensify interactions between humans, wildlife reservoirs, and vectors **(Figure 2)**, thereby promoting spillover and the emergence of new arboviruses (Islam et al., 2025). Urbanisation and peri-urban agriculture contribute to the creation of heterogeneous ecological environments that favour vector proliferation. For example, *Ae. albopictus* thrives in such environments, facilitating arbovirus transmission along urban-rural gradients (Kraemer et al., 2015).

In Europe, several vector species illustrate the interaction between climate, vector biology, and human activity. These include *Ae. albopictus* and *Ae. aegypti* (dengue and chikungunya), *Culex* spp. (West Nile virus), and *I. ricinus* (Lyme disease and tick-borne encephalitis) (Magnavita et al., 2022). OEs are particularly relevant for outdoor workers, including farmers, livestock farmers, foresters, natural space managers, hunters, abattoir workers, veterinarians, healthcare professionals, who operate in environments where vector density is high, activity periods are extended, and new species may become established (Semenza & Suk, 2018).

Figure 2. Urbanisation and habitat change create abundant breeding sites and human hosts. 1: Air transportation can easily spread pathogens and vectors across geographical areas ; 2: urban housing reduces vectors natural habitats and increases human-animal-vector contact, 3,4 & 5: new habitats and host for vectors in urban milieu increase the animal-human-vector contact (Porretta et al., 2013; Zhou et al., 2025)

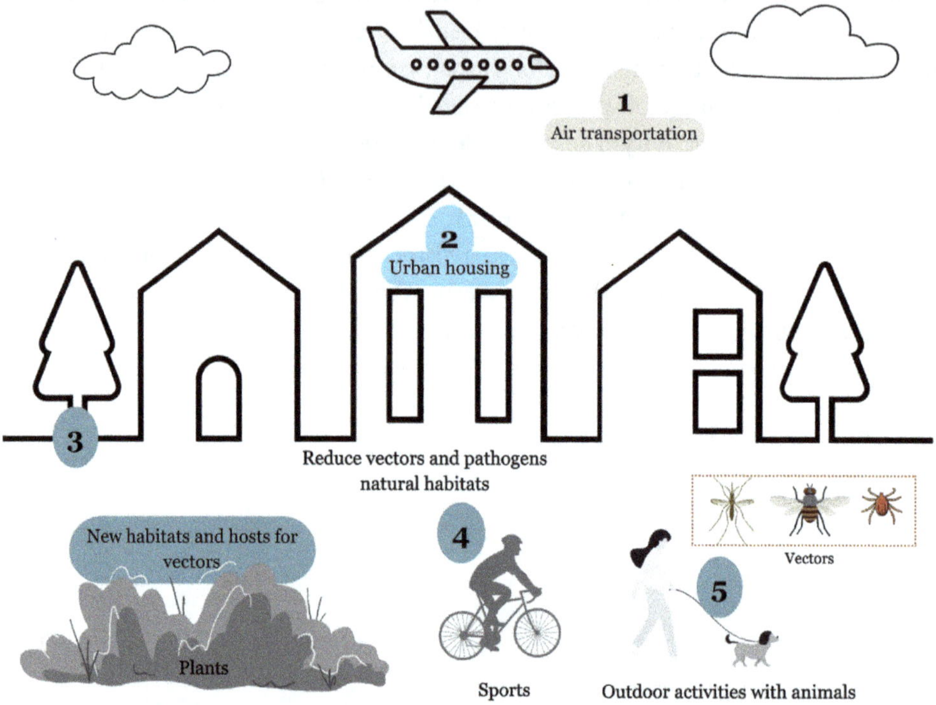

Table 1. Ecological Prerequisites for Vector-Borne Disease Establishment

Component	Necessary Conditions for Establishment
Competent Vector	High abundance, suitable breeding habitat, favourable climate, and adequate host contact rates
Pathogen	Replication capacity in vector and host, horizontal (and sometimes vertical) transmission
Reservoir Hosts	Sufficient density of competent reservoirs, host community amplifying rather than diluting transmission
Climate	Temperature and humidity ranges enabling vector survival, biting, and pathogen development

continued on following page

Table 1. Continued

Component	Necessary Conditions for Establishment
Transmission Cycle	Stable enzootic or urban/peri-urban cycles with frequent vector-host contacts
Environmental and Human Factors	Land use, urbanisation, mobility, socio-economics, and an increasing overlap between vectors, reservoirs, and humans

1.3. The "One Health" Framework

The One Health concept refers to an integrated, intersectoral approach recognising the interdependence of human health, animal health, and ecosystem health, particularly regarding emerging health threats such as zoonoses and VBDs (Cabi, 2025). These diseases are especially sensitive to environmental and climatic perturbations because they occur at the interface of these three domains. Consequently, a One Health approach provides a conceptual and operational framework for understanding and managing such risks (Cabi, 2025). In occupational health, this approach entails integrating environmental determinants (climate, biodiversity, land use), animal health dynamics and working conditions into a unified risk assessment framework. It also supports coordinated surveillance systems linking human epidemiology, veterinary data, and entomological monitoring (Kodama et al., 2023). This integrated perspective facilitates the development of more effective prevention strategies and early warning systems adapted to complex and evolving risk environments (Kodama et al., 2023).

1.4. Occupational Exposure

Occupational exposure (OE) to zoonoses and VBDs is defined as repeated or intense contact during professional activities, with animal reservoirs, vectors, or contaminated environments (Richard & Oppliger, 2015). Such exposure significantly increases the risk of infection compared with the general population (Tokarska-Rodak et al., 2014). It may be direct through animal handling, vector bites, or contact with biological fluids or indirect, through exposure to aerosols, contaminated soil or water, or contaminated surfaces and equipment (Tokarska-Rodak et al., 2014).

In Europe, studies have shown that forestry and agricultural workers exhibit significantly higher seroprevalence rates for certain infections, such as Lyme disease, with odds ratios of approximately 2.3 compared with non-exposed groups (Magnavita et al., 2022; Semenza & Suk, 2018; Tokarska-Rodak et al., 2014). Climate change is expected to further intensify these exposures by increasing vector density, extending activity periods, and expanding geographical range, thereby amplifying occupational risks in both endemic and newly affected areas (Richard & Oppliger, 2015; Rupasinghe et al., 2022).

In sub-Saharan Africa, OE is primarily associated with close interactions between humans, livestock, and wildlife in rural settings. Available data indicate a high diversity of circulating pathogens among vectors and animal hosts, with agricultural and pastoral workers constituting the most exposed groups (Diarra et al., 2023; Djiman et al., 2024). However, data on human infections remain limited and are likely underestimated due to insufficient diagnostic and surveillance capacity, which masks the true burden of OEs (Diarra et al., 2023; Djiman et al., 2024).

In Asia, particularly in South-East Asia, high biodiversity, favourable climatic conditions and intense human–animal–environment interactions contribute to sustained exposure risks. Agricultural workers and other outdoor workers are consistently identified as high-risk groups (Soe et al., 2026). Rapid environmental and socio-economic changes further exacerbate these dynamics, increasing both transmission potential and occupational vulnerability (Msellemu et al., 2024; Soe et al., 2026).

2. CLIMATE CHANGE AND BIOLOGICAL VECTOR DYNAMICS

Biological vectors are not passive carriers ; their physiology, life cycles and behaviour are sensitive to climatic conditions (Rupasinghe et al., 2022). Climate change encompasses not only rising mean temperatures but also altered precipitation patterns, disrupted seasonality, and an increased frequency of extreme weather events such as floods, heatwaves, and droughts (Semenza & Suk, 2018). These environmental changes interact in complex ways and influence both vectors and the pathogens they transmit. Overall, climate change contributes to the expansion of known infectious diseases, facilitates the emergence of new pathogens, and modifies their geographical distribution (Durand-Billaud et al., 2024). In addition, certain pathogens may persist longer in the environment under favourable conditions. For example, leptospires can persist for several weeks in water or moist soil (Durand-Billaud et al., 2024).

Climate change affects several biological parameters of vectors, including their geographical range, mortality rates, activity, life cycles, reproduction, and population density **(Figure 3)** (Durand-Billaud et al., 2024). Key traits such as, survival, fecundity, larval development rates, host-seeking behaviour, and vector competence are closely tied to specific climatic windows, which tend to shift or widen in many regions worldwide (Stufano et al., 2025). These changes have important epidemiological implications, as they alter disease dynamics and increase exposure risks for populations that were previously less affected, including certain occupational groups (Stufano et al., 2025).

Figure 3. Climate change drivers affecting biological vectors, disease transmission dynamics and epidemiological trends, highlighting the need for communication

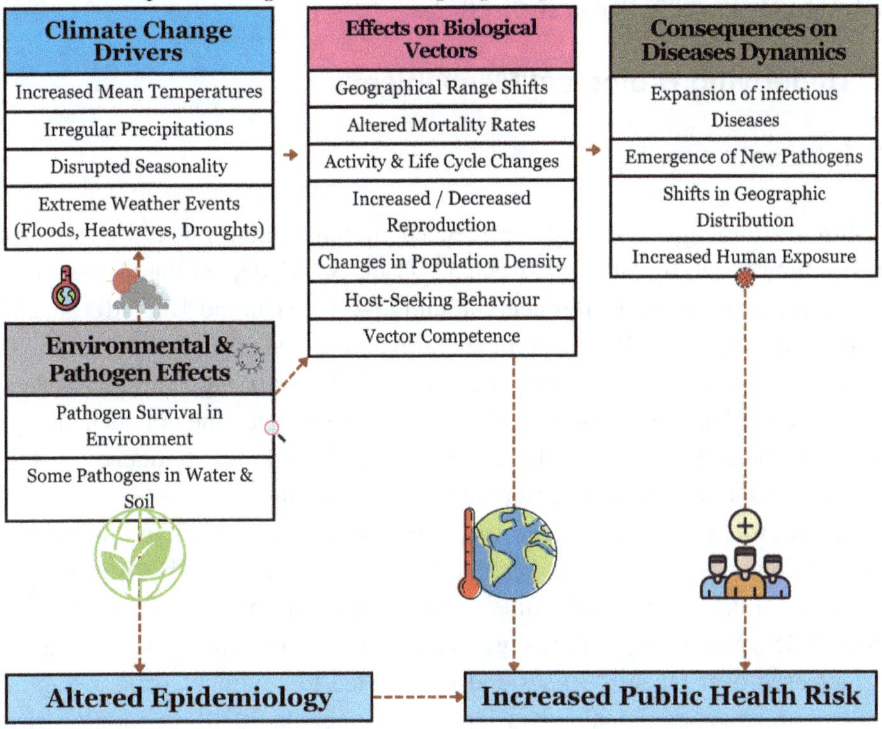

2.1. Geographical Range Shifts: Latitudes and Altitudes

Climate change is driving measurable shifts in vector distributions, both latitudinally and altitudinally. Poleward and upward movements of mosquitoes and ticks have already been documented and are often associated with the emergence of new zoonotic foci (Thomson & Stanberry, 2022).

In South America, for example, the phlebotomine *Lutzomyia flaviscutellata*, a vector of *Leishmania amazonensis*, is projected to expand southward into more densely populated agricultural regions under future climate scenarios (Carvalho et al., 2015). Similarly, in high-altitude regions of Colombia, Ethiopia, and other areas, malaria has been observed at higher elevations during warmer periods, increasing exposure among previously less-affected populations such as mountain farmers (Rocklöv & Dubrow, 2020; Thomson & Stanberry, 2022).

In addition to climatic drivers, land-use changes, including deforestation, agricultural expansion, and urbanisation reduce ecological barriers between vectors, animal hosts, and humans. This reduction in "biogeographical distance" increases contact rates and enhances transmission risks, particularly in occupational settings (Mubemba et al., 2022; Socha et al., 2022).

2.2. Underlying Biological Mechanisms

2.2.1. Mosquitoes

Mosquito biology is strongly influenced by temperature. The upper thermal survival limit generally ranges between 32 °C and 38 °C, depending on the species, while optimal survival typically occurs around 25–28 °C (Durand-Billaud et al., 2024; Rupasinghe et al., 2022). Warmer temperatures and altered precipitation patterns can increase the availability of breeding sites, particularly small stagnant water bodies in urban or peri-urban environments. These conditions shorten the gonotrophic cycle (Durand-Billaud et al., 2024). This could increase blood meal frequency and thus reproductive and transmission potential (Durand-Billaud et al., 2024).

For *Aedes albopictus* and *Aedes aegypti*, rising temperatures in temperate regions facilitate establishment and enable local transmission of arboviruses such as dengue and chikungunya, as documented in several European countries (Semenza & Suk, 2018). Modelling studies suggest that climatic suitability for Aedes aegypti has increased over recent decades and is expected to continue rising, potentially accelerating its geographical expansion (Iwamura et al., 2020). In parallel, global projections indicate that up to one billion additional people could be exposed to Aedes-borne viruses under high-emission scenarios (Ryan et al., 2019; Thomson & Stanberry, 2022). A temperature-dependent transmission model estimates optimal transmission ranges of approximately 21.3–34.0 °C for *Aedes aegypti* and 19.9–29.4 °C for *Aedes albopictus* (Ryan et al., 2019). In Canada, for example, an approximately 10% increase in mosquito-borne diseases has been recorded since the early 2000s (Ludwig et al., 2019; Rakotoarinia, 2025). These changes increase the likelihood that regions previously unsuitable for these vectors may become receptive, leading to the emergence of local transmission in new areas.

2.2.2. Ticks

Tick dynamics are strongly influenced by temperature and humidity. In Europe, *I. ricinus* has shown clear evidence of expansion towards higher altitudes and latitudes, accompanied by changes in seasonal activity patterns (Vonesch et al., 2016).

Warmer conditions have extended the period of tick activity, particularly in spring and autumn, thereby increasing the duration of exposure risk for outdoor workers (Semenza & Suk, 2018).

These changes contribute to the emergence of new risk areas for diseases such as Lyme disease and tick-borne encephalitis, as well as zoonoses like CCHF in the case of *Hyalomma* ticks (Bañuls et al., 2024; Magnavita et al., 2022). Climate change, combined with bird migration and animal trade **(Figure 4)**, facilitates the northward spread of ticks or pathogens from Africa into Europe, raising new occupational health concerns for workers in agriculture, livestock handling, and outdoor environments (Abdelbaset et al., 2023; de Souza & Weaver, 2024).

Figure 4. Global trade and travel move (air, road or maritime transports and migratory birds) vectors and pathogens between continents ; travellers and animals may carry vectors or became infected during or before transit (Bosch et al., 2013; Estrada-Peña et al., 2021; Lee et al., 2013; Nnko et al., 2021).

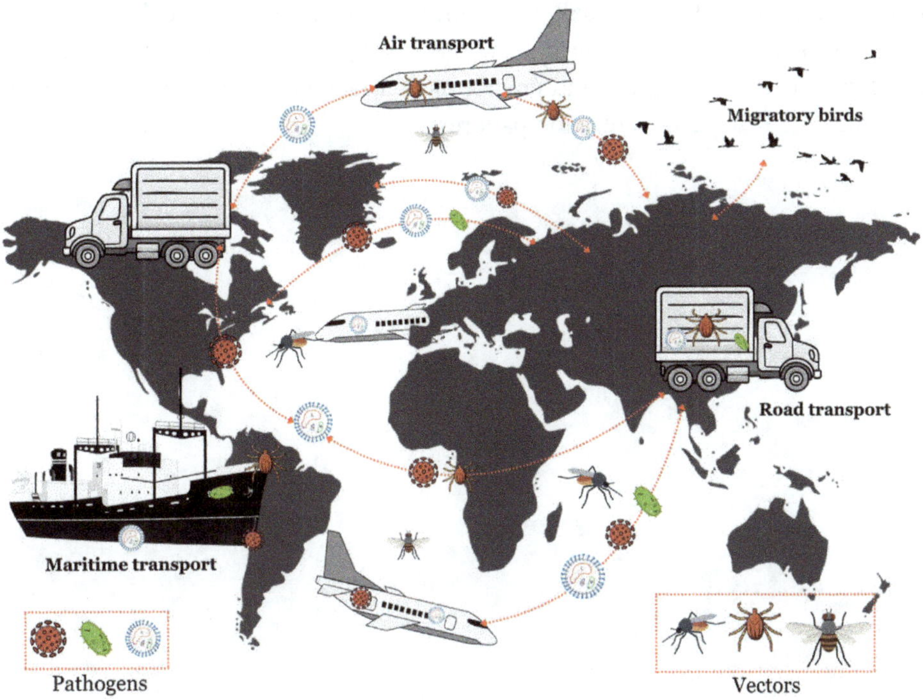

2.2.3. Phlebotomines

Phlebotomine sandflies, vectors of leishmaniasis, are also expanding their geographical range, particularly towards higher altitudes (Carvalho et al., 2015). This expansion is likely driven by milder climatic conditions in previously unsuitable regions, although it remains strongly influenced by habitat characteristics and the availability of animal reservoirs (Carvalho et al., 2015). Increasing temperatures and longer warm seasons may lead to a higher number of vector generations per year and extend the period during which humans are exposed, thereby facilitating colonisation of new areas (Stufano et al., 2025).

2.2.4. The Role of Habitat

Habitat characteristics play a major role in vector survival and development (Lee et al., 2013). Periods of abundant rainfall followed by warm temperatures create favourable conditions for mosquito breeding through the formation of stagnant water bodies (Ryan et al., 2019). However, prolonged droughts may reduce available breeding habitats, illustrating that the effects of climate change are context-dependent and may vary according to local environmental conditions (**Figure 5**) (Lee et al., 2013; Ryan et al., 2019). These interactions highlight that climate change acts in combination with ecological and environmental factors that jointly shape vector dynamics and disease transmission patterns.

Figure 5. Disease transmission dynamics: habitats, climate change and vector ecology — the influence of interactions between habitats and climate on the biological traits of vectors and their effects on disease transmission

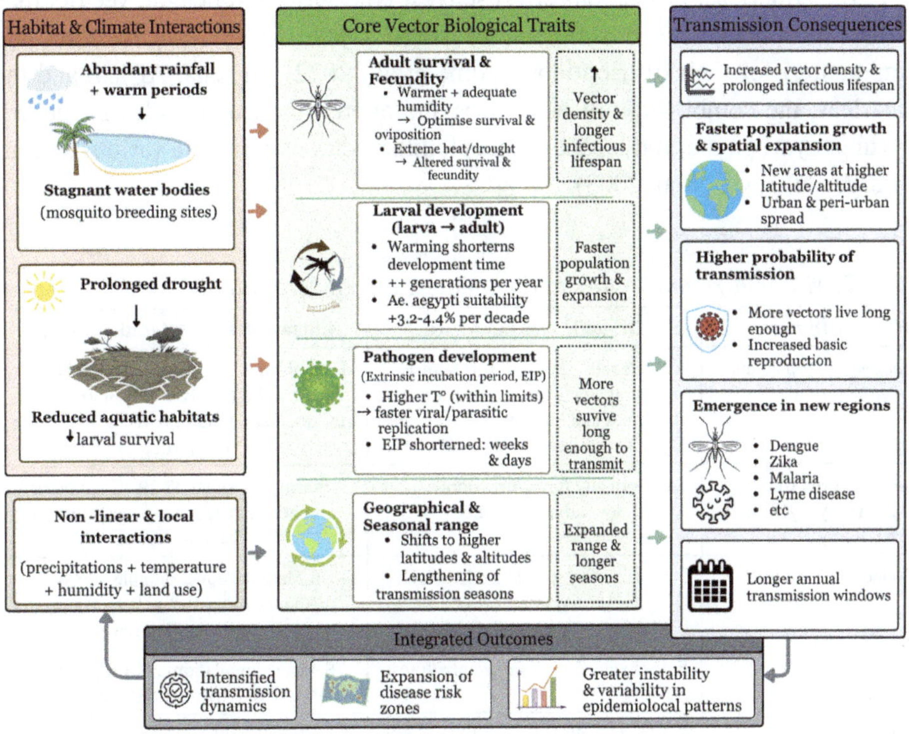

2.3. Regional Focus: Sub-Saharan Africa, Americas, and Asia

Climate change in Africa is projected to raise mean temperatures by 1.5 to 3 °C by 2050, with altered rainfall regimes and more frequent extreme events, affecting multiple vectors (Nnko et al., 2021). Thermal suitability studies show that parts of West Africa and sub-Saharan lowlands may become too hot for optimal malaria transmission but more favourable for Aedes-transmitted arboviruses (dengue, chikungunya, yellow fever, and Zika) (Mordecai et al., 2020; Ryan, 2020). Thus, malaria risk "hotspots" shift to higher altitudes like the Albertine Rift and southern Africa, while *Aedes* virus hotspots expand from West Africa across much of sub-Saharan Africa (Mordecai et al., 2020; Ryan, 2020).

In sub-Saharan Africa, Anopheles spp. malaria vector distribution may shift due to rainfall regime changes and temperature rises, leading to both gains and losses in transmission zones according to regional climate models (Christova et al., 2015;

Egbendewe-Mondzozo et al., 2011). These shifts imply increased arboviral and residual malaria risk for highland farmers, tea and coffee plantation workers, and outdoor urban workers, where risk was previously low (**Table 2**).

In the Americas, projections suggest West Nile virus and dengue vectors may find new habitats, with models emphasising climatic factors (precipitation and temperature) for their distribution (Roome et al., 2022). In Asia, Africa, and the Americas, the combination of heatwaves and altered monsoon cycles influences chikungunya, dengue, and Japanese encephalitis vector dynamics in various regions (Bourgeade & Marchou, 2003).

Table 2. Major African Vectors and Climate-Related Migrations

Vector & Disease	Origin/Ecology in Africa	Climate-Related Migration/Change
Anopheles (human and zoonotic malaria)	Wet zones, irrigated agriculture, rural areas, forests; bites on macaques and humans; foresters and farmers highly exposed.	Low-altitude zones become too hot; risk shifts to highlands. Deforestation and fragmentation increase vector density at forest edges.
Aedes aegypti (dengue, chikungunya, yellow fever, Zika)	Historically forest/peri-domestic; now urban containers.	Warming around 25-29 °C increases climatic suitability; expansion in sub-Saharan Africa.
Tsetse fly (HAT/AAT)	Savanna/forest-wildlife-livestock systems.	In Maasai steppe, suitable area may reduce by 12.9–23.1% by 2050, concentrating risk.
Simuliids (onchocerciasis)	Fast-flowing rivers in savanna/forest.	Savanna populations shifting south to West African forest zones.
Culex spp. (WNV, filariasis)	Wetlands, polluted urban waters, rice fields.	Central belt stable (30°S-30°N) with local suitability variations.
Hard ticks (CCHF, rickettsioses, Lyme disease)	Pastoral landscapes, wildlife-livestock interfaces.	Climate and mobility favour dissemination in Africa and Mediterranean.

Source: (de Souza & Weaver, 2024; Mordecai et al., 2020; Nnko et al., 2021; Randolph & Rogers, 2000; Rocklöv & Dubrow, 2020, 2020; Ryan et al., 2019; Thomson & Stanberry, 2022)

3. OCCUPATIONAL EXPOSURE TO ZOONOTIC AND VECTOR-BORNE DISEASES

OE to zoonotic and vector-borne diseases arises from professional activities that involve direct or indirect contact with animal reservoirs, arthropod vectors, or contaminated environments (Richard & Oppliger, 2015). Such exposure is particularly relevant for workers operating at the human–animal–environment interface, where ecological and climatic changes may intensify transmission risks.

Exposure may occur through multiple pathways, including vector bites, direct contact with infected animals, handling of biological materials, or indirect contact with contaminated soil, water, or surfaces (Tokarska-Rodak et al., 2014). The intensity and frequency of exposure vary according to occupational tasks, environmental conditions, and preventive measures in place.

3.1 High-Risk Occupational Groups

Several occupational groups are consistently identified as being at increased risk of exposure to zoonotic and vector-borne diseases **(Table 3)**. These include agricultural workers, livestock farmers, forestry workers, hunters, veterinarians, slaughterhouse workers, and certain categories of healthcare professionals (Magnavita et al., 2022; Roome et al., 2022; Semenza & Suk, 2018).

These workers frequently operate in environments characterised by high vector density, close contact with animal hosts, and limited protective infrastructure (Msellemu et al., 2024). Outdoor workers are particularly exposed due to prolonged presence in vector habitats and seasonal peaks in vector activity (Roome et al., 2022).

In Europe, studies have shown that forestry and agricultural workers exhibit significantly higher seroprevalence rates for infections such as Lyme disease compared with the general population (Magnavita et al., 2022; Paz, 2020).

Table 3. Summary of vector-borne disease exposure by occupational group, monitoring requirements and preventive measures

Worker group	Main route of exposure	Climate or related exposure factor	Examples of diseases	One Health surveillance needs	Workplace preventive measures
Agricultural workers (arable farming, livestock farming, market gardening, subsistence farming)	Tick and mosquito bites; contact with rodents, contaminated water, aerosols and soil; handling animals	Rising average temperatures and shorter periods of frost	Lyme disease, leptospirosis, tularemia, malaria, Rift Valley Fever, CCHF	Integrated entomological surveillance ; Seroprevalence surveillance among Workers ; veterinary data combined with human health data ; mapping of occupational risks	Appropriate PPE ; body checks following exposure to ticks; training in early detection; management of standing water; animal vaccination; farm biosecurity
Landscape managers and gardeners (urban parks, green spaces on the outskirts of towns, nature conservation technicians)	Tick bites in areas with dense vegetation	Rising temperatures, A longer high-risk season	Lyme disease, tick-borne encephalitis (TBE)	Urban entomological surveillance; symptom-based surveillance in occupational health services; case reporting among park staff	PPE ; post-exposure body check; training in the recognition of erythema migrans; management of vegetation to reduce tick habitats; TBE vaccination (endemic areas)

continued on following page

Table 3. Continued

Worker group	Main route of exposure	Climate or related exposure factor	Examples of diseases	One Health surveillance needs	Workplace preventive measures
Forestry workers and naturalists (foresters, forest managers, ecological researchers, nature guides, outdoor activity leaders)	Tick bites in woodland areas; exposure to wildlife and its reservoirs; remote areas with limited access to healthcare	Latitudinal and altitudinal expansion of *I. ricinus*; prolongation of the nymphal and adult phenology; extended tourist seasons in temperate regions	Lyme disease, TBE, rickettsial diseases, babesiosis	Seroprevalence studies among forestry workers; active vector surveillance; data sharing between human and veterinary epidemiology; early warning systems	PPE (clothing, repellents, boots); training on the correct removal of ticks and early clinical signs; vaccination (TBE, where available); protocols for reporting exposures; access to healthcare in remote areas
Veterinarians, veterinary assistants and professions involving direct contact with animals (clinicians, health inspectors)	Handling of animals with parasites; contact with body fluids, tissues and carcasses; bites and scratches	Milder winters; shift in animal reservoirs linked to climate change	CCHF, brucellosis, leptospirosis, rabies, bartonellosis, rickettsioses, Rift Valley fever, emerging viral zoonoses	Animal sentinel surveillance; coordination between the WOAH, FAO and WHO for the sharing of human and animal epidemiological data; mandatory reporting of occupational infections	High-level PPE when handling animals at risk; animal vaccination; quarantine of imported animals; strict biosecurity protocols; immediate medical consultation following exposure

continued on following page

Table 3. Continued

Worker group	Main route of exposure	Climate or related exposure factor	Examples of diseases	One Health surveillance needs	Workplace preventive measures
Slaughterhouse and meat processing staff	Direct contact with the blood, internal organs and tissues of potentially infected animals; exposure to ticks found on carcasses	Global warming	CCHF, brucellosis, leptospirosis, salmonellosis	Integrated surveillance linking human cases to animal infections; rapid reporting of suspected cases; coordination with veterinary inspection services	Reinforced PPE; strict hygiene standards for work surfaces; pre-slaughter veterinary inspection; specific training on CCHF and zoonoses; isolation protocols in the event of symptoms
Nature guides, ecotourism and seasonal workers (expedition guides, outdoor activity instructors)	Mosquito and tick bites in tropical or temperate natural environments; areas with high biodiversity and limited access to medical care	Extended tourist seasons in temperate regions	Dengue, chikungunya, Lyme disease, rickettsial diseases	Syndromic surveillance in tourist areas; entomological monitoring of *Ae. albopictus*; surveillance of local transmission clusters; feedback to local healthcare workers	Repellents and protective clothing; pre-departure training on local and regional risks; access to preventive treatment and medical care; reporting of exposures; medical repatriation protocols in the event of serious exposure

continued on following page

Table 3. Continued

Worker group	Main route of exposure	Climate or related exposure factor	Examples of diseases	One Health surveillance needs	Workplace preventive measures
Healthcare workers (hospital staff, A&E staff, healthcare workers in endemic areas)	Contact with body fluids from infected patients (nosocomial route); indirect exposure via vectors in healthcare settings in endemic areas	Increase in the number of severe cases (CCHF, Ebola, viral haemorrhagic zoonoses) attributable to the geographical spread of vectors and reservoir hosts	CCHF, Ebola, viral haemorrhagic zoonoses, rickettsiae	Biosafety protocols and active surveillance of nosocomial cases; early warning systems for viral haemorrhagic fevers; data sharing between clinical epidemiology and public health departments	Level 3/4 PPE; ongoing training on standard precautions; protocols for the management of suspected cases; medical monitoring of exposed workers

continued on following page

Table 3. Continued

Worker group	Main route of exposure	Climate or related exposure factor	Examples of diseases	One Health surveillance needs	Workplace preventive measures
Outdoor urban workers (construction workers, municipal workers, informal sector)	Mosquito bites in the vicinity of urban and suburban breeding sites (standing water, containers, construction sites)	Rapid urbanisation leading to an increase in breeding sites; urban heat islands encouraging the proliferation of *Ae. albopictus* and *Ae. aegypti;* heavy rainfall creating new temporary breeding sites	Dengue, chikungunya, West Nile virus, Zika fever	Urban entomological surveillance (*Aedes* density and distribution); symptom-based surveillance ; monitoring of larval breeding sites on construction sites and in public spaces	Management of standing water on construction sites and in workplaces; suitable skin repellents; clothing that covers the skin during peak vector activity times; training for employers and workers; Integrated Vector Management (IVM)
Pastoralists, subsistence farmers (Africa, South-East Asia, the tropical Americas)	Close contact with livestock and wildlife; bites from ticks and mosquitoes in rural/forested areas; consumption of bush meat	Higher altitudes; spread of *Aedes* into urban and peri-urban areas; forest fragmentation increasing contact between humans, wildlife and vectors	Malaria, CCHF, Rift Valley fever, trypanosomiasis (Africa), Japanese encephalitis (Asia), haemorrhagic zoonoses	Integrated One Health surveillance (humans, animals, environment); sentinel surveillance (sentinel herds, environmental indicators); monitoring via wastewater analysis; decentralised diagnostics to address the shortcomings of health systems	PPE suitable for pastoral settings (cost, availability); livestock vaccination; animal quarantine; reducing contact between humans and wildlife; community training and improved access to primary healthcare

Source: (Durand-Billaud et al., 2024; Estrada-Peña et al., 2021; Fanelli & Buonavoglia, 2021; FAO-UNEP-WHO-OIE, 2022; IPCC, 2023; Magnavita et al., 2022; Nnko et al., 2021; Richard & Oppliger, 2015; Semenza & Suk, 2018; Soe et al., 2026; Stufano et al., 2025; Tokarska-Rodak et al., 2014; Vonesch et al., 2016; WHO, 2025a, 2025b)

3.2 Determinants of Occupational Exposure

OE is shaped by a combination of environmental, occupational, and socio-economic factors **(Figure 6).** Environmental determinants include climate conditions, land use, biodiversity, and vector distribution (Gashema et al., 2025). Changes in temperature and precipitation patterns can increase vector abundance, extend activity periods, and facilitate the establishment of new species in previously unaffected areas (Semenza & Suk, 2018).

Occupational determinants relate to the nature of professional activities, including time spent outdoors, type of contact with animals, use of personal protective equipment, and adherence to preventive practices (Roome et al., 2022). Workers in remote or resource-limited settings may face additional challenges in accessing protection and healthcare services. Socio-economic determinants including income level, education, access to information, and working conditions also play a critical role in shaping vulnerability to infection and the ability to implement preventive measures (IPCC, 2023).

These factors often interact, resulting in cumulative and context-specific risks that are difficult to quantify but essential to consider in occupational health assessments.

Figure 6. Occupational Exposures and vulnerabilities to Vector-Borne and Zoonotic diseases in the context of Climate Change: understanding transmission patterns. (±) = positive or negative influence

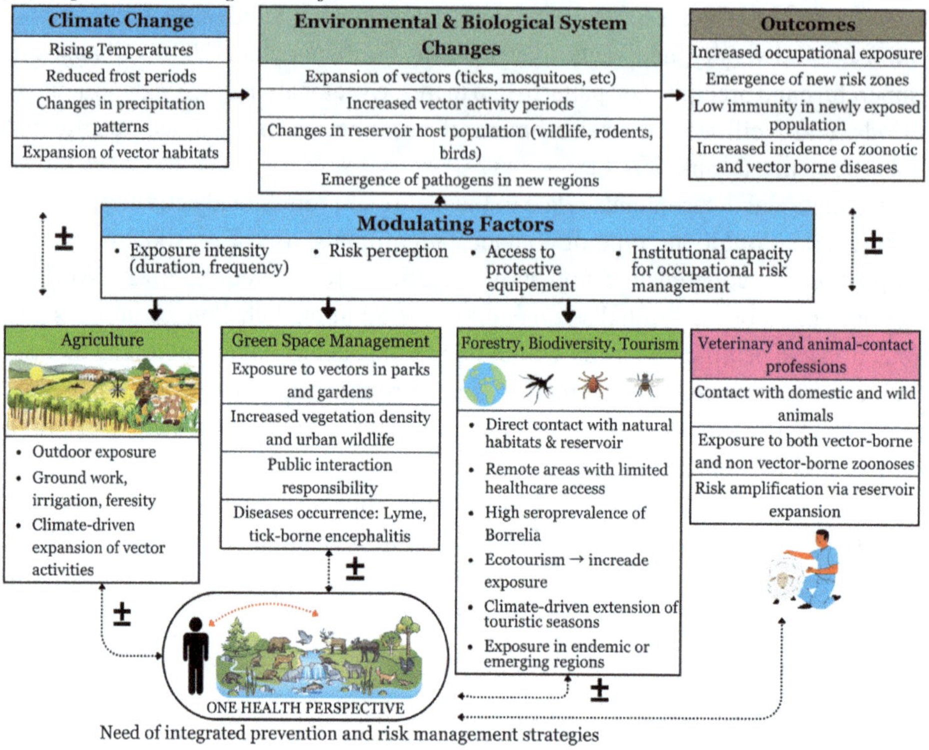

4. IMPACT OF CLIMATE CHANGE ON ZOONOSES AND VECTOR-BORNE DISEASE INCIDENCE AND DISTRIBUTION

VBDs and zoonoses **(Figures 7, 8)** currently represent one of the most dynamic areas of global public health (Lee et al., 2013). The links between climate change and these disease dynamics manifest in incidence, seasonality, and geographical distribution of several key occupational health diseases (Desvars-Larrive et al., 2024). Several representative diseases will be examined, with particular attention to Crimean-Congo haemorrhagic fever, which best illustrates One Health issues.

Figure 7. Climate change is altering the spatial and temporal distribution of vectors, particularly at the edges of their current ranges, leading to an increase in mortality and morbidity associated with vector-borne zoonoses (Damtew et al., 2023; Di Bari et al., 2023; Mora et al., 2022; Scott et al., 2024; WHO, 2017a, 2025b).

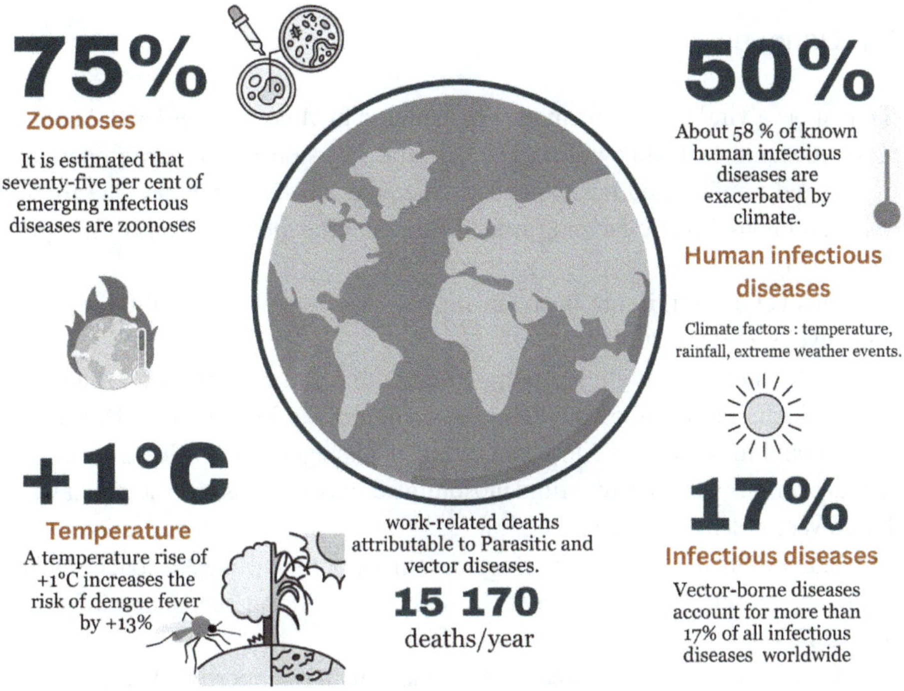

4.1. Mosquito-Related Diseases

The data show with a very high level of confidence that transmission capacity has increased for dengue, malaria and other mosquito-borne diseases and that global warming is expanding the geographical areas conducive to their transmission (IPCC, 2023). Malaria, which is mainly found in Africa, where over 90% of deaths occur, saw its DALYs (Disability-Adjusted Life Years) decrease by 39% between 2007 and 2017 (IPCC, 2023). Its transmission depends heavily on climatic variables such as temperature, but socio-economic factors and health systems could mitigate these effects.

Certain mosquito species can transmit multiple zoonotic viruses: West Nile virus, St. Louis encephalitis virus, Japanese encephalitis virus, and Rift Valley fever virus (Bourgeade & Marchou, 2003; Durand-Billaud et al., 2024; Gahn et al., 2024). Mosquitoes are highly sensitive to weather conditions. A slight temperature rise enables mosquitoes to colonise previously too-cold regions, develop and reproduce faster, leading to more annual generations and thus greater population exposure (Suu-Ire et al., 2025). Wind can aid mosquito dispersal but may also reduce biting capacity (Durand-Billaud et al., 2024).

4.1.1. Dengue

Dengue is a viral disease primarily transmitted by *Aedes aegypti* and *Aedes albopictus* mosquitoes (Damtew et al., 2023). Its transmission depends on temperature, humidity and rainfall, and has expanded since the mid-20th century due to both climatic and non-climatic factors (urbanisation and the globalisation of transport) (IPCC, 2023).

Humans act as amplifying hosts during the viraemic phase, facilitating transmission cycles (WHO, 2025c).

Occupational risk is highest among workers exposed to mosquito-prone environments, including healthcare workers, agricultural workers, field staff, military personnel, and those involved in water or waste management (WHO, 2025c). Exposure is primarily linked to daytime mosquito activity in areas with stagnant water and high vector density (WHO, 2025c).

Clinical presentations range from asymptomatic or mild (fever, headache, aches, rash) to severe dengue, which carries a higher risk during secondary infections (WHO, 2025c).

There is no specific antiviral treatment; management is supportive (WHO, 2025c).

Prevention relies mainly on vector control and personal protection against mosquitoes (protective clothing, repellents, mosquito nets, elimination of breeding sites). A vaccine exists but its use remains restricted to specific contexts (WHO, 2025c).

Climate phenomena such as *El Niño-Southern Oscillation* (ENSO) influence its incidence, with delays of 1-2 months between climatic conditions and case occurrence (IPCC, 2023). Initially confined to tropical and subtropical regions, dengue is currently the most widespread vector-borne viral disease worldwide (Vonesch et al., 2016). *Ae. aegypti* and *Ae. albopictus* evolved as ecological generalists in African forests; *Ae. aegypti* became human-specialised around 5,000 years ago during Sahara desiccation, then spread globally with trade and the Atlantic slave trade (Islam et al., 2025). In Africa, temperature rises around current 25-29 °C levels are projected to increase *Ae. aegypti* transmission suitability, particularly in densely populated low-altitude and peri-urban zones (Ryan et al., 2017; Mordecai et al., 2020).

The incidence of dengue fever has risen sharply over the past few decades. Reported cases increased from 505,430 in 2000 to 14.6 million in 2024 although actual numbers are underestimated due to asymptomatic infections. The disease is now endemic in more than 100 countries (WHO, 2025b).

Estimates suggest approximately 390 million annual infections, including 96 million symptomatic cases, with around 5.6 billion people at risk (WHO, 2025b). From January to July 2025, over 4 million cases and more than 3,000 deaths were reported across 97 countries (WHO, 2025b).

In 2024, more than 12,000 deaths were reported globally, with the Americas accounting for the majority, with over 13 million cases (WHO, 2025b).

In Europe, the establishment *Ae albopictus* has led to local transmission cases, notably in France, Italy, and Spain. In 2024, 308 cases were reported in Europe, in addition to 1,291 cases and 4 deaths in Mayotte and Réunion (WHO, 2025b). Consequently, certain professional groups including tourism and landscaping workers, previously largely spared have been exposed (Semenza & Suk, 2018). From a One Health perspective, dengue is distinctive in that it lacks a non-human reservoir, yet remains strongly driven by environmental conditions and vector ecology (Desvars-Larrive et al., 2024).

4.1.2. West Nile Virus

West Nile virus (WNV) is a *Flavivirus* transmitted by mosquitoes of the *Culex* genus. Its zoonotic cycle primarily involves birds as reservoirs while humans and horses are incidental hosts. The virus has spread widely in North America, Europe, Africa, and the Middle East since the 1990s (ECDC, 2024; Roome et al., 2022).

Humans are dead-end hosts and do not contribute to further transmission (ECDC, 2024).

OE mainly affects workers exposed to mosquito-prone environments (farmers, military personnel, veterinarians and laboratory staff) particularly those in contact with animals, biological fluids or outdoor environments (Bourgeade & Marchou, 2003).

Most infections are asymptomatic. Symptomatic cases may progress to West Nile fever, while severe neuroinvasive forms (encephalitis, meningitis, flaccid paralysis), are rare but potentially disabling (Bourgeade & Marchou, 2003; WHO, 2017b).

There is no specific antiviral treatment; management is supportive (WHO, 2017b). Prevention strategies focus on personal protection against mosquitoes (covering clothing, repellents, mosquito nets), vector control, environmental surveillance and biosafety measures including blood screening (WHO, 2017b).

Specific climatic conditions, such as hot summers and drought periods followed by rains significantly influence *Culex* population and transmission dynamics (Semenza & Suk, 2018). Climatic suitability modelling for *Culex quinquefasciatus* (WNV, filariasis) shows great stability between 30°N and 30°S, covering much of Africa, with some modelled changes in North and Central Africa under future scenarios (Roome et al., 2022).

Warmer temperatures and land-use changes (wetlands, irrigation) increase transmission risk in agricultural and peri-urban ecosystems (Rocklöv & Dubrow, 2020).

Farmers, veterinarians, wildlife managers, and outdoor workers rank among the most at-risk professions. Horses and birds serve as sentinel species in WNV surveillance, illustrating a concrete One Health application (ECDC, 2024).

4.2. Tick-Related Diseases

4.2.1. Lyme Disease

Lyme disease (caused by *Borrelia burgdorferi* sensu lato bacteria) is transmitted by *Ixodes* ticks and is sensitive to temperature and humidity (Randolph et al., 1996; Randolph & Rogers, 2000). Numerous studies report expansion of tick ranges, prolonged active seasons, and increased human exposure (Paz, 2020; Roome et al., 2022). Occupational risk is significantly higher among exposed workers, with a seroprevalence reaching approximately 20% in some European groups (Magnavita et al., 2022).

High-risk professions include foresters, farmers, natural space managers, hunters, veterinarians, military, hunters and nature tourism workers (Roome et al., 2022; Tokarska-Rodak et al., 2014).

The risk is linked to the nature of the work carried out in natural environments, the duration of exposure and the season, with a peak during the period of tick nymph activity between May and September (Magnavita et al., 2022). Transmission requires prolonged tick attachment (>24-36 hours) (Magnavita et al., 2022).

The disease can progress from early forms with erythema migrans and flu-like symptoms to disseminated and late forms, including neurological, joint and chronic complications. Early antibiotic treatment is effective (Magnavita et al., 2022).

Prevention relies on personal protection, tick checks, and early removal (Magnavita et al., 2022).

In a One Health context, wild animals (rodents, deer) serving as reservoirs, along with landscape modifications (forest fragmentation, agricultural practices), influence transmission dynamics in interaction with climate (Bidaisee & Macpherson, 2014).

4.2.2. Crimean-Congo Haemorrhagic Fever (CCHF)

CCHF is a severe viral zoonosis caused by a *Nairovirus* transmitted primarily by *Hyalomma* ticks, with a case fatality rate of 10–40% (WHO, 2025a). The disease is endemic in Africa, the Balkans, the Middle East and certain regions of Asia

(WHO, 2025a). Ticks act both as vectors and reservoirs, while domestic animals serve as amplifying hosts (Boushab et al., 2020). Domestic animals (cattle, sheep, and goats) and certain wildlife can serve as asymptomatic carriers and amplifying hosts, facilitating tick transmission (Belobo et al., 2021). Thus, CCHF constitutes one of the most striking examples of an inherently One Health disease, involving complex interactions between wildlife, domestic animals, vectors, and humans. High-risk groups include livestock farmers, veterinarians, slaughterhouse workers, hunters, and healthcare workers in nosocomial transmission (Belobo et al., 2021; WHO, 2025f).

Transmission to humans occurs through tick bites or contact with the blood or tissues of infected animals, particularly during livestock farming or slaughtering activities (WHO, 2025a). Similarly, human-to-human transmission is possible through contact with biological fluids, including in hospital settings where preventive measures are lacking (WHO, 2025a).

Turkey, Iran, and certain Balkan regions regularly report human cases, with variable incidence rates depending on years and environmental conditions (Bosch et al., 2013; Estrada-Peña et al., 2021; Semenza & Suk, 2018). In France, infected *H. marginatum* ticks were detected in 2023, although no human cases have been reported (Pyrénées-Orientales and Corsica) (Bernard et al., 2024). No validated specific treatment exists; management is supportive. Ribavirin has been used, but evidence remains inconclusive (WHO, 2025a).

Climate change may favour the expansion of *Hyalomma* ticks through milder winters. However, the relative roles of climate, animal movement, and trade remain uncertain (Desvars-Larrive et al., 2024; Estrada-Peña et al., 2021).

4.2.3. Migratory Birds as Long-Distance Vectors of *Hyalomma* Ticks

Migratory birds play a major role in long-distance dissemination of *Hyalomma* ticks. Studies show that up to 2.4% of migratory passerines carry ticks, predominantly *Hyalomma rufipes* (Hoffman et al., 2018). Ticks may remain attached for up to 26 days, allowing intercontinental transport (Estrada-Peña et al., 2021).

Large-scale ringing studies in Spain, Italy, Greece, and Israel showed ~2.4% of over 10,000 northward migratory passerines carried ticks, with *H. rufipes* representing ~78% of collected specimens (Hoffman et al., 2021). Long-distance ground-wintering migrants in open habitats and African wetlands exhibit disproportionately high *H. rufipes* infestation compared to forest/shrub birds (Hoffman et al., 2021). With over one billion birds crossing the Mediterranean annually, extrapolations suggest millions of *Hyalomma* ticks migrate from Africa to Europe each spring (Hoffman et al., 2021).

The CCHF virus genome was detected in *H. rufipes* nymphs on Italian migratory wheatears, with phylogenetic evidence confirming African origin (Mancuso et al., 2019). Complementary studies identified CCHF virus RNA and Alkhurma haemorrhagic fever virus in *Hyalomma* complex ticks on northward migrants (Battisti et al., 2020; Hoffman et al., 2018).

Pathogens detected in transported ticks also include *Rickettsia spp.*, and *Borrelia spp.* (Battisti et al., 2020), as well as *Borrelia burgdorferi s.l.,* and *Anaplasma phagocytophilum* (Palomar et al., 2013; Pedersen et al., 2020).

Climate warming increases the probability that imported ticks complete their life cycle in Europe, enabling local establishment (Battisti et al., 2020; Hoffman et al., 2021; Zhou et al., 2025). Risk assessment requires surveillance of long-distance migrants from African habitats, targeted surveys in "at-risk" Mediterranean and northern regions, and integrated climate-habitat models to predict future establishment patterns (Battisti et al., 2020; Zhou et al., 2025).

4.3. Fly- and Other Vector-Related Diseases

4.3.1. Phlebotomines and Leishmaniasis

Phlebotomines are small blood-sucking insects widely distributed in Mediterranean regions and other tropical and subtropical areas. Their optimal development occurs around 25 °C and 60% relative humidity, conditions that are increasingly met in expanding geographical areas due to climate change (Chalghaf et al., 2018; Durand-Billaud et al., 2024).

They are vectors of several pathogens, including Leishmania spp., responsible for leishmaniasis, as well as *Bartonella, Phlebovirus, and Vesiculovirus* (Chalghaf et al., 2018)

The most frequently transmitted disease is leishmaniasis, which can be cutaneous or visceral. Primary reservoirs are humans and warm-blooded mammals (dogs and foxes) (Chalghaf et al., 2018; WHO, 2023a). An increase in both human and canine cases has been reported in Mediterranean regions, likely associated with rising temperatures facilitating vector expansion (Chalghaf et al., 2018; Gangneux et al., 2015). Leishmaniasis manifests in three main forms: visceral (the most severe and often fatal without treatment), cutaneous (the most common, with ulcerative lesions) and mucocutaneous (affecting the mucous membranes) (WHO, 2023a). Transmission involves multiple animal and human reservoirs (WHO, 2023a). The disease mainly affects poor populations and is associated with socio-economic and environmental factors such as malnutrition, poor housing, migration and climate change (WHO, 2023a).

Approximately 700,000 to 1 million new cases occur each year, although only a fraction of those infected develop the disease (WHO, 2023a).[REMOVED HYPERLINK FIELD]

OE primarily affects workers operating in endemic areas, including farmers, foresters, construction workers, military personnel deployed in rural areas, field researchers, humanitarian workers (Burza et al., 2018).

Exposure occurs mainly during outdoor activities in rural, forest or peri-urban settings, particularly at dusk and at night, when the vectors are most active (Burza et al., 2018; WHO, 2023a). Risk factors include substandard housing, hot and humid environments, proximity to animal reservoirs, and situations involving mobility or temporary accommodation. Prevention relies on personal protection against bites, improving living conditions, reducing night-time exposure and vector control (Burza et al., 2018; WHO, 2023a).

4.3.2. Simuliids and Onchocerciasis

Onchocerciasis (river blindness) is a parasitic disease caused by *Onchocerca volvulus* (*O. bovis, O. gutturosa,* etc in animals) and transmitted by blackflies of the genus *Simulium,* which breed in fast-flowing rivers (WHO, 2025e). The disease is characterised by chronic dermatological manifestations and progressive ocular involvement that may lead to irreversible blindness. It predominantly affects rural populations in sub-Saharan Africa and Yemen, with limited residual foci in Latin America (WHO, 2025e).

In 2024, about 252.3 million people required preventive treatment, and 171.6 million were actually treated across 26 countries and approximately 99% of cases remain concentrated in Africa and Yemen (WHO, 2025e).

OE is particularly significant among individuals working in proximity to vector breeding sites, including farmers, fishermen, forestry workers, and personnel involved in hydraulic infrastructure. Exposure is primarily associated with repeated daytime activities near rivers where vector density is high (Njeshi et al., 2026; WHO, 2025e).

Beyond its health impact, onchocerciasis imposes a substantial socio-economic burden, notably through reduced productivity and loss of income in affected communities (Njeshi et al., 2026; WHO, 2025e).

The main control strategy relies on mass drug administration of ivermectin, which is effective against microfilariae but does not eliminate adult worms, requiring repeated treatment over extended periods (WHO, 2025e).

Environmental and climatic factors influence vector distribution and transmission dynamics. Projections suggest potential shifts in simuliid populations, including a redistribution from savanna to more humid zones in West Africa, which may modify exposure patterns among occupational groups (Thomson & Stanberry, 2022).

However, uncertainties remain regarding the magnitude and spatial consistency of these changes.

4.3.3. Tsetse Flies and Trypanosomiasis

African trypanosomiasis is a vector-borne parasitic disease transmitted by tsetse flies (*Glossina spp.*) and caused by protozoa of the genus *Trypanosoma* (WHO, 2023b). *Trypanosoma brucei gambiense*, responsible for around 92% of reported cases, causes a chronic form of the disease that is widespread in West and Central Africa, whilst *T. brucei rhodesiense*, which accounts for around 8% of cases, causes an acute form found in East and Southern Africa (WHO, 2023b).

This disease is endemic in sub-Saharan Africa and mainly affects rural populations engaged in agriculture, fishing, livestock farming or hunting (WHO, 2023b). Transmission occurs during activities in areas where tsetse flies are present, particularly in humid and vegetated ecosystems (WHO, 2023b).

In addition, animal African trypanosomiasis (AAT) has a major impact on livestock health and agricultural productivity. It affects a wide range of domestic and wild animals (WHO, 2023b). In cattle, the disease is known as *nagana* and constitutes a major obstacle to livestock productivity and rural economic development (WHO, 2023b). The main pathogenic species include *T. congolense, T. vivax, T. brucei brucei* and *Trypanosoma simiae* (WHO, 2023b).

The epidemiology of HAT involves significant animal reservoirs. *T. b. rhodesiense* is a well-established zoonosis, with domestic and wild animals serving as major reservoirs of human infection (WHO, 2023b).

Vectors of human and animal African trypanosomiasis, tsetse flies (*Glossina spp.*), are maintained in wildlife-livestock cycles in savanna and forest mosaics; herders and wildlife staff face particular risk (de Souza & Weaver, 2024; Nnko et al., 2021). In Tanzania's Maasai steppe, current suitable areas for three major species are estimated at 19,225 km^2 (G. m. morsitans), 7,113 km^2 (*G. pallidipes*), and 32,335 km^2 (*G. swynnertoni*) (Nnko et al., 2021). Under mid-century RCP 4.5, habitable areas may reduce by 12.9 - 23.1% across species due to warming and habitat changes (Nnko et al., 2021).

Changes in environmental conditions, including those associated with climate variability, are expected to alter the distribution of suitable habitats for Glossina species (**Figure 8**). Some projections suggest a reduction in overall habitat suitability in certain regions, accompanied by increased concentration of vectors in ecologically favourable areas (Nnko et al., 2021; WHO, 2023b). These shifts may lead to spatial redistribution of risk rather than uniform increases or decreases in transmission.

Figure 8. Impact of climate change on the incidence and distribution of Vector-Borne zoonoses

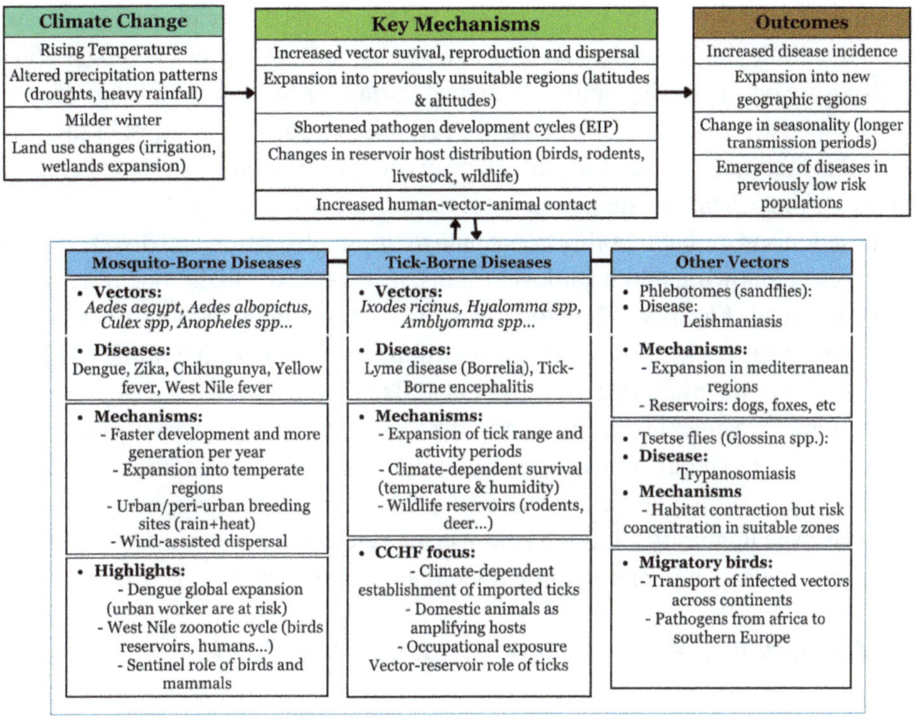

4.4. Deforestation, Urbanisation, and Emerging Zoonoses

Global analyses indicate that the risk of zoonotic disease emergence is highest in tropical regions undergoing rapid land-use change (Wegner et al., 2022). Deforestation, agricultural expansion, and habitat fragmentation increase interactions between humans, wildlife, and domestic animals, thereby facilitating pathogen spillover. Transitional zones between forest and agricultural land represent particularly high-risk interfaces (Kamguem et al., 2025; Wegner et al., 2022).

In Central Africa, deforestation combined with expansion of human settlements has been associated with increased hunting activities and bushmeat consumption, factors linked to the emergence of zoonotic diseases such as Ebola virus disease (Wegner et al., 2022; Wolfe et al., 2005).

OE is especially relevant for individuals in direct contact with animals or biological materials, including healthcare workers, hunters, farmers, forestry workers, and personnel involved in burial practices (WHO, 2025d).

Case fatality rates for emerging zoonoses vary widely depending on the pathogen and outbreak context, with reported ranges between 25% and 90% for certain viral haemorrhagic fevers (WHO, 2025d). The incubation period is 2 to 21 days, and symptoms begin with fever, severe fatigue and pain, which may progress to digestive and neurological disorders and, in some cases, haemorrhages (WHO, 2025d).

Exposure occurs during activities such as caring for the sick, handling infected animals or bush meat, managing bodies, or certain agricultural activities, in environments that facilitate contact with natural reservoirs (fruit-eating bats) or are characterised by fragile health systems and inadequate hygiene conditions (WHO, 2025d). Urbanisation and global mobility contribute to the amplification and dissemination of outbreaks following initial spillover events, increasing the likelihood of regional or global spread **(Figures 1, 2)**. More than 70% of emerging infectious diseases are of zoonotic origin, with a substantial proportion originating from wildlife (Allen et al., 2017; Lawler et al., 2021; Wegner et al., 2022).

Preventive strategies require upstream interventions targeting drivers of emergence, including land-use regulation, control of wildlife trade, and ecosystem conservation (Wegner et al., 2022). Landscape-based One Health approaches emphasise protecting "landscape immunity" through land-use management, ecosystem conservation, indigenous stewardship, and wildlife trade regulation (Lawler et al., 2021).

4.5. Africa's Vulnerability to Zoonoses and VBDs

Africa is disproportionately affected by zoonotic and vector-borne diseases due to a combination of ecological, socio-economic, and health system factors, including high biodiversity, dependence on natural resources, rapid environmental change, and limited healthcare infrastructure (Allen et al., 2017; Baker et al., 2022).

Central Africa's humid tropical forests, with exceptional mammal diversity, are systematically identified as hotspots for emerging zoonoses and viral haemorrhagic fevers (VHFs), where human ecosystem invasion promotes spillover (Baker et al., 2022; Kamguem et al., 2025).

Inadequate diagnostic capacity, underfunded health systems, and limited real-time data sharing hinder rapid epidemic response (Gashema et al., 2025; Sekamatte et al., 2018). In Africa, multiple pathogens, including Rift Valley fever virus, Crimean-Congo haemorrhagic fever virus, and West Nile virus, are known or suspected to circulate in several regions with incomplete epidemiological characterisation (Mubemba et al., 2022). West Nile virus circulates in 28 African countries, but 19 lack basic epidemiological data (Mencattelli et al., 2022).

Forest rural communities, herders, smallholder farmers, livestock keepers, and abattoir workers face high VHF and arboviral risk (Baker et al., 2022; Sekamatte et al., 2018). African One Health approaches emphasise integrated human, animal, and environmental surveillance, using environmental sampling, wastewater monitoring, and decentralised diagnostics. National programmes employ risk mapping, sentinel herds, and environmental indicators for targeted surveillance and early warning systems in a One Health approach (Gashema et al., 2025; Sekamatte et al., 2018).

5. SURVEILLANCE, PREVENTION, AND RISK MANAGEMENT IN OCCUPATIONAL SETTINGS

Climate change challenges traditional sector-based approaches to infectious disease control and highlights the need for integrated strategies that address the interconnected dynamics of human, animal, and environmental health (Roome et al., 2022). The One Health framework provides a conceptual basis for such integration by promoting coordination across sectors and disciplines involved in disease surveillance and prevention (FAO-UNEP-WHO-WOAH, 2022; Rupasinghe et al., 2022).

5.1. Integrated Surveillance in the One Health Framework

Surveillance effectiveness is measured by the ability to collect, analyse, and share data on vectors, animal hosts, and human cases, while identifying links between these components (Fasina et al., 2021). To this end, harmonising data collection systems, using common indicators, and establishing multisectoral governance are essential (Lefrançois et al., 2023).

Human disease epidemiological surveillance must be coordinated with surveillance of infections in domestic and wild animals, as well as entomological vector monitoring (density, seasonal activity, and infection rates) (Cabi, 2025; Desvars-Larrive et al., 2024). Operational leaders recommend combining active and passive surveillance systems—notably syndromic and indicator-based approaches with event-driven surveillance strategies to enable early detection of emerging signals (Kodama et al., 2023). According to the Joint One Health Action Plan (Quadripartite plan), integrating epidemiological, entomological (vector density, present species), veterinary, human (case occurrence), and environmental (climatic data, land use) surveillance is crucial for better targeting interventions, reducing information gaps, and anticipating emergence or re-emergence signals (FAO-UNEP-WHO-WOAH, 2022).

National or regional committees, such as those in France recommending strengthened coordinated vector disease surveillance from a One Health perspective, can play a central coordination role (Lefrançois et al., 2023).

5.2. Primary Prevention Measures in Professional Settings

5.2.1. Personal Protective Equipment and Training

Primary prevention aims to reduce vector and pathogen exposure before disease onset. In a One Health perspective, it must integrate into occupational health, public health, and animal health frameworks. For workers in agriculture, forestry, and natural space management, key measures include:

Personal protective equipment (PPE), which provides a major barrier against vector bites and contact with reservoirs or biological fluids: adapted protective clothing, appropriate footwear, gloves, cutaneous repellents, acaricide or insecticide treatments for textiles, and ocular and respiratory protection devices as needed. Their effectiveness depends on worker compliance (Rupasinghe et al., 2022). Body checks must also be performed after tick exposure (Durand-Billaud et al., 2024).

Training and awareness for workers and supervisors on vector identification, high-risk habitats, early clinical signs (e.g., erythema migrans for Lyme disease or flu-like symptoms after tick or mosquito bite), and appropriate actions (correct tick removal, prompt medical consultation, exposure reporting) are key strategies (Magnavita et al., 2022). Intersectoral training programmes bringing together occupational health professionals, veterinarians, entomologists, and territory managers are encouraged (Lefrançois et al., 2023).

5.2.2. Structural Interventions and Environmental Management

Complementing individual protection, workplace structural interventions can durably reduce vector densities and human-vector-reservoir contact: stagnant water management, targeted vegetation maintenance, and redesign of rest and passage areas to avoid high tick or mosquito density habitats (de la Fuente et al., 2024). Integrated Vector Management (IVM) strategies, inspired by integrated pest management, combine environmental, biological, and chemical measures based on action thresholds and ongoing entomological monitoring (Tiffin et al., 2025). Other measures include animal health inspections, animal vaccination, imported animal quarantine, anti-vector control, farm biosecurity, and strict hygiene practices post-animal or contaminated environment contact **(Figure 9)** (Durand-Billaud et al., 2024).

5.2.3. Legislation

In its 2024 global report on safety and health at work in a changing climate, the International Labour Organization (ILO) stated several legislations applicable in some countries to prevent VBDs at work (Scott et al., 2024). Thus, in Thailand, a medical examination is mandatory for workers exposed to specific hazards (toxic microorganisms such as viruses, bacteria or fungi) associated with their work. In Mexico, an official standard (NOM-032-SSA2-2010) governs the surveillance, prevention and control of VBDs, some of which are recognised as occupational diseases (such as leishmaniasis, etc.). In Barbados, the implementation of vector control programmes in the workplace is mandatory (Scott et al., 2024). Similarly, in Costa Rica, establishments such as beauty salons and gyms must be maintained under strict hygiene conditions to prevent any source of infection or proliferation of vectors. In Finland, regulations require effective control of vectors such as rodents and insects. As regards Africa, in Egypt, Article 210 of the Labour Code requires employers to implement all necessary measures to protect workers from infections caused by biological agents where their work exposes them to such risks(Scott et al., 2024). In Mozambique, companies with more than 30 employees must fit dining areas with mosquito nets where necessary. In Togo, the law prohibits any waste management practices that encourage the proliferation of pests or disease vectors (Scott et al., 2024). Similar legislation exists in Fiji, where land near workplaces must be cleared of vegetation that could harbour mosquitoes (Scott et al., 2024).

5.3. Secondary and Tertiary Prevention

Secondary prevention targets early detection to limit disease progression in workers and reduce secondary transmission (notably nosocomial). Integrating occupational criteria into clinical surveillance enables rapid identification of cases among at-risk groups (farmers, veterinarians, and foresters) (Roome et al., 2022). Using sensitive and specific diagnostic tests adapted to relevant pathogens and vectors (ELISA, PCR, etc.), with attention to situations where clinical symptoms may mimic other infectious diseases (differential diagnosis), is essential (Belij-Rammerstorfer et al., 2022; Fakoorziba et al., 2012).

Tertiary prevention focuses on managing confirmed cases to reduce long-term complications and mortality, and limit secondary transmission risk (e.g., in health-care settings) (Muzammil et al., 2024). Clear treatment and clinical care protocols, along with strict biosecurity practices for healthcare professionals, must be firmly established.

Figure 9. Surveillance, prevention and risk management of Vector-Borne zoonoses at occupational setting in a One Health perpective (±) = positive or negative influence

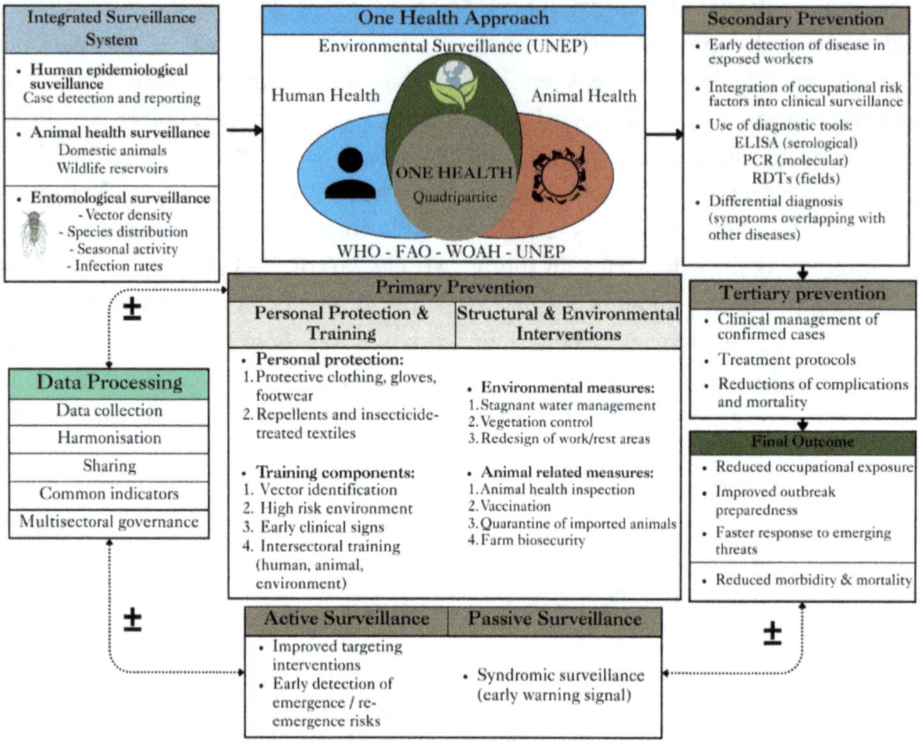

6. METHODOLOGICAL CHALLENGES AND RESEARCH PERSPECTIVES

There is a paucity of profession-stratified quantitative data on these climate-related evolutions in Africa; most work focuses on climatic suitability and general populations rather than workers (de Souza & Weaver, 2024; Thomson & Stanberry, 2022). To implement truly effective surveillance and prevention systems, several methodological, organisational, and scientific challenges must be addressed, and priority research areas identified (Suu-Ire et al., 2025). Indeed, interactions between climate, ecology, production systems, professional practices, and healthcare organisation are complex (Suu-Ire et al., 2025).

6.1. Integrated Approaches and Longitudinal Data

There is a real need for longitudinal studies explicitly linking climatic data (temperatures, precipitation, drought indices), entomological data (densities, infection indices, seasonality), veterinary data (animal infections), and occupational health data (exposures, seroprevalence, incident cases) to better quantify climate effects on professional risk. Many reviews highlight the lack of harmonised occupational zoonosis data and difficulty comparing results between countries or sectors due to methodological heterogeneity and under-reporting (Richard & Oppliger, 2015).

National surveillance systems are generally sectoral, with structural barriers between human and animal data (Gashema et al., 2025). Consequently, the capacity to detect early warning signals before they become epidemics remains limited. Developing algorithms and analytical frameworks inspired by integrated One Health approaches, mapping relationships between zoonotic agents, hosts (biological or reservoirs), vectors, and environmental contexts, would enable cross-referencing entomological, clinical, veterinary, and professional data to identify the most relevant intervention points (Stufano et al., 2025).

6.2. Predictive Models and Risk Assessment

Predictive risk models integrating climatic, ecological, and professional variables remain rare but represent a promising research area for anticipating new risk zones and refining workplace risk assessments (Roome et al., 2022). For CCHF, for example, developed models demonstrate the value of combining climate projections, host data (birds and livestock), and vector presence data to generate spatiotemporal risk maps (Desvars-Larrive et al., 2024).

From an occupational health perspective, these models should incorporate data on professional activity distribution ; working hours ; exposure patterns (day or night shifts, vegetation or water contact intensity, PPE use) ; and inequalities in prevention measure access. Integrating these models into risk assessment procedures and prevention plans would enable anticipating training, PPE, and workstation adaptation needs (Desvars-Larrive et al., 2024). It is also necessary to evaluate prevention measure effectiveness under real working conditions, particularly integrated control strategies and One Health training programmes, as well as document benefits and potential adverse effects (on biodiversity or chemical use) (Bidaisee & Macpherson, 2014; Tiffin et al., 2025).

SOLUTIONS AND RECOMMENDATIONS

In light of the controversies and shortcomings identified, several potential solutions are worth considering at different levels.

To disentangle the respective contributions of climate and non-climatic factors, the development of multivariate predictive models that simultaneously incorporate climatic, ecological, socio-economic and occupational data is a research priority. Causal modelling approaches, combining remote sensing data, entomological monitoring and epidemiological records, would enable a more rigorous quantification of the relative weight of each determinant and guide preventive policies on an evidence-based foundation (Desvars-Larrive et al., 2024; Rupasinghe et al., 2022).

To address geographical and methodological gaps in occupational health, investment is needed in seroprevalence studies stratified by occupation in under-documented regions, particularly in sub-Saharan Africa and South-East Asia. Harmonising surveillance protocols and establishing common indicators across countries would facilitate international comparisons and reveal the true burden of disease among exposed workers (Msellemu et al., 2024; Richard & Oppliger, 2015). The establishment of occupational exposure registries, linked to occupational disease reporting systems, represents a concrete and achievable step in the short term.

Finally, for the One Health approach to move beyond mere rhetoric, dedicated funding mechanisms, shared data platforms across the human, veterinary and environmental sectors, and cross-sectoral training for practitioners in the field are essential (FAO-UNEP-WHO-OIE, 2022; Gashema et al., 2025). The experience of countries with national One Health coordination committees, such as France with its enhanced vector surveillance system, offers a model that can be replicated and adapted to other national contexts (Lefrançois et al., 2023).

FUTURE RESEARCH DIRECTIONS

Recent scientific advances open up several promising avenues of research for better characterising and anticipating occupational risks associated with vector-borne and zoonotic diseases in the context of climate change.

Firstly, the development of real-time surveillance systems, combining artificial intelligence, satellite data and connected environmental sensors, represents a major opportunity for the early detection of signs of vector emergence in areas of high occupational exposure (Stufano et al., 2025). These tools could feed into early warning platforms tailored to the specific needs of occupational health managers.

Furthermore, approaches involving environmental metagenomics and wastewater surveillance offer unprecedented opportunities to detect the circulation of zoonotic pathogens before human cases emerge, including in settings with limited diagnostic resources (Gashema et al., 2025).

Research into the real-world effectiveness of integrated prevention strategies particularly Integrated Vector Management programmes and One Health training in real-world working conditions remains insufficient and is a priority area (Tiffin et al., 2025). Finally, longitudinal studies explicitly linking climate change, vector dynamics and workers' health data over several decades remain essential to consolidate the epidemiological foundations of this still-emerging field.

CONCLUSION

Climate change reshapes the ecological conditions governing zoonotic and VBDs, with measurable effects on vector distribution, transmission dynamics and exposure patterns. These changes are not isolated, but interact with changes in land use, socio-economic factors and the capacity of health systems, resulting in heterogeneous and context-dependent outcomes.

Occupational groups working at the human-animals-environment interface remain disproportionately exposed. However, current knowledge is limited by fragmented surveillance systems, insufficient occupation-specific data and the variability of methodological approaches across regions.

The findings strengthen the relevance of integrated One Health strategies combining human, animal and environmental data. Reinforcing surveillance systems, improving data integration and developing context-appropriate prevention measures may be essential for better characterising and managing evolving risks.

Further research is needed to clarify the relative contribution of climatic and non-climatic factors, and to improve predictive approaches capable of informing targeted occupational health interventions.

REFERENCES

Abdelbaset, A. E., Kwak, M. L., Nonaka, N., & Nakao, R. (2023). Human-biting ticks and zoonotic tick-borne pathogens in North Africa: Diversity, distribution, and trans-Mediterranean public health challenges. *One Health*, *16*, 100547. DOI: 10.1016/j.onehlt.2023.100547 PMID: 37363219

Ahmed, A., Ali, Y., Salim, B., Dietrich, I., & Zinsstag, J. (2022). Epidemics of Crimean-Congo Hemorrhagic Fever (CCHF) in Sudan between 2010 and 2020. *Microorganisms*, *10*(5), 5. Advance online publication. DOI: 10.3390/microorganisms10050928 PMID: 35630372

Allen, T., Murray, K. A., Zambrana-Torrelio, C., Morse, S. S., Rondinini, C., Di Marco, M., Breit, N., Olival, K. J., & Daszak, P. (2017). Global hotspots and correlates of emerging zoonotic diseases. *Nature Communications*, *8*(1), 1124. DOI: 10.1038/s41467-017-00923-8 PMID: 29066781

Baker, R. E., Mahmud, A. S., Miller, I. F., Rajeev, M., Rasambainarivo, F., Rice, B. L., Takahashi, S., Tatem, A. J., Wagner, C. E., Wang, L.-F., Wesolowski, A., & Metcalf, C. J. E. (2022). Infectious disease in an era of global change. *Nature Reviews. Microbiology*, *20*(4), 193–205. DOI: 10.1038/s41579-021-00639-z PMID: 34646006

Bañuls, P., Galán-Relaño, Á., Romero-Salmoral, A., Gómez-Gascón, L., Tarradas, C., Astorga Márquez, R. J., Luque, I., & Huerta, B. (2024). Knowledge, perception and attitude about Crimean-Congo Haemorrhagic Fever (CCHF) among veterinary students and professionals of Spain. *Comparative Immunology, Microbiology and Infectious Diseases*, *115*, 102260. DOI: 10.1016/j.cimid.2024.102260 PMID: 39476521

Battisti, E., Urach, K., Hodžić, A., Fusani, L., Hufnagl, P., Felsberger, G., Ferroglio, E., & Duscher, G. G. (2020). *Zoonotic Pathogens in Ticks from Migratory Birds, Italy—Volume 26, Number 12—December 2020—Emerging Infectious Diseases journal—CDC*. DOI: 10.3201/eid2612.181686

Belij-Rammerstorfer, S., Limon, G., Maze, E. A., Hannant, K., Hughes, E., Tchakarova, S. R., Alexandrov, T., Mmbaga, B. T., Willett, B., Booth, G., Lyons, N. A., Baker, N., Thomas, K. M., Wright, D., Saunders, J., Browning, C., Wilsden, G., Carroll, M., Hewson, R., & Ludi, A. B. (2022). Development of anti-Crimean-Congo hemorrhagic fever virus Gc and NP-specific ELISA for detection of antibodies in domestic animal sera. *Frontiers in Veterinary Science*, *9*, 913046. Advance online publication. DOI: 10.3389/fvets.2022.913046 PMID: 36090164

Belobo, J. T. E., Kenmoe, S., Kengne-Nde, C., Emoh, C. P. D., Bowo-Ngandji, A., Tchatchouang, S., Wobessi, J. N. S., Mikangue, C. A. M., Tazokong, H. R., Bebey, S. R. K., Noura, E. A., Ka'e, A. C., Simo, R. E. G., Modiyinji, A. F., Ngongang, D. T., Che, E., Kenfack, S., Nzukui, N. D., Adjia, N. A., & Njouom, R. (2021). Worldwide epidemiology of Crimean-Congo Hemorrhagic Fever Virus in humans, ticks and other animal species, a systematic review and meta-analysis. *PLoS Neglected Tropical Diseases, 15*(4), e0009299. DOI: 10.1371/journal.pntd.0009299 PMID: 33886556

Bernard, C., Joly Kukla, C., Rakotoarivony, I., Duhayon, M., Stachurski, F., Huber, K., Giupponi, C., Zortman, I., Holzmuller, P., Pollet, T., Jeanneau, M., Mercey, A., Vachiery, N., Lefrançois, T., Garros, C., Michaud, V., Comtet, L., Despois, L., Pourquier, P., & Vial, L. (2024). Detection of Crimean–Congo haemorrhagic fever virus in Hyalomma marginatum ticks, southern France, May 2022 and April 2023. *Eurosurveillance, 29*(6). Advance online publication. DOI: 10.2807/1560-7917.ES.2024.29.6.2400023 PMID: 38333936

Bidaisee, S., & Macpherson, C. N. L. (2014). Zoonoses and One Health: A Review of the Literature. *Journal of Parasitology Research, 874345*, 1–8. Advance online publication. DOI: 10.1155/2014/874345 PMID: 24634782

Bosch, J., Muñoz, M. J., Martínez, M., de la Torre, A., & Estrada-Peña, A. (2013). Vector-Borne Pathogen Spread Through Ticks on Migratory Birds: A Probabilistic Spatial Risk Model for South-Western Europe. *Transboundary and Emerging Diseases, 60*(5), 403–415. DOI: 10.1111/j.1865-1682.2012.01358.x PMID: 22781365

Boulanger, N., Aran, D., Maul, A., Camara, B. I., Barthel, C., Zaffino, M., Lett, M.-C., Schnitzler, A., & Bauda, P. (2024). Multiple factors affecting Ixodes ricinus ticks and associated pathogens in European temperate ecosystems (northeastern France). *Scientific Reports, 14*(1), 9391. DOI: 10.1038/s41598-024-59867-x PMID: 38658696

Bourgeade, A., & Marchou, B. (2003). Fièvre jaune, dengue, encéphalite japonaise et virose West Nile, 4 arboviroses majeures. *Médecine et Maladies Infectieuses, 33*(8), 385–395. DOI: 10.1016/S0399-077X(03)00220-8

Boushab, B. M., Kelly, M., Kébé, H., Bollahi, M. A., & Basco, L. K. (2020). Crimean-Congo Hemorrhagic Fever, Mauritania. *Emerging Infectious Diseases, 26*(4), 817–818. DOI: 10.3201/eid2604.191292 PMID: 32187505

Burza, S., Croft, S. L., & Boelaert, M. (2018). Leishmaniasis. *Lancet, 392*(10151), 951–970. DOI: 10.1016/S0140-6736(18)31204-2 PMID: 30126638

Cabi. (2025). Research into zoonotic disease risks requires a One Health approach. *CABI.Org*. https://www.cabi.org/news-article/research-into-zoonotic-disease-risks-requires-a-one-health-approach/

Carvalho, B. M., Rangel, E. F., Ready, P. D., & Vale, M. M. (2015). Ecological Niche Modelling Predicts Southward Expansion of Lutzomyia (Nyssomyia) flaviscutellata (Diptera: Psychodidae: Phlebotominae), Vector of Leishmania (Leishmania) amazonensis in South America, under Climate Change. *PLoS One*, *10*(11), e0143282. DOI: 10.1371/journal.pone.0143282 PMID: 26619186

Chalghaf, B., Chemkhi, J., Mayala, B., Harrabi, M., Benie, G. B., Michael, E., & Ben Salah, A. (2018). Ecological niche modeling predicting the potential distribution of Leishmania vectors in the Mediterranean basin: Impact of climate change. *Parasites & Vectors*, *11*(1), 461. DOI: 10.1186/s13071-018-3019-x PMID: 30092826

Christova, I., Petrov, A., Papa, A., Vutchev, D., Kalvatchev, N., Vatev, N., & Stoycheva, M. (2015). Fatal Outcome of Coinfection of Crimean-Congo Hemorrhagic Fever and Malaria. *Japanese Journal of Infectious Diseases*, *68*(2), 131–134. DOI: 10.7883/yoken.JJID.2014.106 PMID: 25420643

Damtew, Y. T., Tong, M., Varghese, B. M., Anikeeva, O., Hansen, A., Dear, K., Zhang, Y., Morgan, G., Driscoll, T., Capon, T., & Bi, P. (2023). Effects of high temperatures and heatwaves on dengue fever: A systematic review and meta-analysis. *EBioMedicine*, *91*, 104582. DOI: 10.1016/j.ebiom.2023.104582 PMID: 37088034

de la Fuente, J., Ghosh, S., Lempereur, L., Garrison, A., Sprong, H., Lopez-Camacho, C., Maritz-Olivier, C., Contreras, M., Moraga-Fernández, A., & Bente, D. A. (2024). Interventions for the control of Crimean-Congo hemorrhagic fever and tick vectors. *NPJ Vaccines*, *9*(1), 1–10. DOI: 10.1038/s41541-024-00970-5 PMID: 39353956

de Souza, W. M., & Weaver, S. C. (2024). Effects of climate change and human activities on vector-borne diseases. *Nature Reviews. Microbiology*, *22*(8), 476–491. DOI: 10.1038/s41579-024-01026-0 PMID: 38486116

Desvars-Larrive, A., Vogl, A. E., Puspitarani, G. A., Yang, L., Joachim, A., & Käsbohrer, A. (2024). A One Health framework for exploring zoonotic interactions demonstrated through a case study. *Nature Communications*, *15*(1), 5650. DOI: 10.1038/s41467-024-49967-7 PMID: 39009576

Di Bari, C., Venkateswaran, N., Fastl, C., Gabriël, S., Grace, D., Havelaar, A. H., Huntington, B., Patterson, G. T., Rushton, J., Speybroeck, N., Torgerson, P., Pigott, D. M., & Devleesschauwer, B. (2023). The global burden of neglected zoonotic diseases: Current state of evidence. *One Health*, *17*, 100595. DOI: 10.1016/j.onehlt.2023.100595 PMID: 37545541

Diarra, A. Z., Kelly, P., Davoust, B., & Parola, P. (2023). Tick-Borne Diseases of Humans and Animals in West Africa. *Pathogens (Basel, Switzerland)*, *12*(11), 1276. DOI: 10.3390/pathogens12111276 PMID: 38003741

Djiman, T. A., Biguezoton, A. S., & Saegerman, C. (2024). Tick-Borne Diseases in Sub-Saharan Africa: A Systematic Review of Pathogens, Research Focus, and Implications for Public Health. *Pathogens (Basel, Switzerland), 13*(8), 697. DOI: 10.3390/pathogens13080697 PMID: 39204297

Durand-Billaud, E. BERTHOL, N., BIJAOUI, A., BOUSELHAM, M., PERIS, E., & BOUSLAMA, M. (2024). *Maladies zoonotiques et vectorielles en milieu professionnel: Impacts du changement climatique - Article - INRS*. https://www.inrs.fr/media.html?refINRS=TC%20181

ECDC. (2024, June 10). *Historical data on local transmission in Europe for West Nile virus*. https://www.ecdc.europa.eu/en/west-nile-fever/surveillance-and-disease-data/historical

Egbendewe-Mondzozo, A., Musumba, M., McCarl, B. A., & Wu, X. (2011). Climate Change and Vector-borne Diseases: An Economic Impact Analysis of Malaria in Africa. *International Journal of Environmental Research and Public Health, 8*(3), 913–930. DOI: 10.3390/ijerph8030913 PMID: 21556186

Ergonul, O., & Whitehouse, C. A. (2007). *Crimean-Congo Hemorrhagic Fever: A Global Perspective*. Springer Science & Business Media. DOI: 10.1007/978-1-4020-6106-6

Estrada-Peña, A. (2025). *Hyalomma marginatum* (the Mediterranean Hyalomma). *Trends in Parasitology, 41*(9), 822–823. DOI: 10.1016/j.pt.2025.06.007 PMID: 40617740

Estrada-Peña, A., D'Amico, G., & Fernández-Ruiz, N. (2021). Modelling the potential spread of Hyalomma marginatum ticks in Europe by migratory birds. *International Journal for Parasitology, 51*(1), 1–11. DOI: 10.1016/j.ijpara.2020.08.004 PMID: 32991918

Fakoorziba, M. R., Golmohammadi, P., Moradzadeh, R., Moemenbellah-Fard, M. D., Azizi, K., Davari, B., Alipour, H., Ahmadnia, S., & Chinikar, S. (2012). Reverse Transcription PCR-Based Detection of Crimean-Congo Hemorrhagic Fever Virus Isolated from Ticks of Domestic Ruminants in Kurdistan Province of Iran. *Vector Borne and Zoonotic Diseases, 12*(9), 794–799. DOI: 10.1089/vbz.2011.0743 PMID: 22651389

Fanelli, A., & Buonavoglia, D. (2021). Risk of Crimean Congo haemorrhagic fever virus (CCHFV) introduction and spread in CCHF-free countries in southern and Western Europe: A semi-quantitative risk assessment. *One Health, 13*, 100290. DOI: 10.1016/j.onehlt.2021.100290 PMID: 34307823

FAO-UNEP-WHO-OIE. (2022). *One Health Joint Plan of Action, 2022–2026*. FAO; UNEP; WHO; World Organisation for Animal Health (WOAH) (founded as OIE); DOI: 10.4060/cc2289en

Fasina, F. O., Fasanmi, O. G., Makonnen, Y. J., Bebay, C., Bett, B., & Roesel, K. (2021). The one health landscape in Sub-Saharan African countries. *One Health*, *13*, 100325. DOI: 10.1016/j.onehlt.2021.100325 PMID: 34584927

Gahn, M. C. B., Diouf, G., Cissé, N., Ciss, M., Bordier, M., Ndiaye, M., Bakhoum, M. T., Djiba, M. L., Brown, C., Faburay, B., Fall, A. G., & Lo, M. M. (2024). Large-Scale Serological Survey of Crimean-Congo Hemorrhagic Fever Virus and Rift Valley Fever Virus in Small Ruminants in Senegal. *Pathogens (Basel, Switzerland)*, *13*(8), 8. Advance online publication. DOI: 10.3390/pathogens13080689 PMID: 39204289

Gale, P., Stephenson, B., Brouwer, A., Martinez, M., de la Torre, A., Bosch, J., Foley-Fisher, M., Bonilauri, P., Lindström, A., Ulrich, R. G., de Vos, C. J., Scremin, M., Liu, Z., Kelly, L., & Muñoz, M. J. (2012). Impact of climate change on risk of incursion of Crimean-Congo haemorrhagic fever virus in livestock in Europe through migratory birds. *Journal of Applied Microbiology*, *112*(2), 246–257. DOI: 10.1111/j.1365-2672.2011.05203.x PMID: 22118269

Gangneux, J.-P., Belaz, S., & Robert-Gangneux, F. (2015). Mise au point et actualités sur la leishmaniose viscérale méditerranéenne. *Journal des Anti-Infectieux*, *17*(1), 25–28. DOI: 10.1016/j.antinf.2014.12.005

Gashema, P., Sesonga, P., Iradukunda, P. G., Muvunyi, R., Mugisha, J. C., Nday-isenga, J., Musafiri, T., Habimana, R., Bigirimana, R., Kabanda, A., Gashegu, M., Gahamanyi, N., Izudi, J., Siddig, E. E., Ngabonziza, J. C. S., Ahmed, A., Dzinamarira, T., Mutesa, L., & Muvunyi, C. M. (2025). Enhancing Global Health Security in Sub-Saharan Africa: The case for integrated One Health surveillance against zoonotic diseases and environmental threats. *One Health*, *21*, 101136. DOI: 10.1016/j.onehlt.2025.101136 PMID: 40687600

Gottdenker, N. L., Streicker, D. G., Faust, C. L., & Carroll, C. R. (2014). Anthropogenic Land Use Change and Infectious Diseases: A Review of the Evidence. *EcoHealth*, *11*(4), 619–632. DOI: 10.1007/s10393-014-0941-z PMID: 24854248

Haagsma, J. A., Tariq, L., Heederik, D. J., & Havelaar, A. H. (2012). Infectious disease risks associated with occupational exposure: A systematic review of the literature. *Occupational and Environmental Medicine*, *69*(2), 140–146. DOI: 10.1136/oemed-2011-100068 PMID: 22006935

Hoffman, T., Carra, L. G., Öhagen, P., Fransson, T., Barboutis, C., Piacentini, D., Figuerola, J., Kiat, Y., Onrubia, A., Jaenson, T. G. T., Nilsson, K., Lundkvist, Å., & Olsen, B. (2021). Association between guilds of birds in the African-Western Palaearctic region and the tick species *Hyalomma rufipes,* one of the main vectors of Crimean-Congo hemorrhagic fever virus. *One Health, 13*, 100349. DOI: 10.1016/j.onehlt.2021.100349 PMID: 34825045

Hoffman, T., Lindeborg, M., Barboutis, C., Erciyas-Yavuz, K., Evander, M., Fransson, T., Figuerola, J., Jaenson, T. G. T., Kiat, Y., Lindgren, P.-E., Lundkvist, Å., Mohamed, N., Moutailler, S., Nyström, F., Olsen, B., & Salaneck, E. (2018). *Alkhurma Hemorrhagic Fever Virus RNA in Hyalomma rufipes Ticks Infesting Migratory Birds, Europe and Asia Minor—Volume 24, Number 5—May 2018—Emerging Infectious Diseases journal—CDC.* DOI: 10.3201/eid2405.171369

IPCC. ntergovernmental P. O. C. C. (2023). *Climate Change 2022 – Impacts, Adaptation and Vulnerability: Working Group II Contribution to the Sixth Assessment Report of the Intergovernmental Panel on Climate Change* (1st ed.). Cambridge University Press. DOI: 10.1017/9781009325844

Islam, M. A., Begum, S., Solaiman, M., Zohora, F.-T., Datta, S. K., Ahmed, K. A., & Ahmed, M. S. (2025). Complete mitogenomes of two major dengue vectors Aedes aegypti and Aedes albopictus from Bangladesh: Insights from comparative genomics with global mitogenome diversity and phylogenetics. *PLoS One*, *20*(9), e0333693. DOI: 10.1371/journal.pone.0333693 PMID: 41026721

Iwamura, T., Guzman-Holst, A., & Murray, K. A. (2020). Accelerating invasion potential of disease vector Aedes aegypti under climate change. *Nature Communications*, *11*(1), 2130. DOI: 10.1038/s41467-020-16010-4 PMID: 32358588

Kamguem, I. S., Kirschvink, N., Wade, A., & Linard, C. (2025). Determinants of viral haemorrhagic fever risk in Africa's tropical moist forests: A scoping review of spatial, socio-economic, and environmental factors. *PLoS Neglected Tropical Diseases*, *19*(1), e0012817. DOI: 10.1371/journal.pntd.0012817 PMID: 39820141

Kilpatrick, A. M., & Randolph, S. E. (2012). Drivers, dynamics, and control of emerging vector-borne zoonotic diseases. *Lancet*, *380*(9857), 1946–1955. DOI: 10.1016/S0140-6736(12)61151-9 PMID: 23200503

Kodama, C., El Rifay, A. S., Badra, R., Abu Salbi, R., Abubakar, A., & Kayali, G. (2023). Operationalizing One Health: Strategic guidance for prevention and control of emerging and re-emerging vector-borne and zoonotic diseases in the Eastern Mediterranean Region. *Frontiers in Tropical Diseases*, *4*, 1237688. Advance online publication. DOI: 10.3389/fitd.2023.1237688

Kraemer, M. U., Sinka, M. E., Duda, K. A., Mylne, A. Q., Shearer, F. M., Barker, C. M., Moore, C. G., Carvalho, R. G., Coelho, G. E., Van Bortel, W., Hendrickx, G., Schaffner, F., Elyazar, I. R., Teng, H.-J., Brady, O. J., Messina, J. P., Pigott, D. M., Scott, T. W., Smith, D. L., & Hay, S. I. (2015). The global distribution of the arbovirus vectors Aedes aegypti and Ae. Albopictus. *eLife*, *4*, e08347. DOI: 10.7554/eLife.08347 PMID: 26126267

Kuehnert, P. A., Stefan, C. P., Badger, C. V., & Ricks, K. M. (2021). Crimean-Congo Hemorrhagic Fever Virus (CCHFV): A Silent but Widespread Threat. *Current Tropical Medicine Reports*, *8*(2), 141–147. DOI: 10.1007/s40475-021-00235-4 PMID: 33747715

Lawler, O. K., Allan, H. L., Baxter, P. W. J., Castagnino, R., Tor, M. C., Dann, L. E., Hungerford, J., Karmacharya, D., Lloyd, T. J., López-Jara, M. J., Massie, G. N., Novera, J., Rogers, A. M., & Kark, S. (2021). The COVID-19 pandemic is intricately linked to biodiversity loss and ecosystem health. *The Lancet. Planetary Health*, *5*(11), e840–e850. DOI: 10.1016/S2542-5196(21)00258-8 PMID: 34774124

Lee, S. H., Nam, K. W., Jeong, J. Y., Yoo, S. J., Koh, Y.-S., Lee, S., Heo, S. T., Seong, S.-Y., & Lee, K. H. (2013). The Effects of Climate Change and Globalization on Mosquito Vectors: Evidence from Jeju Island, South Korea on the Potential for Asian Tiger Mosquito (Aedes albopictus) Influxes and Survival from Vietnam Rather Than Japan. *PLoS One*, *8*(7), e68512. DOI: 10.1371/journal.pone.0068512 PMID: 23894312

Lefrançois, T., Lina, B., & Autran, B. (2023). One Health approach at the heart of the French Committee for monitoring and anticipating health risks. *Nature Communications*, *14*(1), 7540. DOI: 10.1038/s41467-023-43089-2 PMID: 37990001

Lesne, J. (2021). L'émergence de maladies infectieuses d'origine zoonotique: Complexité écologique et responsabilités socio-économiques. *Environnement, Risques & Santé : ERS*, *20*(3), 244–257. DOI: 10.1684/ers.2021.1560

Ludwig, A., Zheng, H., Vrbova, L., Drebot, M., Iranpour, M., & Lindsay, L. (2019). Increased risk of endemic mosquito-borne diseases in Canada due to climate change. *Canada Communicable Disease Report = Relevé des Maladies Transmissibles au Canada*, *45*(4), 91–97. DOI: 10.14745/ccdr.v45i04a03 PMID: 31285698

Magnavita, N., Capitanelli, I., Ilesanmi, O., & Chirico, F. (2022). Occupational Lyme Disease: A Systematic Review and Meta-Analysis. *Diagnostics (Basel)*, *12*(2), 296. DOI: 10.3390/diagnostics12020296 PMID: 35204387

Mancuso, E., Toma, L., Polci, A., d'Alessio, S. G., Luca, M. D., Orsini, M., Domenico, M. D., Marcacci, M., Mancini, G., Spina, F., Goffredo, M., & Monaco, F. (2019). *Crimean-Congo Hemorrhagic Fever Virus Genome in Tick from Migratory Bird, Italy—Volume 25, Number 7—July 2019—Emerging Infectious Diseases journal—CDC*. DOI: 10.3201/eid2507.181345

Mencattelli, G., Ndione, M. H. D., Rosà, R., Marini, G., Diagne, C. T., Diagne, M. M., Fall, G., Faye, O., Diallo, M., Faye, O., Savini, G., & Rizzoli, A. (2022). Epidemiology of West Nile virus in Africa: An underestimated threat. *PLoS Neglected Tropical Diseases*, *16*(1), e0010075. DOI: 10.1371/journal.pntd.0010075 PMID: 35007285

Mora, C., McKenzie, T., Gaw, I. M., Dean, J. M., von Hammerstein, H., Knudson, T. A., Setter, R. O., Smith, C. Z., Webster, K. M., Patz, J. A., & Franklin, E. C. (2022). Over half of known human pathogenic diseases can be aggravated by climate change. *Nature Climate Change*, *12*(9), 869–875. DOI: 10.1038/s41558-022-01426-1 PMID: 35968032

Mordecai, E. A., Ryan, S. J., Caldwell, J. M., Shah, M. M., & LaBeaud, A. D. (2020). Climate change could shift disease burden from malaria to arboviruses in Africa. *The Lancet. Planetary Health*, *4*(9), e416–e423. DOI: 10.1016/S2542-5196(20)30178-9 PMID: 32918887

Msellemu, D., Tanner, M., Yadav, R., & Moore, S. J. (2024). Occupational exposure to malaria, leishmaniasis and arbovirus vectors in endemic regions: A systematic review. *Current Research in Parasitology & Vector-borne Diseases*, *6*, 100185. DOI: 10.1016/j.crpvbd.2024.100185 PMID: 39027087

Mubemba, B., Mburu, M. M., Changula, K., Muleya, W., Moonga, L. C., Chambaro, H. M., Kajihara, M., Qiu, Y., Orba, Y., Hayashida, K., Sutcliffe, C. G., Norris, D. E., Thuma, P. E., Ndubani, P., Chitanga, S., Sawa, H., Takada, A., & Simulundu, E. (2022). Current knowledge of vector-borne zoonotic pathogens in Zambia: A clarion call to scaling-up "One Health" research in the wake of emerging and re-emerging infectious diseases. *PLoS Neglected Tropical Diseases*, *16*(2), e0010193. DOI: 10.1371/journal.pntd.0010193 PMID: 35120135

Muzammil, K., Rayyani, S., Abbas Sahib, A., Gholizadeh, O., Naji Sameer, H., Jwad Kazem, T., Badran Mohammed, H., Ghafouri Kalajahi, H., Zainul, R., & Yasamineh, S. (2024). Recent Advances in Crimean-Congo Hemorrhagic Fever Virus Detection, Treatment, and Vaccination: Overview of Current Status and Challenges. *Biological Procedures Online*, *26*(1), 20. DOI: 10.1186/s12575-024-00244-3 PMID: 38926669

Njeshi, C. N., Robertson, A. P., & Martin, R. J. (2026). Onchocerciasis overview: Pathogenesis, clinical spectrum, and therapeutic challenges of a neglected tropical disease. *Frontiers in Parasitology*, *5*, 1690973. Advance online publication. DOI: 10.3389/fpara.2026.1690973 PMID: 41908972

Nnko, H. J., Gwakisa, P. S., Ngonyoka, A., Sindato, C., & Estes, A. B. (2021). Potential impacts of climate change on geographical distribution of three primary vectors of African Trypanosomiasis in Tanzania's Maasai Steppe: G. m. morsitans, G. pallidipes and G. swynnertoni. *PLoS Neglected Tropical Diseases*, *15*(2), e0009081. DOI: 10.1371/journal.pntd.0009081 PMID: 33571190

Oliveira, S., Rocha, J., Sousa, C. A., & Capinha, C. (2021). Wide and increasing suitability for Aedes albopictus in Europe is congruent across distribution models. *Scientific Reports*, *11*(1), 9916. DOI: 10.1038/s41598-021-89096-5 PMID: 33972597

Palomar, A. M., Portillo, A., Santibáñez, P., Mazuelas, D., Arizaga, J., Crespo, A., Gutiérrez, Ó., Cuadrado, J. F., & Oteo, J. A. (2013). Crimean-Congo Hemorrhagic Fever Virus in Ticks from Migratory Birds, Morocco1. *Emerging Infectious Diseases*, *19*(2), 260–263. DOI: 10.3201/eid1902.121193 PMID: 23347801

Paz, S. (2020). Climate change impacts on vector-borne diseases in Europe: Risks, predictions and actions. *The Lancet Regional Health. Europe*, *1*, 100017. DOI: 10.1016/j.lanepe.2020.100017 PMID: 35104838

Pedersen, B. N., Jenkins, A., & Kjelland, V. (2020). Tick-borne pathogens in Ixodes ricinus ticks collected from migratory birds in southern Norway. *PLoS One*, *15*(4), e0230579. DOI: 10.1371/journal.pone.0230579 PMID: 32271774

Porretta, D., Mastrantonio, V., Amendolia, S., Gaiarsa, S., Epis, S., Genchi, C., Bandi, C., Otranto, D., & Urbanelli, S. (2013). Effects of global changes on the climatic niche of the tick Ixodes ricinus inferred by species distribution modelling. *Parasites & Vectors*, *6*(1), 271. DOI: 10.1186/1756-3305-6-271 PMID: 24330500

Rakotoarinia, M. R. (2025). *Effets combinés des changements climatiques et des changements d'utilisation des sols sur les populations de moustiques en Ontario.* https://hdl.handle.net/1866/41127

Randolph, S. E., Gern, L., & Nuttall, P. A. (1996). Co-feeding ticks: Epidemiological significance for tick-borne pathogen transmission. *Parasitology Today (Personal Ed.)*, *12*(12), 472–479. DOI: 10.1016/S0169-4758(96)10072-7 PMID: 15275266

Randolph, S. E., & Rogers, D. J. (2000). Fragile transmission cycles of tick-borne encephalitis virus may be disrupted by predicted climate change. *Proceedings. Biological Sciences*, *267*(1454), 1741–1744. DOI: 10.1098/rspb.2000.1204 PMID: 12233771

Richard, S., & Oppliger, A. (2015). Zoonotic occupational diseases in forestry workers – Lyme borreliosis, tularemia and leptospirosis in Europe. *Annals of Agricultural and Environmental Medicine : AAEM*, *22*(1), 43–50. DOI: 10.5604/12321966.1141368 PMID: 25780827

Rocklöv, J., & Dubrow, R. (2020). Climate change: An enduring challenge for vector-borne disease prevention and control. *Nature Immunology*, *21*(5), 479–483. DOI: 10.1038/s41590-020-0648-y PMID: 32313242

Roome, A., Gouli, S., Yodsuwan, R., Victory, J., Collins, C., Jenkins, P., Scribani, M., Krupa, N., Freilich, D., & Gadomski, A. (2022). Tick magnets: The occupational risk of tick-borne disease exposure in forestry workers in New York. *Health Science Reports*, *5*(2), e509. DOI: 10.1002/hsr2.509 PMID: 35229053

Rüegg, S. R., McMahon, B. J., Häsler, B., Esposito, R., Nielsen, L. R., Ifejika Speranza, C., Ehlinger, T., Peyre, M., Aragrande, M., Zinsstag, J., Davies, P., Mihalca, A. D., Buttigieg, S. C., Rushton, J., Carmo, L. P., De Meneghi, D., Canali, M., Filippitzi, M. E., Goutard, F. L., & Lindberg, A. (2017). A Blueprint to Evaluate One Health. *Frontiers in Public Health*, *5*. Advance online publication. DOI: 10.3389/fpubh.2017.00020 PMID: 28261580

Rupasinghe, R., Chomel, B. B., & Martínez-López, B. (2022). Climate change and zoonoses: A review of the current status, knowledge gaps, and future trends. *Acta Tropica*, *226*, 106225. DOI: 10.1016/j.actatropica.2021.106225 PMID: 34758355

Ryan, S. J., Carlson, C. J., Mordecai, E. A., & Johnson, L. R. (2019). Global expansion and redistribution of Aedes-borne virus transmission risk with climate change. *PLoS Neglected Tropical Diseases*, *13*(3), e0007213. DOI: 10.1371/journal.pntd.0007213 PMID: 30921321

Scott, N., Graczyk, H., & Papandrea, D. (2024). *Ensuring safety and health at work in a changing climate: Global report—International Labour Organization* (No. 995361942702676; p. 132). https://s3.eu-central-1.amazonaws.com/eu-st01.ext.exlibrisgroup.com/41ILO_INST/storage/alma/5C/8E/93/00/A8/1F/1F/D3/44/0E/55/11/5E/A1/2F/FF/wcms_922850.pdf?response-content-type=application%2Fpdf&X-Amz-Algorithm=AWS4-HMAC-SHA256&X-Amz-Date=20260504T071608Z&X-Amz-SignedHeaders=host&X-Amz-Credential=AKIAJUYRSPBSIJGGDX6A%2F20260504%2Feu-central-1%2Fs3%2Faws4_request&X-Amz-Expires=119&X-Amz-Signature=5edfaf9b8da827053cdb37cea40de514c7ceacb85ce9d2cd66312f495bbd6922

Sekamatte, M., Krishnasamy, V., Bulage, L., Kihembo, C., Nantima, N., Monje, F., Ndumu, D., Sentumbwe, J., Mbolanyi, B., Aruho, R., Kaboyo, W., Mutonga, D., Basler, C., Paige, S., & Behravesh, C. B. (2018). Multisectoral prioritization of zoonotic diseases in Uganda, 2017: A One Health perspective. *PLoS One*, *13*(5), e0196799. DOI: 10.1371/journal.pone.0196799 PMID: 29715287

Semenza, J. C., & Suk, J. E. (2018). Vector-borne diseases and climate change: A European perspective. *FEMS Microbiology Letters*, *365*(2), fnx244. Advance online publication. DOI: 10.1093/femsle/fnx244 PMID: 29149298

Socha, W., Kwasnik, M., Larska, M., Rola, J., & Rozek, W. (2022). Vector-Borne Viral Diseases as a Current Threat for Human and Animal Health—One Health Perspective. *Journal of Clinical Medicine*, *11*(11), 3026. Advance online publication. DOI: 10.3390/jcm11113026 PMID: 35683413

Soe, B. K., Huertas-Lopez, A., Sato, H., & Adisakwattana, P. (2026). Ticks and tick-borne pathogens with zoonotic importance in animals in Southeast Asia: A decade of evidence from systematic review and meta-analysis. *Acta Tropica*, *274*, 107974. DOI: 10.1016/j.actatropica.2026.107974 PMID: 41529770

Spengler, J. R., Bergeron, É., & Rollin, P. E. (2016). Seroepidemiological Studies of Crimean-Congo Hemorrhagic Fever Virus in Domestic and Wild Animals. *PLoS Neglected Tropical Diseases*, *10*(1), e0004210. DOI: 10.1371/journal.pntd.0004210 PMID: 26741652

Stufano, A., Schino, V., Plantone, D., & Lucchese, G. (2025). Occupational zoonoses, neurological diseases, and public health: A one health approach. *Infectious Medicine*, *4*(2), 100184. DOI: 10.1016/j.imj.2025.100184 PMID: 40530081

Suu-Ire, R. D., Gyebi, G., Baum, S., Abugri, H. A., Breen, R. W. B., Karesh, W. B., & Machalaba, C. (2025). The effects of climate change on zoonotic disease risks in Ghana: Reviewing the state of knowledge from a One Health perspective. *PAMJ - One Health : Pan African Medical Journal, 18*(9). Advance online publication. DOI: 10.11604/pamj-oh.2025.18.9.46717

Taylor, L. H., Latham, S. M., & woolhouse, M. E. J. (2001). Risk factors for human disease emergence. *Philosophical Transactions of the Royal Society of London. Series B, Biological Sciences, 356*(1411), 983–989. DOI: 10.1098/rstb.2001.0888 PMID: 11516376

Thomson, M. C., & Stanberry, L. R. (2022). Climate Change and Vectorborne Diseases. *The New England Journal of Medicine, 387*(21), 1969–1978. DOI: 10.1056/NEJMra2200092 PMID: 36416768

Tiffin, H. S., Gordon, J. R., & Poh, K. C. (2025). One Health, many approaches: Integrated vector management strategies support One Health goals. *Frontiers in Insect Science, 5*, 1549348. DOI: 10.3389/finsc.2025.1549348 PMID: 40530168

Tokarska-Rodak, M., Plewik, D., Kozioł-Montewka, M., Szepeluk, A., & Paszkiewicz, J. (2014).. . *Medycyna Pracy, 65*(1), 109–117. DOI: 10.13075/mp.5893.2014.017 PMID: 24834698

Vonesch, N., Binazzi, A., Bonafede, M., Melis, P., Ruggieri, A., Iavicoli, S., & Tomao, P. (2019). Emerging zoonotic viral infections of occupational health importance. *Pathogens and Disease, 77*(2), ftz018. Advance online publication. DOI: 10.1093/femspd/ftz018 PMID: 30916772

Vonesch, N., D'Ovidio, M. C., Melis, P., Remoli, M. E., Ciufolini, M. G., & Tomao, P. (2016). Climate change, vector-borne diseases and working population. *Annali dell'Istituto Superiore di Sanità, 52*(3), 397–405. DOI: 10.4415/ANN_16_03_11 PMID: 27698298

Wegner, G. I., Murray, K. A., Springmann, M., Muller, A., Sokolow, S. H., Saylors, K., & Morens, D. M. (2022). Averting wildlife-borne infectious disease epidemics requires a focus on socio-ecological drivers and a redesign of the global food system. *EClinicalMedicine, 47*, 101386. Advance online publication. DOI: 10.1016/j.eclinm.2022.101386 PMID: 35465645

WHO. (2017a). *Global vector control response 2017–2030*. https://www.who.int/publications/i/item/9789241512978

WHO. (2017b). *West Nile virus*. https://www.who.int/news-room/fact-sheets/detail/west-nile-virus

WHO. (2023a). *Leishmaniasis*. https://www.who.int/news-room/fact-sheets/detail/leishmaniasis

WHO. (2023b). *Trypanosomiasis, human African (sleeping sickness)*. https://www.who.int/news-room/fact-sheets/detail/trypanosomiasis-human-african-(sleeping-sickness)

WHO. (2025a). *Crimean-Congo haemorrhagic fever*. https://www.who.int/news-room/fact-sheets/detail/crimean-congo-haemorrhagic-fever

WHO. (2025b). *Dengue*. https://www.who.int/news-room/fact-sheets/detail/dengue-and-severe-dengue

WHO. (2025c). *Dengue*. https://www.who.int/news-room/fact-sheets/detail/dengue-and-severe-dengue

WHO. (2025d). *Ebola disease*. https://www.who.int/news-room/fact-sheets/detail/ebola-disease

WHO. (2025e). *Onchocerciasis (river blindness)*. https://www.who.int/news-room/fact-sheets/detail/onchocerciasis

WHO. (2025f). *Public health advice on Crimean-Congo haemorrhagic fever*. https://www.who.int/europe/news-room/questions-and-answers/item/public-health-advice-on-crimean-congo-haemorrhagic-fever

Wolfe, N. D., Daszak, P., Kilpatrick, A. M., & Burke, D. S. (2005). *Bushmeat Hunting, Deforestation, and Prediction of Zoonotic Disease—Volume 11, Number 12—December 2005—Emerging Infectious Diseases journal—CDC*. DOI: 10.3201/eid1112.040789

Zhou, R., Sprong, H., Liu, Q., Krafft, T., & Estrada-Peña, A. (2025). Mapping the potential suitable habitats for Hyalomma rufipes (Acari: Ixodidae) in Africa and Western Eurasia. *PLoS Neglected Tropical Diseases*, *19*(3), e0012923. DOI: 10.1371/journal.pntd.0012923 PMID: 40100878

Zinsstag, J., Kaiser-Grolimund, A., Heitz-Tokpa, K., Sreedharan, R., Lubroth, J., Caya, F., Stone, M., Brown, H., Bonfoh, B., Dobell, E., Morgan, D., Homaira, N., Kock, R., Hattendorf, J., Crump, L., Mauti, S., del Rio Vilas, V., Saikat, S., Zumla, A., & de la Rocque, S. (2023). Advancing One human–animal–environment Health for global health security: What does the evidence say? *Lancet*, *401*(10376), 591–604. DOI: 10.1016/S0140-6736(22)01595-1 PMID: 36682371

Chapter 10
Understanding Climate Change Through Visualization and Its Environmental Impacts

D. Rajalakshmi
http://orcid.org/0000-0003-4417-155X
SRM Institute of Science and Technology, India

A. Rehash Rushmi Pavitra
http://orcid.org/0000-0002-1220-1497
SRM Institute of Science and Technology, India

V. Vijayalakshmi
SRM Institute of Science and Technology, India

R. Radha
http://orcid.org/0000-0001-9822-0949
SRM Institute of Science and Technology, India

S. Sharanya
SRM Institute of Science and Technology, India

ABSTRACT

In Recent days climate change is one of the major global concerns. The intent of implementing this proposed Tableau-based visualization system is to aid in bridging the gap between detailed scientific knowledge of climate change and the general public's ability to understand such scientific information by converting highly complex data sets related to climate change into visually manageable data narratives.

DOI: 10.4018/979-8-3373-8470-2.ch010

Key indicators that will be addressed through the proposed Tableau-based visualization system include all climate change actions and the cascading effects of these phenomena on many ecosystems and human populations. Furthermore, it illustrates interactive or dynamic visualizations creates considerably higher communicative values than static representations, allowing the user to investigate temporal trends, spatial distributions and scenario projections in real-time. Thus, the findings of this chapter support visualization as a catalyst for increased levels of climate literacy, support for better-informed policy-making and the increase of public engagement with climate action.

INTRODUCTION TO CLIMATE CHANGE

Climate change is one of the biggest problems in earth and economy right now. It has huge effects on the environment that will last for a long time. Climate change is now viewed by many as one of the most significant environmental challenges facing the world today, with implications that are likely to have a direct impact on both natural systems and human societies for generations to come. As the speed at which our climate continues to change increases, there is a growing need to communicate climate change science effectively to many different types of audience members. Significant and long-term changes in Earth's temperature patterns brought on mostly by human activity and natural processes are known as climate change. Greenhouse gases (GHGs), such as carbon dioxide (CO_2), methane (CH_4), and nitrous oxide (N_2O), have being released at a quicker rate than they were a century ago due to human activities such as burning fossil fuels, deforestation, and industrial operations (Lun et al., 2021). These gases trapping heat in the atmosphere causes global warming by increasing the temperature of the earth. The earth is being affected by climate change in many ways, such as rising sea levels, severe weather, and the loss of species. It also influences people's health, the availability of food, and economic growth, especially in poor countries. Governments, businesses, citizens, and people must all work together to stop climate change. It needs new policies and solutions to lower carbon emissions and adjust to the changing environment (Tang et al., 2021).

Following is the few consequences of climate change on the entire planet and the ways in which it influences various elements of our lives:

1. **Increasing temperatures and sea levels:** Global temperatures have increased by 1.1C on average since the start of the industrial era and scientists predict that, it will be climb an additional 0.5C in the coming decade. The melting of glaciers and ice caps leads to increase the levels of Sea and its effect the low-lying areas and coastal regions (Pervin & Gan, 2021). The Maldives is an island republic

in the Indian Ocean, that is the risk from rising the sea levels in the next future decades.

Figure 1. Introduction to Climate Change and its effects

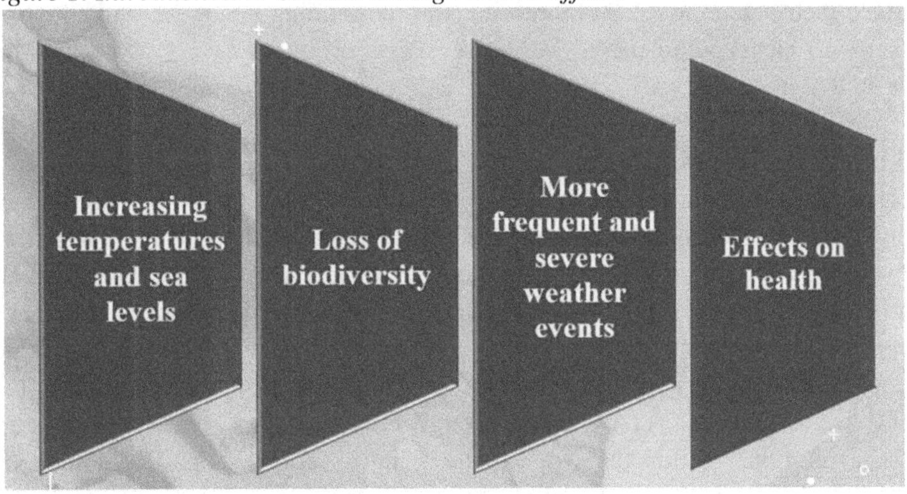

2. **Loss of biodiversity:** biodiversity loss is becoming worse due to climate change. Ecosystems are already struggling to adjust to new average temperatures and precipitation patterns, and this problem is only becoming worse (Cai et al., 2020). Many people's livelihoods are directly impacted by the extinction of certain plant and animal species.
3. **More often and worse weather conditions:** Hurricanes, floods, and droughts are becoming more often due to this climate change (Jain et al., 2020). The effects of these disasters on local populations are catastrophic, especially in areas that are already struggling, such as Southeast Asia and Sub-Saharan Africa. Example 2018 floods in Kerala, 2023 floods in Chennai, Thoothukudi and wayanad landslide in India. They were responsible for over few hundreds of deaths and the displacement of over few million people.
4. **Effects on health:** Children, the elderly, and the poor are among the most susceptible groups to the health consequences of climate change. Malaria, dengue fever, Chikungunya and other tropical diseases might spread, while crop failures can cause starvation. More number of people died due to the heatwave, mostly the elderly and those with existing health problems.

Addressing climate change necessitates action at all levels and is a complicated and pressing subject. The implementation of legislation aimed at lowering emissions and promoting low-carbon technologies must be the responsibility of governments. By implementing sustainable practices and making investments in renewable energy, businesses may also make a significant contribution. Finally, individuals may help reduce their carbon footprint by driving less, consuming less meat, and adopting energy-efficient equipment.

LITERATURE WORK

The climate change poses a societal and technological challenge apart from government and industry sectors. Because its solely responsible for individual efforts to overcome these kinds of climate actions like global warming, acid rain etc. The escalating threat of global warming and climate change has been a significant concern globally over the past two decades. Following are the completion of the First Assessment Report (AR) by the Intergovernmental Panel on Climate Change (IPCC) in 1990. Within the scope of the United Nations scope Convention on Climate Change (UNFCCC), climate debates have gained significant importance, but frequently accompanied by controversy. Since 1997, from the Kyoto Protocol to the Katowice COP24 climate conference in 2018, the primary focus across all these forums has been the reduction of greenhouse gases that contribute to climate change. The Kyoto Protocol (1997) stipulated a decrease of carbon emissions by 5.2% below 1990 levels over the period from 2008 to 2012. It was enacted in 2005. Carbon dioxide (CO_2) was attributed to 58.8% of greenhouse gas emissions, among many environmental contaminants contributing to climate change (The World Bank, 2019). Climate change is undeniably imminent, and the globe is now seeing its adverse effects. The World Health Organization (WHO, 2021) estimates that climate change is expected to contribute to approximately 250,000 additional deaths per year between 2030 and 2050, attributable to factors such as malnutrition, malaria, diarrhoea, and heat stress; it should be noted, however, that isolating climate change as a sole cause of mortality involves considerable methodological uncertainty. Growth techniques that exacerbate issues of food insecurity, environmental degradation, national instability, and social discontent contribute to unsustainable growth. Consequently, development goals that enhance food security, social protection, and mitigate climate change are of paramount significance for scholars and public policymakers, since they facilitate sustainable development. Developing nations encounter a significant difficulty in economic growth due to the intensified issues of population expansion and dependence on non-fossil fuel energy sources, which exacerbate environmental concerns.

It is important to acknowledge that according to the Environmental Kuznets Curve (EKC), although there is evidence of decreasing energy consumption during the development process, it is uncertain whether further development will lead to an increase or decrease in CO_2 emissions in the long term for many countries still in the developing stage. Examining carbon emission patterns in conjunction with economic growth is crucial from a policy standpoint. The EKC premise suggests a 'inverted-U' association between per heads GDP and per heads carbon emissions.

The theory states that economic progress incurs environmental worsening during the initial steps of development. The EKC theory, proliferated by Grossman and Krueger (1991) after Simon Kuznets' methods that links the income difference with economic change, it joins economic progress to environmental worsening and is often known as the Environmental Kuznets Curve (EKC) theory. It represents a dynamic revolution process that directly associates with environmental degradation in the early stages of economic development, expected to invert after reaching a specific level of advancement.The unique reiteration of the EPI is extended from the Environmental Sustainability Index (ESI) (1999-2005), formed by a consortium of the Yale Center. It evaluates 180 countries according to 32 performance metrics in 11 categories, encompassing environmental health (EH) and ecosystem vitality (EV).

There is a growing body of evidence on climate-related models and policy frameworks, yet there is a significant gap in this area of research concerning the use of interactive dashboard-based visualisation tools (e.g., Tableau) to communicate climate datasets with multiple variables to non-specialists. Most other work in this topic area has either focused on scientific aspects of climate modelling or general public education campaigns without providing a systematic demonstration of how to design commercial visualisation platforms that allow for the visualisation of uncertainty, change over time and relationships between different variables in such a way that is both policy relevant and publicly accessible. This chapter provides an example of how to close the above noted gap, by providing an application of a Tableau-based climate visualisation system demonstrating the application of a multi-variable, multi-country dataset.

If the pace of global warming continues to rise at the same rate as it is already, it is likely that it will hit 1.5 degrees Celsius between the years 2030 and 2052 (IPCC Special Report, 2018). Warming caused by anthropogenic emissions from the pre-industrial period to the present day will continue to create more long-term changes in the climate system, such as an increase in sea level, along with the related consequences. This warming will continue for hundreds to millennia. While it is anticipated that limiting global warming to 1.5 degrees Celsius rather than 2 degrees Celsius will lower rises in ocean temperature, it is also anticipated that this will lessen accompanying increases in ocean acidity and losses in ocean oxygen levels. Consequently, it is anticipated that limiting global warming to 1.5 degrees Celsius will lessen threats to marine biodiversity, fisheries, and ecosystems, as well as their functions and benefits to humans. This is demonstrated by the recent changes that have occurred in the ecosystems of warm-water coral reefs and Arctic Sea ice for example. CO2 emissions are forecast to decrease by around 25% by the year 2030 in most routes (10% - 30% interquartile range), and they are expected to reach net zero around the year 2070 (2065 - 2080 interquartile range). This is necessary in order to keep the global temperature to below 2 degrees Celsius. According to the intergovernmental panel on climate change (IPCC) Special Report (2018), routes that limit global warming to 1.5 degrees Celsius demonstrate deep reductions that are comparable to those in pathways that limit warming to 2 degrees Celsius.

According to the Intergovernmental Panel on Climate Change (IPCC), the expansion of economic development, rising income, a growing rate of urbanization, and rising per capita energy consumption have all contributed to greater access to energy in emerging nations. The consequences are exacerbated by the release of gases that are caused by a variety of human activities that are a direct result of urbanization and industrialization. In the many emerging countries that are now driven by energy, end-use Oil, natural gas, and coal are the primary sources of energy that are used to power "energy services," either directly or indirectly. Fossil fuels have historically accounted for a substantial share of India's primary energy supply, though this proportion has been gradually declining as the country scales up its renewable energy capacity, particularly in solar and wind power (IEA, 2023). Nonetheless, the reliance on coal and other fossil fuels for electricity generation continues to contribute significantly to carbon dioxide (CO2) emissions from the energy sector. The problem of excessive carbon dioxide emissions from the burning of fossil fuels, especially coal, in the energy sector of the country has become even more severe (The World Bank Report, 2019). Projections regarding the future energy mix vary across scenarios and should be interpreted with caution, as policy interventions, technological change, and investment patterns can significantly alter outcomes. If energy consumption grows in accordance with forecasts made by the International Energy Agency (IEA) and many other organizations, the world will consume approximately forty percent more energy than it uses today. The degree to which a nation is vulnerable to climate change will mostly be determined by its economic standing and its capacity for energy use.

IMPORTANT IDEAS AND FUNDAMENTAL CONCEPTS OF CLIMATE CHANGE

1. **Greenhouse Effect:** This natural phenomenon maintains the Earth's temperature in a constant level. It occurs when the gases are more in the atmosphere's referred as greenhouse gases, trap solar heat in the Earth's atmosphere. The average temperature of the Earth would be roughly -18°C, without the greenhouse effect. which would be too cold for living things to survive. But because of human activities like burning fossil fuels, the amount of greenhouse gases in the atmosphere has been increased, increasing the greenhouse effect and raising temperatures.

Figure 2. Fundamental concepts of Climate Change

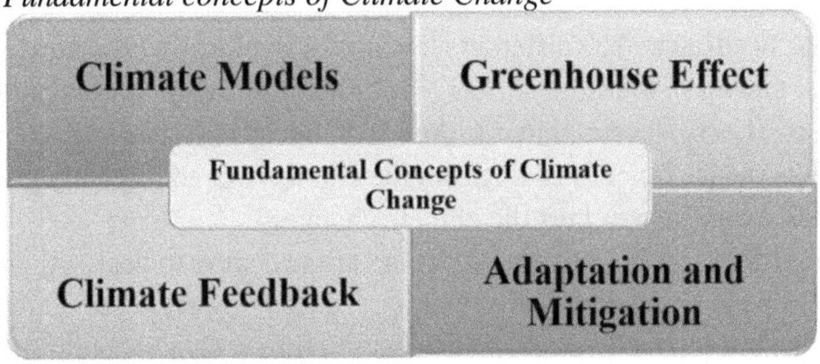

2. **Climate Models:** Through computer simulations its predict how the climate will change in the future is called as climate models. These climate models predict the future climate patterns by various factors such as ocean currents, solar radiation, and greenhouse gas emissions. It provides the important information about how climate change will affect the environment.

Figure 3. Climate Models based on Land Average Temperature

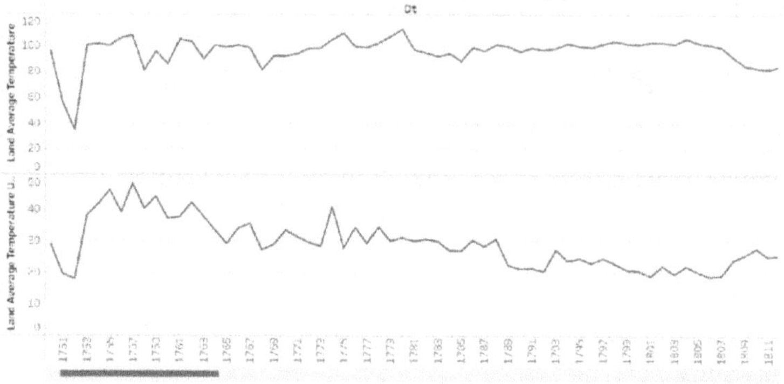

Source: Created by the authors using Tableau Desktop and Open-source climate datasets

3. **Climate Feedback:** The different elements of the Earth's climate system work together can make the climate changes neither improving or worsening. For example, it reflects less sunlight into the space. It leads to increase the temperature and melts the ice quickly. Changes in the cloud cover additionally provides the feedback either help or hurt the climate change.
4. **Modification and Improvement:** There are two ways to deal with the climate change: one is by changing it or making it better. "Modification" refers to the steps taken to lessen the negative effects of climate change, building sea walls to protect against rising sea levels. Another one is Mitigation, it refers to actions that lower greenhouse gas emissions and slow down climate change by using renewable energy sources.

Climate change is a complicated problem, need to know about its causes, effects, and possible solutions. To understand this significant by examining essential concepts and ideas, including the greenhouse effect, climate models, climate feedback, and adaptation and mitigation strategies.

CLIMATE CHANGE AND EFFECTS ON THE ATMOSPHERE OF THE EARTH:

The atmosphere of Earth that protects us from the sun's dangerous rays and saves the temperature of our planet within a range for people to live in the environment. However, people do burn of fossil fuels and cutting down trees, have made the increase of greenhouse gas emissions.

1. Depletion of Ozone Layer:

The ozone layer is a layer of gas that protects the earth. It absorbs harmful UV rays that is produced by the sun. The ozone layer has been damaged by human activities, like using chlorofluorocarbons (CFCs) in aerosols and refrigerants. Because of this depletion, more people are getting cataracts, skin cancer, and other health problems. The ozone layer has become thinner, which has let more UV radiation reach the surface of the Earth. This has triggered the temperature of the earth to change. Ozone depletion triggers more UV light to reach the Earth's surface. While there are other similarities, both ozone depletion and climate change are caused by human activities resulting in the interaction between ozone and climate. Ozone depletion is primarily driven by the release of halogenated chemicals such as CFCs that result in increased UV radiation reaching the earth's surface. Climate change is mainly caused by the accumulation of greenhouse gases that trap infrared radiation causing global temperatures to rise. Although there is an interaction between the two phenomena, for example where certain greenhouse gases have an influence on stratospheric chemistry, the phenomena are created by two very different mechanisms, i.e. should not be considered interchangeable when discussing policies regarding either problem or the scientific research related to both.

2. Acid Rain:

Acid rain is initiated by human activities that release acidic pollutants into the air, like sulfur dioxide and nitrogen oxides. Acid rain damages the crops, forests, and rivers, that's is the one of the reasons why biodiversity is decreasing, and more environmental problems are getting worse. It also makes the ozone layer dangerous and causes more greenhouse gases to be emitted into the air. Acid rain is another atmospheric phenomenon that is associated directly with climate change yet is distinct from it. Acid rain results from the emissions of sulfur dioxide (SO2) and nitrogen oxides (NOx), which are released into the atmosphere mainly from burning fossil fuels and industrial processes. The gaseous products of these processes react with the atmosphere to form sulphuric and nitric acids. There are some similarities between acid rain and climate change in that both involve fossil fuel combustion, but the two phenomena are not linked directly; rather, acid rain has primarily localized effects by damaging ecological systems in soil, bodies of water, and plants. Although there is a common belief that acid rain damages the ozone layer, this needs to be clarified because while acidic aerosols can alter stratospheric chemistry when they exist at certain altitude and atmospheric conditions, they do not represent the primary cause of ozone depletion. While climatologists, scientists studying ozone depletion, and researchers studying acid rain recognize that climate change, ozone depletion, and acid rain; all share a common cause, they are all different issues that require independent analysis and implementation of effective policies.

Figure 4. Acidic Pollutants in the Atmosphere

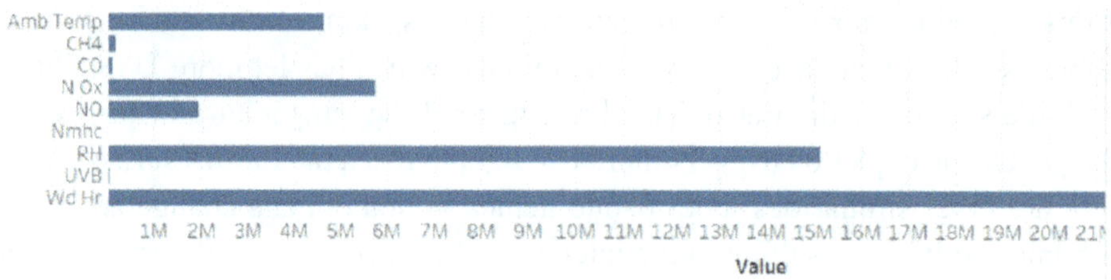

Source: Created by the authors using Tableau Desktop and Open-source climate datasets

3. Emission of greenhouse gases:

Examples of greenhouse gases are Carbon dioxide, methane, and nitrous oxides. These gases help preserve heat in the atmosphere of earth, which leads to increase the temperature of the atmosphere. The increase in greenhouse gases, which are mostly caused by the human activities. When it comes to greenhouse gas emissions, the biggest contributors are burning of fossil fuels for energy, destruction of forests, and agricultural practices. Another large contributor to emissions of greenhouse gases is the transportation sector, which includes automobiles, trucks, and airplanes among other modes of transportation.

Figure 5. Sunlight warms earth's surface

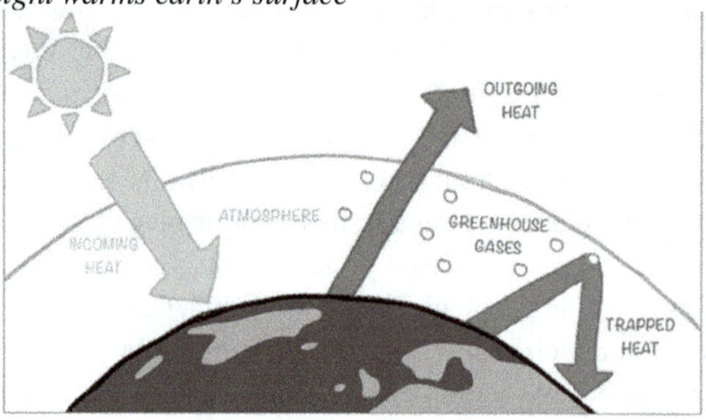

4. Strategies to Reduce the Impact:

Reducing emissions of greenhouse gases through a range of different ways is necessary to lessen the impact of climate change will affect the atmosphere of earth. The transition to renewable energy sources, improvements in energy efficiency, and reductions in deforestation and emissions from agriculture are some of the methods that are under consideration. It utilizes the carbon capture and storage technology, to capture and preserve the carbon dioxide (CO_2) emissions that are produced by power plants and other industrial operations.

VISUALIZATIONS OF CLIMATE CHANGE AND INDIGENOUS KNOWLEDGE:

If you are using current visualization tools with Indigenous ecological knowledge (IEK), understand how the world is changing in a more complete way.

In Canada, people have used participatory planning, GIS, and narrative maps to combine oral histories with geographical data in projects like the "First Nations Climate Initiative." Ford et al. (2016) assert that this policy applies traditional storytelling to facilitate the development of systematic and policy responses.

Working with Indigenous groups helps long-term local interpretations to records and backs up the area's plans for distributing with climate change.

ROLE OF VISUALIZATION IN CLIMATE CHANGE

Data visualizations for climate change, it transforms difficult to learn datasets into simpler form, visualization helps the people to understand the structure clearly and analyze the impact of climate change. Representing complex data into visualization forms, it gives more clear about the data it leads to better understanding, take better decision making and meaningful solution. Climate change is a difficult process that affects various sectors, such as meteorology, ecology, economics, and public health. Even the well-trained professionals also facing lot of challenges why because large dataset and understanding of climate related data. This data is converted into accessible, engaging formats via visualization, facilitating comprehension of patterns, trends, and correlations for many audiences. Dashboard visualizations serve as instruments for policymakers, amalgamating environmental, economic, and social data to enhance decision-making processes.

Infographics and interactive tools make climate issues more accessible and actionable for the general populace. Visualization tools promote cooperation by offering a communal platform for study and discourse. Digital tools and interactive dashboards examine climatic data in real-time, develop scenarios based on diverse assumptions and identify the solutions and localized effects. Google Earth Engine with ArcGIS enable researchers and activists to collaboratively map deforestation and flood danger, thereby fostering a global, community-oriented approach to climate action. A thousand of years, line graph shows historical warning trends, while area charts show temperature changes. The rising of sea level it indicates that the different regions of country is in danger zone based on that the Maps of geographic information systems (GIS). Sankey maps show how carbon emissions move from one industry to another, it highlights the important areas where action needs to be needed. "Climate Change Visualization" displaying climate change data and patterns both clear and useful to a wide range of people.

THE ROLE OF CLIMATE COMMUNICATION IN PUBLIC AWARENESS

The importance of communication is to deliver the climate change data clearly to the end users, to change their minds, and start acting differently. Traditional climate science studies are long and complicated, which makes it hard for people who aren't experts to understand how bad and urgent climate problems are. Visualization fills in this gap by turning complicated data into stories that are easy to understand and interesting to read. Visualizations help with thinking by using the dual-coding theory, which says that hearing and seeing things together help with remembering and understanding things (Paivio, 1991). Often, time-lapse images that make you feel something, like how mountains melt or wildfires spread over time, have a bigger effect than words alone. Visual proof, such as satellite images and trends in deforestation, can also be a strong counter-narrative to climate misinformation and rejection (Leiserowitz et al., 2020). Global groups like the IPCC, NASA, and WWF use pictures a lot to make people aware of climate policies, stress how important they are, and get people to support them. No longer is talking about climate change just about giving people facts; it's also about getting them to move on an emotional and intellectual level.

Different types of data are used to represent the climate data visualizations:

1. **Temporal Data**: Temporal data is an information that is collected, observed, or changed throughout time. It is a main tool for analysing procedures that are continuously moving because it can discover patterns, trends, or events that are associated to time. Many fields, such as climate science, economics, healthcare, and the social sciences, use temporal data a lot

2. **Spatial Data:** Spatial data is also known as geospatial data. This approach to show information about the things surrounded on Earth and how it becomes shaped. It explains the geological features, such as land patterns, climate patterns, cities, vegetation can be traced in space. Spatial data is very helpful for both reviewing climate change and monitoring the environment.
3. **Categorical Data:** Data can be categorized into different categories. Most of the time it can be represented as not a number but sometimes it can be represented as a number (for example, numbered as 1, 2, 3). Categorical data can be used to represent the data visualization more clear and perform analysis, to compare and analyses distributions, frequencies, and relationships.
4. **Quantitative Data:** Information that is numerical in nature and represents quantities that may be measured is referred to as quantitative data. It is vital for extensive data analysis and scientific investigations because of its ability to be tallied, compared, or evaluated numerically. This ability is what distinguishes it from other data types. This category of data is utilized extensively in a variety of sectors, including engineering, economics, healthcare, and climate research.

OPEN-SOURCE DATASETS AND COLLABORATIVE PLATFORMS

Its reliable, open-source climate data, researchers and citizens who want to do something about climate change have transformed the way they work. Some important platforms are:

- NASA Earth Data: This site provides weather and terrain data from throughout the world, as well as satellite photographs.
- NOAA Climate Data Online (CDO): This site enables you examine weather, rain, and storm data from the past and virtually real time.
- Use Google Earth Engine to see a lot of satellite photos that show topics like water levels, urban development, and the loss of forests.
- Global Forest Watch utilizes satellite data to illustrate where trees are being chopped down and where fires are happening right now.

GitHub repositories and Kaggle datasets let data scientists, organizers, and students produce and share pictures. These kinds of open tools help everyone understand climate science better and allow individuals of all skill levels come up with fresh ideas (Gorelick et al., 2017).

PROPOSED SYSTEM: VISUALIZE THE CLIMATE CHANGE USING TABLEAU

Introduction to Tableau:

This chapter presents a proposed Tableau-based visualization tool for the purpose of communicating climate change scientific information through the use of geospatial mapping, infographic style and by using interactive and data-driven dashboard types of visualizations. The proposed Tableau-based visualization system include: increasing surface temperatures, changing precipitation patterns, melting glaciers, rising sea levels and the cascading effects of these phenomena on many ecosystems and human populations. Worksheets and dashboards are the two most important components of Tableau. Worksheets contain individual visualizations, and dashboards may contain one or more worksheets to display the data. Additionally, you can merge worksheets and dashboards into narratives to share specific insights with the audience in a presentation environment. Finally, you arrange all the worksheets, dashboards, and stories into workbooks, accessible through the Tableau desktop application, server, reader, or desktop version of the Tableau mobile app.

Dataset Description and Variable Selection:

The dataset used to illustrate the proposed Tableau-based visualization system in this chapter consists of a multi-country, multi-variable climate dataset covering the period from 2000 to 2023. The dataset contains information for 16 countries representing a range of geographic locations and levels of development. These countries include, but are not limited to: India, USA, China, Brazil, Germany, South Africa, Australia, and many more. The dataset consists of ten variables per country/year observation: average surface temperature (°C), CO_2 emissions per capita (tonnes), sea-level rise (mm), annual rainfall (mm), population, percentage of renewable energy use, number of extreme weather events, and forest area as a percentage of total land area. Each of these variables has been selected specifically to demonstrate the inter-related nature of climate change as described in the IPCC's framework for the assessment of climate change: temperature change is the primary indicator of global warming, CO_2 emissions are the principal driving force of global warming, sea-level rise and number of extreme weather events are the indicators of physical impacts of climate change, the percentage of renewable energy used is the indicator of mitigation progress, and forest area is the proxy for carbon sequestration. The inclusion of the population variable enables per capita normalization and comparison between countries that differ in size. It is important to note that the dataset is used in this chapter to demonstrate dashboard design principles and the communication capabilities of Tableau, and is not intended to be a significant contribution to the scientific understanding of climate change.

Dashboard Design and Chart Type Rationale:

Developing a Tableau dashboard involves creating four interconnected worksheets that make up one entire workbook. The choice of visualization was based upon the type of variable(s) being depicted, and the corresponding analysis.

The first worksheet utilizes a Dual-Axis Time Series Line chart to display the Average Temperature and CO2 Emission (per capita) for each country over time. With both variables being continuous and temporally ordered, a line chart would be suitable because the Dual-Axis format allows direct visual comparison between two different measurements that may be represented on separate scales without a loss of precision. In addition, there are country filters in place for individual countries, allowing users to view the individual trends of each country while also allowing users to observe other countries with respect to the change in temperature that corresponds to a higher-county's CO2 emissions in the data.

The second worksheet contains a geo-based map of the world created in Tableau that uses "country" as a geographic element and either average temperature or greenhouse gas (GHG) emissions per person, respectively, as a means of representing an average by the colour of the country displayed on it. The blue-to-red diverging colour scale makes it easy to see the different levels of each factor, where red represents the highest level for average temperature or average GHG emissions. Using this type of map allows users to see climatic data for an entire country at once, while also allowing users to compare the climate of different countries (if viewed together on the same map) and see significant differences between countries.

As an example of how the data represent significant variation even within the same country, there were temperatures from approximately 7.7 degrees Celsius (°C) to 31°C throughout the various years (between 2006 and 2010) and throughout the different areas of India, as recorded in the data, and GHG emissions per person during that time period ranged from approximately 3.8 tonnes to 19 tonnes.

In the Third Worksheet, a scatterplot is implemented as a means of displaying CO2 emissions per capita on the x-axis, average temperature on the y-axis, the size of the bubbles representing the country's share of renewable energy, and colour denoting which country it belongs to. The scatterplot is the best method for representing the correlational structure between emissions and temperature for countries around the world. It also allows us to examine the effect of renewable energy use on these two variables over time. In order to contextualise the points shown in the scatterplot, we have included a trend line with a 95% confidence interval to provide a statistical reference point, and in doing so, we have included an example of how to display analytical uncertainty within the dashboard.

The Fourth Worksheet displays a grouped bar chart comparing the proportion of renewable energy to the number of extreme weather events for each country during a user-selected year. Grouped bar charts were chosen for this data representation due to the fact that both of the variables are categorical (country specific) in nature; thus, the best means of analysing each country is by using a side-by-side format. In this way, the Fourth Worksheet can provide direct support to the policy-relevant question of whether countries that have invested more heavily in renewable energy are more likely to experience different frequencies of extreme weather events. However, when interpreting this cross-sectional data, due to the many variables influencing both renewable energy use and extreme weather events, we strongly recommend that you be cautious in your interpretation of the data.

Representation of Uncertainty:

Responsible climate data visualization includes communicating uncertainty. The proposed Tableau dashboard addresses uncertainty in a variety of ways. For example, on the trend line in the scatter plot, confidence intervals (shown by the bands around the trend line) are used to help users assess the statistical reliability of the observed relationship between temperature and emissions. Additionally, tooltips are configured to show the number of observations underlying each aggregated data point. This will indicate to users when the number of observations is small and therefore when they need to be cautious interpreting results. Last, there is a dashboard annotation layer that explains that the dataset used for illustration only and therefore should not be used as the basis for making causal inferences from cross-sectional country comparisons. The above design choices were made in accordance with ethical climate visualization best practices and follow a principle of providing informational transparency rather than using rhetorical persuasion (Druckman & Lupia, 2017).

KEY OBSERVATIONS AND INTERPRETIVE CONCLUSIONS FROM THE DASHBOARD

In analyzing the dataset using the proposed Tableau dashboard, we find some illustrative results. First, the time series visualizations display that consistently high per-capita CO_2 emissions countries (USA, Australia and Canada) have, on average, higher temperatures over the period of study than those countries that are below the global average, as represented in the dataset. Second, the scatter plot indicates that those countries with a higher share of renewable energy (greater bubble size) do not necessarily also show a decrease in per-capita emissions for the period covered by this dataset, indicating that for a number of economies, the transition to renewable energy is still in its infancy and emissions reductions occur after the addition of renewable capacity. Third, the grouped bar chart demonstrates significant variability in extreme weather event occurrences per country and per year, and shows no simple or linear relationship between renewable energy share and occurrences of such events, reflecting the complexity of the relationship between climatic impacts and their various causes. Collectively, these findings can only be described as illustrative, not conclusive, and should be interpreted with respect to the dataset's scope and the limitations of cross-country observational studies. The dashboard's primary value lies not in generating novel empirical findings but in demonstrating how a Tableau-based system can render multi-dimensional climate data explorable, interpretable, and communicable to non-specialist users.

Figure 6. Tableau desktop workspace environment

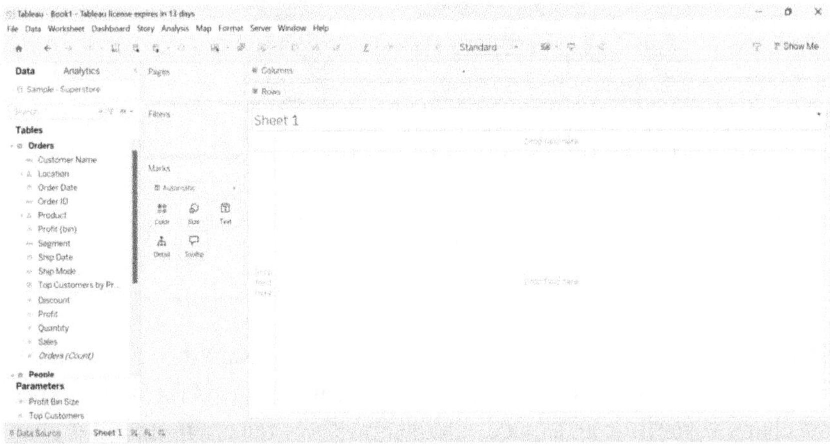

Source: Created by the authors using Tableau Desktop and Open-source climate datasets

Visualization Techniques:

Visualization tools are essential for comprehending and conveying climate change data. They facilitate the conversion of intricate statistics into understandable insights, so simplifying the interpretation of trends, the identification of patterns, and the formulation of informed judgments.

1. **Time Series Chart**: A Time series chart is the basic data visualization method that presents data points in chronological order, usually with the x-axis denoting

time and the y-axis indicating the variable of interest. This graphic is essential for discerning trends, patterns, and relationships throughout time, particularly in subjects like climate science, economics, stock markets, and any area where temporal data is crucial for study.

Figure 7. Global land temperature and uncertainty in the state

Source: Created by the authors using Tableau Desktop and Open-source climate datasets

2. **Heatmaps**: A heatmap uses color gradients to represent the strong values in a two-dimensional space or area. Heatmaps are mostly used for showing patterns that change over time and space, and it shows how the environment is changing rapidly. It includes changes in temperature, rainfall, air quality, rising sea levels and other environmental factors.
3. **Geospatial maps:** Geospatial maps are useful for showing climate change data because they show how different environmental factors, like temperature, sea level rise, rainfall, deforestation, and carbon emissions, are circulated out over diverse regions. This map uses spatial data, which data is linked to the specific geographic coordinates, to show how climate change is modifying different parts of the regions. These Geospatial maps merge the geographic and environmental data to visualize how climate-related events are extend out over time and space. It leads to lot of advantages to the end user perspectives like taking correct decisions, placing new policies, and increasing public awareness of that climate data.

Figure 8. Climate changes in India

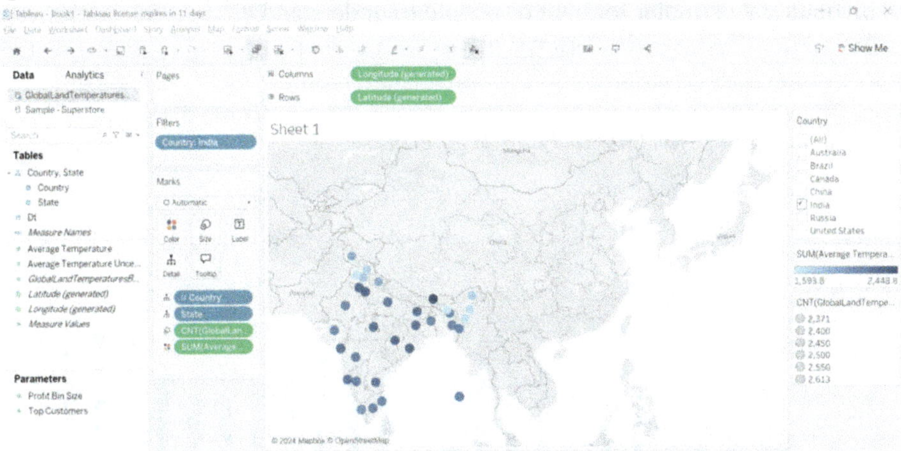

Source: Created by the authors using Tableau Desktop and Open-source climate datasets

4. **Scatter Plot:** A scatter plot is a more powerful and commonly used statistical tool for visualizing data that shows the relationship between two continuous variables. On a two-dimensional graph, each data point is shown as a dot (or point), and the location of each dot is based on the values of two variables. The x-axis (horizontal) shows one of the variables, and the y-axis (vertical) shows the other.

In the background of climate change data, scatter plots are predominantly effective for examining connections among various environmental variables, including temperature and carbon emissions, rainfall and scarcity existence, or sea level rise and global warming. They help scientists, policymakers, and the public understand how one climatic variable interacts or associates with another, as well as giving them ideas about patterns and causal links.

5. **Bar charts and column charts:** This Bar charts and column charts are the best chart to show the data in groups and compare with different groups of data. People use these bar chart and column chart a lot, to show climate change data because it clearly represents the data between different groups such as geographical locations and different types of climate variables. Both bar chart and column chart represent the data in terms of rectangular bars. The main difference between these two charts is bar charts use horizontal bars and column chart use vertical bars. These charts are important tools in the field of climate

change because they show trends, compare climate-related variables, and make difficult data simpler for a lot of people to understand.

Figure 9. Bar Chart for Global Land Temperature

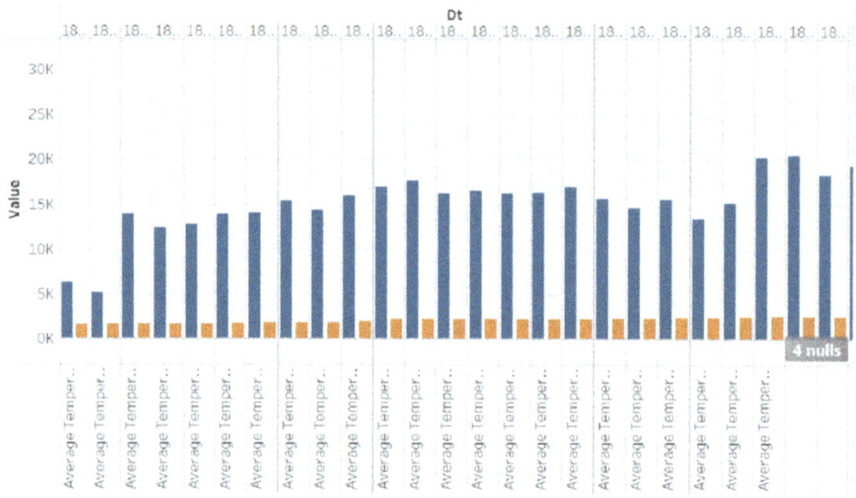

Source: Created by the authors using Tableau Desktop and Open-source climate datasets

6.Pie Charts: The circular/round graphs are called as pie charts. It shows the data as parts of a whole. This chart illustrates the distribution of greenhouse gas emissions, energy use by source or how different sectors contribute to climate effects. Pie charts are circular in shape it divides the data into slices. Each slice represents the different categories of data relevant to climate information.

Figure 10. Climate Changes in the overall country

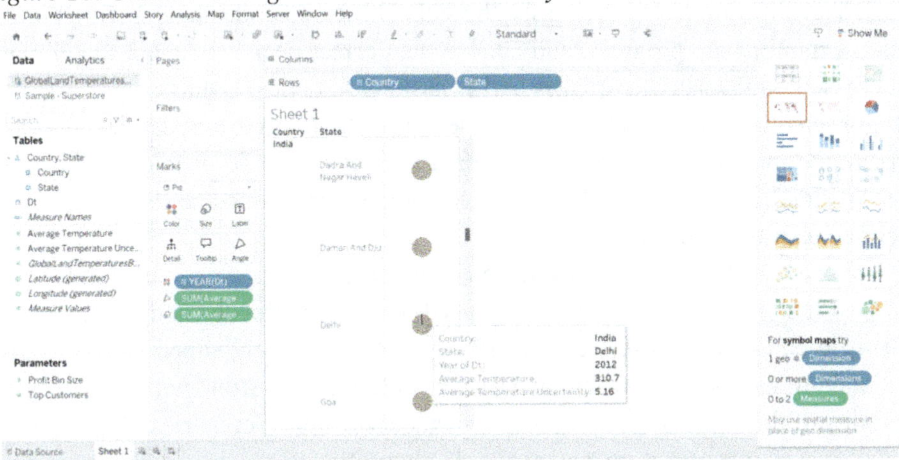

Source: Created by the authors using Tableau Desktop and Open-source climate datasets

New Trends in Climate Visualizations AI-driven

AI is improving the climate models and photos, to give more accurate and real time information to all kinds of people to easily understand the climate change effects. For more accurate prediction use deep learning models.

- NLP-to-Visualization Technologies: It shows the data in terms of visualization patterns.
- Generative AI: It generates the videos that depict how different conditions may trigger to rise the sea levels or cultivated crops to change.
- AI in Satellite Images: Microsoft's AI for Earth and IBM's Green Horizons use data from satellites periodically monitor the air quality and plant development (Rolnick et al., 2019).

These tools help the individual users to understand the climate data and take clear decisions.

DATA VISUALIZATION FOR CLIMATE SOLUTIONS

Climate change is one of the biggest problems in world. Before giving solution to that problem first understand the problem clearly then only suggest proper solution for that. Visualization plays a vital role to give easy understanding of data even though if it's a complex data sets it provides more insightful information to the end users. Visualization techniques assist the stakeholders to understand difficult interactions, monitor the progress, and choose the correct decisions about the climate data, based on the satellite images and emissions statistics. Data visualization is a way for scientists, policymakers, businesses and individuals to turn data into actions. Traditional models lack in prediction, data shortages etc. To overcome this issue use Artificial Intelligence (AI) is to solve complex problems and its predict more accurate results in the climate data.

Ensure that fossil fuels remain in the ground: The burning and combustion of fossil fuels, such as coal, oil, and gas will make the climate change worse. The scientific consensus, as reflected in IPCC reports, strongly supports a managed and rapid transition away from unabated fossil fuel combustion, with the pace and modality of this transition necessarily varying according to countries' development status, energy infrastructure, and available alternatives.

Invest in renewable energy: Instead of burning the fossil fuels, use the primary energy sources to clean and gain the renewable energy. Solar, wind, wave, tidal, and geothermal technologies comprise this category.

Transition to sustainable transportation: Aircraft, ships, and automobiles use fossil fuels. Reducing the use of cars, transitioning to electrified vehicles, and minimizing air travel will not only mitigate climate change but also reduce air pollution.

Help us maintain the warmth of our residences. It is a waste of money and a miserable experience in the winter for homes to be draughty and chilly. By insulating walls and roofs and transitioning from oil or gas furnaces to heat pumps, the government can assist households in heating their homes in an environmentally friendly manner.

Enhance natural ecosystems to increase carbon absorption: The natural world is quite effective at cleaning up our emissions. The planting of trees in the appropriate locations or the implementation of "rewilding" programs that return land to its natural state are both excellent places to start. This is because plants that engage in photosynthesis absorb carbon dioxide as they develop, thereby storing it in the soil's tissues.

Reduce the number of things that people consume: The decisions regarding transportation, clothes, food, and other aspects of our lifestyle all have varied effects on the environment. Often, corporations in the fashion and technology industries intentionally produce more than what is truly necessary. While reducing these foods is difficult, it is worth it. Reducing total consumption in more prosperous nations can aid in reducing the strain on the planet.

Preserve forests such as the Amazon: Forests are essential in the war against climate change, and their preservation is a critical climate solution. The destruction of forests on an industrial scale result in the destruction of gigantic trees that have the potential to absorb significant quantities of carbon. Nevertheless, companies deforest forests to facilitate the development of soya, palm oil, or animal husbandry. Governments can implement better legislation to prevent these practices.

Preserve the oceans: The oceans are responsible for the absorption of significant quantities of carbon dioxide from the atmosphere, which contributes to the stability of our climate. However, deep-sea mining, oil and gas extraction, and overfishing endanger numerous species. Ultimately, safeguarding the oceans and the life within them is a means of safeguarding ourselves from the effects of climate change.

Minimize the use of plastic: The process of extracting, refining, and converting oil into plastic (or polyester for garments) is surprisingly carbon intensive. Oil serves as the raw material to produce plastics. Incineration of a significant amount of plastic contributes to emissions, as it does not decompose rapidly in nature. The demand for plastic is increasing at an unprecedented rate, and by 2050, the creation and disposal of plastics will consume 17% of the global carbon budget. The Paris agreement requires us to maintain this emissions threshold.

Enhancing agriculture complements vegan diets: Individuals can significantly contribute to mitigating climate change by either reducing their consumption of meat and dairy products or adopting a completely vegan diet. This is currently one of the most effective real-time approaches available. Businesses and food merchants can enhance agricultural practices and boost the production of plant-based products to assist individuals transitioning.

ETHICAL CONSIDERATIONS IN CLIMATE VISUALIZATION

As climate images become more popular and have more of an impact, it is even more important to stick to moral standards. Using the wrong axis scale, only showing some data, or using colours that are too strong can change the meaning of scientific facts. Even if they don't mean to, these kinds of lies can hurt people's trust. It's also very important that things are easy to get to. Charts should be easy to read for people who are colour blind, have trouble reading, or can't get online. It is wrong to picture weak groups, like climate refugees or Indigenous people, without their permission and knowledge of their culture (Druckman & Lupia, 2017). Also, images should find a balance between what people need and what they hope for. If you think about all the bad things that could happen too much, you might get eco-anxiety and not want to get involved. Ethical design gives people the right information so they can act without being too much.

POLICY MAKING AND VISUALIZATION

Visualizations are a key part of making climate policy because they help leaders quickly understand hard problems. Dashboards put together information about health, the economy, and the environment so that people can make decisions based on facts.

- Risk maps help cities and towns get ready for storms, heat waves, and droughts, among other things.
- Scenario planning tools show what could happen if certain policy decisions are made.
- Cost-benefit charts show how much more money green energy makes than fossil fuels.

Hsu et al. (2014) show how the governments of Germany and Singapore use Tableau, ArcGIS, and Power BI to keep track of how well they are meeting their climate goals. Policies are more successful because it interacts the user to promotes openness and responsible to the people community.

CLIMATE CHANGE BEST PRACTICES USING VISUALIZING DATA

Avoid complexity: Use the clear image, include the additional features of that data points. To focus the content clearer and more precise.

Accurate interpretation: It relies on providing context, which includes baselines, reference points, and explanations of the data source.

Use of Colors: Choose the color correctly. Green color indicates the normal routine operation, Red color indicates the warning or caution to indicate through alarm notifications.

Interactivity: It is very difficult to provide the user details with interactive tools that allow them to different aspects especially in the complex datasets.

Storytelling: To help people grasp the human effect of climate change, use visualizations to present a captivating tale that ties facts to real-world implications.

CLIMATE EDUCATION THROUGH VISUALIZATION

A key aspect of SDG 13 (Climate Action) is teaching people about climate change. When people can see things, they learn better because it helps them grasp how things are connected.

- Use AR/VR applications like “Climate Quest” or “Tree VR” to understand how climate change truly impacts the globe.
- Story maps for students in grades K–12 that combine stories with real-world facts.
- Games that employ eco-points and carbon monitoring to urge people to change the way they act.

NASA's Earth Now and UNEP's Ozone App are two mobile apps that provide students real-time, changing data that they can use to study about pollution, global warming, and the ozone layer. Individuals learn better from the beginning when they use visual tools to educate. This produces the next generation of individuals who will fight for the environment.

CONCLUSION

In recent years, climate change has become a crucial worldwide risk factor, exerting enduring effects on the environment and civilizations. This research highlights the need of representing climate data using geographical maps, infographics, and interactive models, therefore connecting intricate scientific knowledge with public comprehension. By converting scientific data into engaging visual narratives, these visualizations adeptly communicate the urgent concerns of increasing global temperatures, changing weather patterns, glacier melt, rising the sea levels, and their effects on the ecosystems and societies. Instead of using static representations use the dynamic visualizations to allow the users to understand the complex climate data pattens to improve the comprehension and engagement in real time data sets. These kinds of case studies to focus the importance of innovative visual strategies to create the climate awareness, meaningful discussion leads to inspire the people to work together to solve the problems of climatic change.

Future Enhancements

Now a days the innovative technologies are implemented by the AI and Machine Learning (ML) functionalities to represent the flow of climatic trends, adaptive visualizations and customize the end user experience. Additionally use the Augmented Reality (AR) and Virtual Reality (VR) for real time simulations perspectives, to enable the user interaction to visualize the changes of climate changes in the real world. It combines to deliver the real time climate changes and actions to perform to take a quick decision-making and emergency alert actions. After taking a correct and strong decision, it allows the users to interact with the visual data patterns and sharing of information to the intended users. These information's are communicated to the educational sectors to create awareness of the climate changes and its consequences.

REFERENCES

Cai, J., Zhu, J., Dai, Q., Yang, Q., & Zhang, S. (2020). Sensitivity of a weather research and forecasting model to downscaling schemes in ensemble rainfall estimation. *Meteorological Applications*, *27*(1), e1806. DOI: 10.1002/met.1806

Druckman, J. N., & Lupia, A. (2017). Visual Communication in Politics. *Annual Review of Political Science*, *20*, 369–388.

Ford, J. D., Cameron, L., Rubis, J., Maillet, M., Nakashima, D., Willox, A. C., & Pearce, T. (2016). Including Indigenous Knowledge and Experience in IPCC Assessment Reports. *Nature Climate Change*, *6*(4), 349–353. DOI: 10.1038/nclimate2954

Gorelick, N., Hancher, M., Dixon, M., Ilyushchenko, S., Thau, D., & Moore, R. (2017). Google Earth Engine: Planetary-scale geospatial analysis for everyone. *Remote Sensing of Environment*, *202*, 18–27. DOI: 10.1016/j.rse.2017.06.031

Hsu, A., Emerson, J. W., Levy, M. A., de Sherbinin, A., Johnson, L., & Malik, O. (2014). *The 2014 Environmental Performance Index*. Yale Center for Environmental Law & Policy.

IPCC. Climate Change (2020). Synthesis Report. Cambridge University Press, Cambridge and New York

Jain, M., & Nagpal, A. (2020). Analysing the nexus of environmental performance and human well-being: An empirical study of selected emerging economies. *International Journal of Green Economics*, *14*(3), 266–277. DOI: 10.1504/IJGE.2020.111470

Leiserowitz, A., Maibach, E., Rosenthal, S., Kotcher, J., Ballew, M., Goldberg, M., & Gustafson, A. (2020). *Climate Change in the American Mind: April 2020*. Yale University and George Mason University.

Lun, Y., Liu, L., Cheng, L., Li, X., Li, H., & Xu, Z. (2021). Assessment of GCMs simulation performance for precipitation and temperature from CMIP5 to CMIP6 over the Tibetan Plateau. *International Journal of Climatology*, *41*(7), 3994–4018. DOI: 10.1002/joc.7055

Moser, S. C. (2010). Communicating climate change: History, challenges, process and future directions. *Wiley Interdisciplinary Reviews: Climate Change*, *1*(1), 31–53. DOI: 10.1002/wcc.11

Paivio, A. (1991). Dual Coding Theory: Retrospect and Current Status. *Canadian Journal of Psychology*, *45*(3), 255–27. DOI: 10.1037/h0084295

Pervin, L., & Gan, T. Y. (2021). Sensitivity of physical parameterization schemes in WRF model for dynamic downscaling of climatic variables over the MRB. *Journal of Water and Climate Change*, *12*(4), 1043–1058. DOI: 10.2166/wcc.2020.036

Rolnick, D., Donti, P. L., Kaack, L. H., Kochanski, K., Lacoste, A., Sankaran, K., . . . Bengio, Y. (2019). Tackling Climate Change with Machine Learning. arXiv preprint arXiv:1906.05433.

Tang, B., Hu, W., & Duan, A. (2021). Assessment of extreme precipitation indices over Indochina and South China in CMIP6 models. *Journal of Climate*, *34*(18), 7507–7524. DOI: 10.1175/JCLI-D-20-0948.1

Compilation of References

Abballe, A., Felip, E. D., Dellatte, E., Iacovella, N., Marra, V., Miniero, R., Valentini, S., & Ingelido, A. M. (2026). Polybrominated Diphenyl Ether (PBDE) Serum Concentrations in Italian Women of Reproductive Age. *Toxics*, *14*(1), 72. Advance online publication. DOI: 10.3390/toxics14010072 PMID: 41600621

Abdelbaset, A. E., Kwak, M. L., Nonaka, N., & Nakao, R. (2023). Human-biting ticks and zoonotic tick-borne pathogens in North Africa: Diversity, distribution, and trans-Mediterranean public health challenges. *One Health*, *16*, 100547. DOI: 10.1016/j.onehlt.2023.100547 PMID: 37363219

Acharya, B., Behera, A., Moharana, S., Prajapati, B. G., & Behera, S. (2025). Nanoparticle-Mediated Embryotoxicity: Mechanisms of Chemical Toxicity and Implications for Biological Development. *Chemical Research in Toxicology*, *38*(4), 521–541. DOI: 10.1021/acs.chemrestox.4c00472 PMID: 40105412

Adamopoulos, I. (2025). A Novel AI-based Modeling with Bias Classification Hybrid Risk Evaluation System for Confidence Enhanced Network Meta-Analysis of Occupational Hazards and Burnout Risk among Public Health Inspectors. Mesopotamian Journal of Artificial Intelligence in Healthcare. 219–233. DOI: 10.58496/MJAIH/2025/021

Adamopoulos, I., Lamnisos, D., Syrou, N., Boustras, G. (2022c). Training needs and quality of public health inspectors in Greece during the COVID-19 pandemic. European Journal of Public Health 2022, 32, ckac131.373. DOI: 10.1093/eurpub/ckac131.373

Adamopoulos, I., Syrou, N. (2024). Climate Change, Air Pollution, African Dust Impacts on Public Health and Sustainability in Europe, European Journal of Public Health, Volume 34, ckae144.1374. DOI: 10.1093/eurpub/ckae144.1374

Adamopoulos, I., Syrou, N., Lamnisos, D., Boustras, G. (2023). Cross-sectional nationwide study in occupational safety & health: Inspection of job risks context, burn out syndrome and job satisfaction of public health inspectors in the period of the COVID-19 pandemic in Greece. Safety Science, Elsevier, 2023, 158, 105960. . 105960DOI: 10.1016/j.ssci.2022

Adamopoulos, I., Syrou, N., Mpourazanis, G., Constantinidis, T. C., & Dounias, G. (2025a). The Association of the Global Climate Crisis with Environmental Risks and the Impact of Heat Stress on Occupational Safety, Health, and Hygiene. Med. Sci. Forum, 33, 2. DOI: 10.3390/msf2025033002

Adamopoulos, I., Lamnisos, D., Syrou, N., & Boustras, G. (2022a). Public health and work safety pilot study: Inspection of job risks, burn out syndrome and job satisfaction of public health inspectors in Greece. *Safety Science*, *2022*(147), 105592. DOI: 10.1016/j.ssci.2021.105592

Adamopoulos, I., Lamnisos, D., Syrou, N., & Boustras, G. (2022b). Inspection of job risks, burn out syndrome and job satisfaction of Greek public health inspectors. *Safety and Health at Work*, *2022*(13), S294. DOI: 10.1016/j.shaw.2021.12.1670

Adamopoulos, I., Syrou, N., Lamnisos, D., Valamontes, A., Karantonis, J. T., Tsirkas, P., & Dounias, G. (2026). *Classifying and mitigating occupational risks for public health inspectors in the context of the global climate crisis*. European Journal of Sustainable Development Research., DOI: 10.29333/ejosdr/17273

Adamopoulos, I., Valamontes, A., Tsirkas, P., & Dounias, G. (2025b). Predicting Workplace Hazard, Stress and Burnout Among Public Health Inspectors: An AI-Driven Analysis in the Context of Climate Change. *European Journal of Investigation in Health, Psychology and Education*, *15*(5), 65. DOI: 10.3390/ejihpe15050065 PMID: 40422294

Adam-Poupart, A., Labrèche, F., Smargiassi, A., Duguay, P., Busque, M. A., Gaucher, C., & Fournier, M. (2015). Climate change and occupational health and safety in a temperate climate: Potential impacts on vector-borne diseases. *International Journal of Environmental Research and Public Health*, *12*(10), 13232–13250.

Ahmed, A., Ali, Y., Salim, B., Dietrich, I., & Zinsstag, J. (2022). Epidemics of Crimean-Congo Hemorrhagic Fever (CCHF) in Sudan between 2010 and 2020. *Microorganisms*, *10*(5), 5. Advance online publication. DOI: 10.3390/microorganisms10050928 PMID: 35630372

Alahmad, B., Kessler, W., Alwadi, Y., Schwartz, J., Wagner, G. R., & Michaels, D. (2025). A nationwide analysis of heat and workplace injuries in the United States. *Environmental Health : A Global Access Science Source*, *24*(1), 65. DOI: 10.1186/s12940-025-01231-1 PMID: 41047384

Albrecht, G. (2005). Solastalgia: A new concept in health and identity. *PAN: Philosophy Activism Nature*, *3*, 41–55.

Alfaro-Serrano, D., Balantrapu, T., Chaurey, R., Goicoechea, A., & Verhoogen, E. (2021). Interventions to Promote Technology Adoption in Firms: A Systematic Review. *Campbell Systematic Reviews*, *17*(4), e1181. DOI: 10.1002/cl2.1181 PMID: 36950339

Allen, T., Murray, K. A., Zambrana-Torrelio, C., Morse, S. S., Rondinini, C., Di Marco, M., Breit, N., Olival, K. J., & Daszak, P. (2017). Global hotspots and correlates of emerging zoonotic diseases. *Nature Communications*, *8*(1), 1124. DOI: 10.1038/s41467-017-00923-8 PMID: 29066781

Altizer, S., Ostfeld, R. S., Johnson, P. T., Kutz, S., & Harvell, C. D. (2013). Climate change and infectious diseases: From evidence to a predictive framework. *Science*, *341*(6145), 514–519. DOI: 10.1126/science.1239401 PMID: 23908230

American College of Obstetricians and Gynecologists. (2018). Optimizing postpartum care. *ACOG Committee Opinion No. 736. Obstetrics and Gynecology*, *131*(5), e140–e150. https://www.acog.org/clinical/clinical-guidance/committee-opinion/articles/2018/05/optimizing-postpartum-care. DOI: 10.1097/AOG.0000000000002633 PMID: 29683911

American Psychological Association. (2019). About EE and why it matters. Retrieved from https://naaee.org/about-us/about-ee-and-why-it-matters

Amini, M., Ghavamabadi, L., Rangkooy, H., & Dehaghi, B. (2021). Climate change and its effects on farm workers. *Environmental Health Engineering and Management*, *8*(3), 179–185. DOI: 10.34172/EHEM.2021.21

Amoadu, M., Ansah, E. W., Sarfo, J. O., & Hormenu, T. (2023). Impact of climate change and heat stress on workers' health and productivity: A scoping review. *The Journal of Climate Change and Health*, *12*, 100249. Advance online publication. DOI: 10.1016/j.joclim.2023.100249

Ananthakumar, B., & Raghavan, R. (2026). Social and rural determinants of glycemic control in type 2 diabetes mellitus. *The Indian Journal of Medical Research*, *163*(2), 231–236. DOI: 10.25259/IJMR_1884_2025 PMID: 41949130

Anderer, S. (2025). Beyond Genes: Human Exposome Project to Tackle External Drivers of Disease. *Journal of the American Medical Association*, *334*(1), 7–8. DOI: 10.1001/jama.2025.8666 PMID: 40445605

Anderson, G. B., & Bell, M. L. (2011). Heat waves in the United States: Mortality risk during heat waves and effect modification by heat wave characteristics in 43 U.S. communities. *Environmental Health Perspectives*, *119*(2), 210–218. DOI: 10.1289/ehp.1002313 PMID: 21084239

Andonotopo, W., Bachnas, M. A., Dewantiningrum, J., Pramono, M. B. A., Bernolian, N., Yeni, C. M., Wiradnyana, A. A. G. P., Sanjaya, I. N. H., Akbar, M. I. A., Darmawan, E., Sulistyowati, S., Stanojevic, M., & Kurjak, A. (2025). The fetal exposome and Preterm Birth: A systematic synthesis of environmental exposures and multi-omics evidence. *Journal of Perinatal Medicine*. Advance online publication. DOI: 10.1515/jpm-2025-0231 PMID: 41242981

Angol, B. (2024). Comparison between WBGT app prototype and WBGT monitor to assess heat stress risk in groundskeeping in an eastern North Carolina university setting.

Ansah, E. W., Ankomah-Appiah, E., Amoadu, M., & Sarfo, J. O. (2021). Climate change, health and safety of workers in developing economies: A scoping review. *The Journal of Climate Change and Health*, *3*, 100034. Advance online publication. DOI: 10.1016/j.joclim.2021.100034

Antoniou, E. E., & Otter, R. (2024). Phthalate exposure and neurotoxicity in children: A systematic review and meta-analysis. *International Journal of Public Health*, *69*, 1606802. DOI: 10.3389/ijph.2024.1606802 PMID: 38590582

Antu, U. B., Roy, T. K., Kulsum, T. I., Mitu, P. R., Ismail, Z., Arifin, M., Datta, M., Hossain, S. A., Islam, M. S., Mahiddin, N. A., Al Bakky, A., Hossin, S., Islam, S., & Idris, A. M. (2025). Role of humic acid for climate change adaptation measures to boost up sustainable agriculture and soil health: A potential review. *International Journal of Biological Macromolecules*, *313*, 144043. DOI: 10.1016/j.ijbiomac.2025.144043 PMID: 40350124

Applebaum, K. M., Graham, J., Gray, G. M., LaPuma, P., McCormick, S. A., Northcross, A., & Perry, M. J. (2016). An overview of occupational risks from climate change. *Current Environmental Health Reports*, *3*(1), 13–22. DOI: 10.1007/s40572-016-0081-4 PMID: 26842343

Assis, M., Barcella, R., Padilha, J., Pohl, H., & Krug, S. (2020). Health problems in agricultural workers occupationally exposed to pesticides. *Revista Brasileira de Medicina do Trabalho : Publicação Oficial da Associaação Nacional de Medicina do Trabalho-ANAMT, 18*(03), 352–363. DOI: 10.47626/1679-4435-2020-532 PMID: 33597986

Badham, J., Kee, F., & Hunter, R. F. (2018). Simulating Network Intervention Strategies: Implications for Adoption of Behaviour. *Network Science, 6*(2), 265–280. DOI: 10.1017/nws.2018.4

Baek, S., & Choi, E. H. (2025). Climate change health risks and workplace protective strategies for construction workers. *Scientific Reports, 16*(1), 3423. DOI: 10.1038/s41598-025-33471-z PMID: 41422155

Baethge, C., Goldbeck-Wood, S., & Mertens, S. (2019). SANRA—A scale for the quality assessment of narrative review articles. *Research Integrity and Peer Review, 4*(1), 5. Advance online publication. DOI: 10.1186/s41073-019-0064-8 PMID: 30962953

Bagavathiannan, M., & Davis, A. (2018). An ecological perspective on managing weeds during the great selection for herbicide resistance. *Pest Management Science, 74*(10), 2277–2286. DOI: 10.1002/ps.4920 PMID: 29569406

Bai, J., Johar, S. B., Mohammad, N. B., & Xia, J. (2026). Traditional village protection and sustainable development: Knowledge graph construction and hotspot evolution. *Sustainable Futures : An Applied Journal of Technology, Environment and Society, 11*, 101577. DOI: 10.1016/j.sftr.2025.101577

Baker, R. E., Mahmud, A. S., Miller, I. F., Rajeev, M., Rasambainarivo, F., Rice, B. L., Takahashi, S., Tatem, A. J., Wagner, C. E., Wang, L.-F., Wesolowski, A., & Metcalf, C. J. E. (2022). Infectious disease in an era of global change. *Nature Reviews. Microbiology, 20*(4), 193–205. DOI: 10.1038/s41579-021-00639-z PMID: 34646006

Bakshi, A., Doren, A., Berens, A., Jacob, C., Omari, A., Foreman, A., & Chew, G. (2025). Sociodemographic and housing factors for flood and mold damage in the Greater New Orleans Area after Hurricane Ida. *Disaster Medicine and Public Health Preparedness, 19*, 19. DOI: 10.1017/dmp.2025.13 PMID: 40556425

Balducci, S., D'Errico, V., Haxhi, J., Sacchetti, M., Orlando, G., Cardelli, P., Vitale, M., Bollanti, L., Conti, F., Zanuso, S., Nicolucci, A., & Pugliese, G.Balducci, Stefano, Valeria D'Errico, Jonida Haxhi, Massimo Sacchetti, Giorgio Orlando, Patrizia Cardelli, Martina Vitale, Lucilla Bollanti, Francesco Conti, Silvano Zanuso, Antonio Nicolucci, Giuseppe Pugliese, and for the Italian Diabetes and Exercise Study 2. (2017). Effect of a Behavioral Intervention Strategy for Adoption and Maintenance of a Physically Active Lifestyle: The Italian Diabetes and Exercise Study 2 (IDES_2). *Diabetes Care*, *40*(11), 1444–1452. DOI: 10.2337/dc17-0594 PMID: 28821576

Bañuls, P., Galán-Relaño, Á., Romero-Salmoral, A., Gómez-Gascón, L., Tarradas, C., Astorga Márquez, R. J., Luque, I., & Huerta, B. (2024). Knowledge, perception and attitude about Crimean-Congo Haemorrhagic Fever (CCHF) among veterinary students and professionals of Spain. *Comparative Immunology, Microbiology and Infectious Diseases*, *115*, 102260. DOI: 10.1016/j.cimid.2024.102260 PMID: 39476521

Bao, M., Nobile, F., Koureas, M., Bottini, I., Fomenko, E., Brimicombe, C., Tirado, V., Mouchtouri, V., Hadjichristodoulou, C., Ljungman, P., Chersich, M., Jackson, D., Luchters, S., Stephansson, O., Stafoggia, M., & de Bont, J. (2026). Trimester-specific exposure to multiple heat indicators and adverse birth outcomes across four European countries. *Environmental Research*, *296*, 124092. DOI: 10.1016/j.envres.2026.124092 PMID: 41740699

Barn, P., Keefe, A., Slot, N., Jardine, K., Ziembicki, S., Telfer, J., Palmer, A. L., & Peters, C. (2020). Canada Should Move Toward Adopting Harmonized Evidence-Based OELs to Consistently and Adequately Protect Workers. *Annals of Work Exposures and Health*, *65*(4), 367–372. DOI: 10.1093/annweh/wxaa110 PMID: 33336241

Barouki, R. (2024). A toxicological perspective on climate change and the exposome. *Frontiers in Public Health*, *12*, 1361274. Advance online publication. DOI: 10.3389/fpubh.2024.1361274 PMID: 38651121

Basu, R., Malig, B., & Ostro, B. (2010). High ambient temperature and the risk of preterm delivery. *American Journal of Epidemiology*, *172*(10), 1108–1117. DOI: 10.1093/aje/kwq170 PMID: 20889619

Battaglia, M., Pasetti, E., Bianchi, L., & Frey, M. (2020). Innovation in managing sustainability: A tentative integration of accounting for employee health and safety. *Sinergie Italian Journal of Management*, *38*(1), 73–101.

Battisti, E., Urach, K., Hodžić, A., Fusani, L., Hufnagl, P., Felsberger, G., Ferroglio, E., & Duscher, G. G. (2020). *Zoonotic Pathogens in Ticks from Migratory Birds, Italy—Volume 26, Number 12—December 2020—Emerging Infectious Diseases journal—CDC*. DOI: 10.3201/eid2612.181686

Beck, N., Balanay, J. A. G., & Johnson, T. (2018). Assessment of occupational exposure to heat stress and solar ultraviolet radiation among groundskeepers in an eastern North Carolina university setting. *Journal of Occupational and Environmental Hygiene*, *15*(2), 105–116. DOI: 10.1080/15459624.2017.1392530 PMID: 29090983

Bekkar, B., Pacheco, S., Basu, R., & DeNicola, N. (2020). Association of air pollution and heat exposure with preterm birth, low birth weight, and stillbirth in the US: A systematic review. *JAMA Network Open*, *3*(6), e208243. DOI: 10.1001/jamanetworkopen.2020.8243 PMID: 32556259

Belij-Rammerstorfer, S., Limon, G., Maze, E. A., Hannant, K., Hughes, E., Tchakarova, S. R., Alexandrov, T., Mmbaga, B. T., Willett, B., Booth, G., Lyons, N. A., Baker, N., Thomas, K. M., Wright, D., Saunders, J., Browning, C., Wilsden, G., Carroll, M., Hewson, R., & Ludi, A. B. (2022). Development of anti-Crimean-Congo hemorrhagic fever virus Gc and NP-specific ELISA for detection of antibodies in domestic animal sera. *Frontiers in Veterinary Science*, *9*, 913046. Advance online publication. DOI: 10.3389/fvets.2022.913046 PMID: 36090164

Belobo, J. T. E., Kenmoe, S., Kengne-Nde, C., Emoh, C. P. D., Bowo-Ngandji, A., Tchatchouang, S., Wobessi, J. N. S., Mikangue, C. A. M., Tazokong, H. R., Bebey, S. R. K., Noura, E. A., Ka'e, A. C., Simo, R. E. G., Modiyinji, A. F., Ngongang, D. T., Che, E., Kenfack, S., Nzukui, N. D., Adjia, N. A., & Njouom, R. (2021). Worldwide epidemiology of Crimean-Congo Hemorrhagic Fever Virus in humans, ticks and other animal species, a systematic review and meta-analysis. *PLoS Neglected Tropical Diseases*, *15*(4), e0009299. DOI: 10.1371/journal.pntd.0009299 PMID: 33886556

Benedek, D. M., Fullerton, C., & Ursano, R. J. (2007). First responders: Mental health consequences of natural and human-made disasters for public health and public safety workers. *Annual Review of Public Health*, *28*(1), 55–68. DOI: 10.1146/annurev.publhealth.28.021406.144037 PMID: 17367284

Bennett, C. M., & McMichael, A. J. (2010). Non-heat related impacts of climate change on working populations. *Global Health Action*, *3*(1), 5640. DOI: 10.3402/gha.v3i0.5640 PMID: 21191440

Berger, S. E., Ordway, M. R., Schoneveld, E., Lucchini, M., Thakur, S., Anders, T., Natale, L., & Barnett, N. (2023). The impact of extreme summer temperatures in the United Kingdom on infant sleep: Implications for learning and development. *Scientific Reports*, *13*(1), 10061. DOI: 10.1038/s41598-023-37111-2 PMID: 37344536

Berisha, V., Hondula, D. M., Roach, M., White-Newsome, J. L., McCormick, S., Marnin, M. L., & Jolliet, O. (2017). Assessing adaptation strategies for extreme heat: A public health evaluation of cooling centers in Maricopa County, Arizona. *Weather, Climate, and Society*, *9*(1), 71–80. DOI: 10.1175/WCAS-D-16-0033.1

Bernard, C., Joly Kukla, C., Rakotoarivony, I., Duhayon, M., Stachurski, F., Huber, K., Giupponi, C., Zortman, I., Holzmuller, P., Pollet, T., Jeanneau, M., Mercey, A., Vachiery, N., Lefrançois, T., Garros, C., Michaud, V., Comtet, L., Despois, L., Pourquier, P., & Vial, L. (2024). Detection of Crimean–Congo haemorrhagic fever virus in Hyalomma marginatum ticks, southern France, May 2022 and April 2023. *Eurosurveillance*, *29*(6). Advance online publication. DOI: 10.2807/1560-7917.ES.2024.29.6.2400023 PMID: 38333936

Bernstein, A. S., Sun, S., Weinberger, K. R., Spangler, K. R., Sheffield, P. E., & Wellenius, G. A. (2022). Warm season and emergency department visits to U.S. children's hospitals. *Environmental Health Perspectives*, *130*(1), 017001. DOI: 10.1289/EHP8083 PMID: 35044241

Berry, H. L., Bowen, K., & Kjellstrom, T. (2010). Climate change and mental health: A causal pathways framework. *International Journal of Public Health*, *55*(2), 123–132. DOI: 10.1007/s00038-009-0112-0 PMID: 20033251

Bethune, S. (2019). Gen Z more likely to report mental health concerns. *Monitor on Psychology*, *45*(3), 226–234.

Bhati, N., & Sheth, A. (2026). Occupational heat stress and adaptation among outdoor workers: A narrative review of global evidence and policy responses. *International Journal of Biometeorology*, *70*(1), 19. DOI: 10.1007/s00484-025-03101-4 PMID: 41501481

Bhuvana, G. N., Umesh, S. M., & Raghavan, R. (2026, January). (2025). Expanding the frontiers of Schistosomiasis Treatment: The promise of Nanoparticle-Based Therapies. *Journal of Vector Borne Diseases*, *63*(1), 16–26. Advance online publication. DOI: 10.4103/jvbd.jvbd_180_23 PMID: 40662683

Biabani, A., Falahati, M., Alimohammadi, I., Zokaei, M., Jalilian, H., Dehghani, A., & Majareh, M. N. (2021). Validation of heat pressure assessment (HPA) method and WBGT index in Iranian South Oil Company. *Journal of Health and Safety at Work*, *11*(1), 106–116.

Bianco, G., Espinoza-Chávez, R. M., Ashigbie, P. G., Junio, H., Borhani, C., Miles-Richardson, S., & Spector, J. (2024). Projected impact of climate change on human health in low- and middle-income countries: A systematic review. *BMJ Global Health*, *8*(Suppl 3), e015550. Advance online publication. DOI: 10.1136/bmjgh-2024-015550 PMID: 39357915

Bidaisee, S., & Macpherson, C. N. L. (2014). Zoonoses and One Health: A Review of the Literature. *Journal of Parasitology Research*, *874345*, 1–8. Advance online publication. DOI: 10.1155/2014/874345 PMID: 24634782

Biradar, S. P., & Raghavan, R. (2026). The Paternal Link to Congenital Heart Disease: Phthalate Exposure Alters Sperm Gene Expression and Increases Offspring Cardiac Risk. *The Multifactorial Etiology of Congenital Heart Disease: Insights from Teratology, Genetics, and Exposomics*, 99–130. DOI: 10.4018/979-8-3373-3735-7.ch004

Black, C., Tesfaigzi, Y., Bassein, J. A., & Miller, L. A. (2017). Wildfire smoke exposure and human health: Significant gaps in research for a growing public health issue. *Environmental Toxicology and Pharmacology*, *55*, 186–195. DOI: 10.1016/j.etap.2017.08.022 PMID: 28892756

Böhm, S., Misoczky, M. C., & Moog, S. (2012). Greening Capitalism? A Marxist Critique of Carbon Markets. *Organization Studies*, *33*(11), 1617–1638. DOI: 10.1177/0170840612463326

Borba, V., Carrera-Bastos, P., Hernández, A. L., & Shoenfeld, Y. (2026). Climate Change and the Exposome in Autoimmune Rheumatic Diseases. *Rheumatic Diseases Clinics of North America*, *52*(1), 13–31. DOI: 10.1016/j.rdc.2025.09.004 PMID: 41265939

Borromeo, A. S., Manaloto, A. M., Vicedo, R. V., Antonio, R., Cura, J. D., & Wider, W. (2025). The rise of virtual care in nursing practice: A global bibliometric review. *Journal of Nursing Regulation*, *16*(3), 141–149. DOI: 10.1016/j.jnr.2025.08.013

Bosch, J., Muñoz, M. J., Martínez, M., de la Torre, A., & Estrada-Peña, A. (2013). Vector-Borne Pathogen Spread Through Ticks on Migratory Birds: A Probabilistic Spatial Risk Model for South-Western Europe. *Transboundary and Emerging Diseases*, *60*(5), 403–415. DOI: 10.1111/j.1865-1682.2012.01358.x PMID: 22781365

Bouchama, A., & Knochel, J. P. (2002). Heat stroke. *The New England Journal of Medicine*, *346*(25), 1978–1988. DOI: 10.1056/NEJMra011089 PMID: 12075060

Boulanger, N., Aran, D., Maul, A., Camara, B. I., Barthel, C., Zaffino, M., Lett, M.-C., Schnitzler, A., & Bauda, P. (2024). Multiple factors affecting Ixodes ricinus ticks and associated pathogens in European temperate ecosystems (northeastern France). *Scientific Reports, 14*(1), 9391. DOI: 10.1038/s41598-024-59867-x PMID: 38658696

Bourgeade, A., & Marchou, B. (2003). Fièvre jaune, dengue, encéphalite japonaise et virose West Nile, 4 arboviroses majeures. *Médecine et Maladies Infectieuses, 33*(8), 385–395. DOI: 10.1016/S0399-077X(03)00220-8

Boushab, B. M., Kelly, M., Kébé, H., Bollahi, M. A., & Basco, L. K. (2020). Crimean-Congo Hemorrhagic Fever, Mauritania. *Emerging Infectious Diseases, 26*(4), 817–818. DOI: 10.3201/eid2604.191292 PMID: 32187505

Bowers, B. J. (2021). Sustainability reports based on GRI 403 occupational safety and health 2018 guidelines.

Bowleg, L. (2012). The problem with the phrase women and minorities: Intersectionality-an important theoretical framework for public health. *American Journal of Public Health, 102*(7), 1267–1273. DOI: 10.2105/AJPH.2012.300750 PMID: 22594719

Bowler, D. E., Buyung-Ali, L., Knight, T. M., & Pullin, A. S. (2010). Urban greening to cool towns and cities: A systematic review of the empirical evidence. *Landscape and Urban Planning, 97*(3), 147–155. DOI: 10.1016/j.landurbplan.2010.05.006

Braglia, M., Paco, F., Frosolini, M., Gabbrielli, R., & Marrazzini, L. (2023). A Novel Tool for Preliminary Risk Assessment of Climate Change on Workers' Health and Safety in Outdoor Worksites, 1514–1521.

Braveman, P., & Gruskin, S. (2003). Defining equity in health. *Journal of Epidemiology and Community Health, 57*(4), 254–258. DOI: 10.1136/jech.57.4.254 PMID: 12646539

Broadbent, A. M., Declet-Barreto, J., Krayenhoff, E. S., Harlan, S. L., & Georgescu, M. (2022). Targeted implementation of cool roofs for equitable urban adaptation to extreme heat. *The Science of the Total Environment, 811*, 151326. DOI: 10.1016/j.scitotenv.2021.151326 PMID: 34757097

Brooks, S. K., & Greenberg, N. (2023). Climate Change Effects on Mental Health: Are There Workplace Implications? *Occupational Medicine (Oxford, England), 73*(3), 133–137. DOI: 10.1093/occmed/kqac100 PMID: 36170162

Broster, J., Boutsalis, P., Gill, G., & Preston, C. (2022). The extent of herbicide resistance in *Lolium rigidum* Gaud. (annual ryegrass) across south-eastern Australia as determined from random surveys. *Australian Journal of Agricultural Research, 73*(11), 1308–1317.

Broster, J., Jalaludin, A., Widderick, M., Chambers, A., & Walsh, M. (2023). Herbicide Resistance in Summer Annual Weeds of Australia's Northern Grains Region. *Agronomy (Basel), 13*(7), 1862. DOI: 10.3390/agronomy13071862

Brown, J., Ahmed, N., Biel, M., Patchen, L., Rethy, J., Thomas, A., & Arem, H. (2023). Considerations in implementation of social risk factor screening and referral in maternal and infant care in Washington, DC: A qualitative study. *PLoS One, 18*(4), e0283815. Advance online publication. DOI: 10.1371/journal.pone.0283815 PMID: 37053233

Burrows, K., Denckla, C. A., Hahn, J., Schiff, J. E., Okuzono, S. S., Randriamady, H., Mita, C., Kubzansky, L. D., Koenen, K. C., & Lowe, S. R. (2024). A systematic review of the effects of chronic, slow-onset climate change on mental health. *Nature Mental Health, 2*(2), 228–243. DOI: 10.1038/s44220-023-00170-5 PMID: 41098556

Burza, S., Croft, S. L., & Boelaert, M. (2018). Leishmaniasis. *Lancet, 392*(10151), 951–970. DOI: 10.1016/S0140-6736(18)31204-2 PMID: 30126638

Bustos, D., Guedes, J. C., Baptista, J. S., Vaz, M. P., Costa, J. T., & Fernandes, R. J. (2021). Applicability of physiological monitoring systems within occupational groups: A systematic review. *Sensors (Basel), 21*(21), 7249. DOI: 10.3390/s21217249 PMID: 34770556

Butler-Dawson, J., Krisher, L., Yoder, H., Dally, M., Sorensen, C., Johnson, R., Asensio, C., Cruz, A., Johnson, E. C., Carlton, E. J., Tenney, L., Asturias, E. J., & Newman, L. (2019). Evaluation of heat stress and cumulative incidence of acute kidney injury in sugarcane workers in Guatemala. *International Archives of Occupational and Environmental Health, 92*(7), 977–990. DOI: 10.1007/s00420-019-01426-3 PMID: 30997573

Cabi. (2025). Research into zoonotic disease risks requires a One Health approach. *CABI.Org*. https://www.cabi.org/news-article/research-into-zoonotic-disease-risks-requires-a-one-health-approach/

Cai, J., Zhu, J., Dai, Q., Yang, Q., & Zhang, S. (2020). Sensitivity of a weather research and forecasting model to downscaling schemes in ensemble rainfall estimation. *Meteorological Applications, 27*(1), e1806. DOI: 10.1002/met.1806

Campbell, L. P., Luther, C., Moo-Llanes, D., Ramsey, J. M., Danis-Lozano, R., & Peterson, A. T. (2015). Climate change influences on global distributions of dengue and chikungunya virus vectors. *Philosophical Transactions of the Royal Society B*, 370(1665), 20140135.

Carlson, C. J., Albery, G. F., Merow, C., Trisos, C. H., Zipfel, C. M., Eskew, E. A., Olival, K. J., Ross, N., & Bansal, S. (2022). Climate change increases cross-species viral transmission risk. *Nature*, *607*(7918), 555–562. DOI: 10.1038/s41586-022-04788-w PMID: 35483403

Carroll, S. R., Garba, I., Figueroa-Rodríguez, O. L., Holbrook, J., Lovett, R., Materechera, S., Parsons, M., Raseroka, K., Rodriguez-Lonebear, D., Rowe, R., Sara, R., Walker, J. D., Anderson, J., & Hudson, M. (2020). The CARE principles for Indigenous data governance. *Data Science Journal*, *19*, 43. https://datascience.codata.org/articles/dsj-2020-043. DOI: 10.5334/dsj-2020-043

Carter, A. W., Zaitchik, B. F., Gohlke, J. M., Wang, S., & Richardson, M. B. (2020). Methods for estimating wet bulb globe temperature from remote and low-cost data: A comparative study in Central Alabama. *GeoHealth*, *4*(5), e2019GH000231.

Carvalho, B. M., Rangel, E. F., Ready, P. D., & Vale, M. M. (2015). Ecological Niche Modelling Predicts Southward Expansion of Lutzomyia (Nyssomyia) flaviscutellata (Diptera: Psychodidae: Phlebotominae), Vector of Leishmania (Leishmania) amazonensis in South America, under Climate Change. *PLoS One*, *10*(11), e0143282. DOI: 10.1371/journal.pone.0143282 PMID: 26619186

Cascio, W. E. (2018). Wildland fire smoke and human health. *The Science of the Total Environment*, *624*, 586–595. DOI: 10.1016/j.scitotenv.2017.12.086 PMID: 29272827

Casey, J. A., Fukurai, M., Hernández, D., Balsari, S., & Kiang, M. V. (2020). Power outages and community health: A narrative review. *Current Environmental Health Reports*, *7*(4), 371–383. DOI: 10.1007/s40572-020-00295-0 PMID: 33179170

Cavalli, L. S. (2025). Incorporating occupational health and safety into One Health approaches to aquaculture. *Journal of Agromedicine*, *30*(2), 214–220. DOI: 10.1080/1059924X.2024.2446256 PMID: 39720893

Cazé, A., Lacerda, A., Lüders, D., Conto, J., Marques, J., & Leroux, T. (2018). Perception of the Quality of Life of Tobacco Growers Exposed to Pesticides: Emphasis on Health, Hearing, and Working Conditions. *International Archives of Otorhinolaryngology*, 23(01), 050–059.

Cerceo, E., Saxer, K., Grossman, L., Shapley-Quinn, K., & Feldman-Winter, L. (2024). The climate crisis and breastfeeding: Opportunities for resilience. *Journal of Human Lactation : Official Journal of International Lactation Consultant Association, 40*(1), 33–50. DOI: 10.1177/08903344231216726 PMID: 38158719

Chalghaf, B., Chemkhi, J., Mayala, B., Harrabi, M., Benie, G. B., Michael, E., & Ben Salah, A. (2018). Ecological niche modeling predicting the potential distribution of Leishmania vectors in the Mediterranean basin: Impact of climate change. *Parasites & Vectors, 11*(1), 461. DOI: 10.1186/s13071-018-3019-x PMID: 30092826

Charlson, F., Ali, S., Benmarhnia, T., Pearl, M., Massazza, A., Augustinavicius, J., & Scott, J. G. (2021). Climate Change and Mental Health: A Scoping Review. *International Journal of Environmental Research and Public Health, 18*(9), 4486. Advance online publication. DOI: 10.3390/ijerph18094486 PMID: 33922573

Cheng, V., Sayre, E. C., Amiri, N., Cragg, J. J., Harrison, M., Proulx, L., & De Vera, M. A. (2026). Perinatal use of targeted synthetic DMARDs and maternal-neonatal outcomes in autoimmune disease: A population-based cohort study. *Journal of Autoimmunity, 159*, 103538. DOI: 10.1016/j.jaut.2026.103538 PMID: 41780201

Chen, K. Y., Chan, H. C., & Chan, C. M. (2025). Are dexamethasone implants a safe and effective breakthrough in uveitis treatment? A systematic review and meta-analysis. *Steroids, 223*, 109675. DOI: 10.1016/j.steroids.2025.109675 PMID: 40829733

Cherappurath, N., Thoompenthodi, M. A., Elayaraja, M., Pelasseri, M., Perumbalath, S., Muriyedath, A., & Parambath, S. P. (2025). Anterior Cruciate Ligament Injury in Football (Soccer) Players: A Bibliometric Review of Global Sports Orthopaedic Literature. *Journal of Orthopaedic Reports*, •••, 100834. DOI: 10.1016/j.jorep.2025.100834

Chersich, M. F., Pham, M. D., Areal, A., Haghighi, M. M., Manyuchi, A., Swift, C. P., Wernecke, B., Robinson, M., Hetem, R., Boeckmann, M., & Hajat, S.Climate Change and Heat-Health Study Group. (2020). Associations between high temperatures in pregnancy and risk of preterm birth, low birth weight, and stillbirths: Systematic review and meta-analysis. *BMJ (Clinical Research Ed.), 371*, m3811. DOI: 10.1136/bmj.m3811 PMID: 33148618

Chow, W. T. L., Brennan, D., & Brazel, A. J. (2012). Urban heat island research in Phoenix, Arizona: Theoretical contributions and policy applications. *Bulletin of the American Meteorological Society, 93*(4), 517–530. DOI: 10.1175/BAMS-D-11-00011.1

Christina, A. L., & Claire, V. M. (2025). A general population KAP survey: Supporting the design of interventions and tiger mosquito control efforts in response to arbovirus emergence in non-endemic areas. *One Health, 21*, 101247. DOI: 10.1016/j.onehlt.2025.101247 PMID: 41292613

Christova, I., Petrov, A., Papa, A., Vutchev, D., Kalvatchev, N., Vatev, N., & Stoycheva, M. (2015). Fatal Outcome of Coinfection of Crimean-Congo Hemorrhagic Fever and Malaria. *Japanese Journal of Infectious Diseases*, *68*(2), 131–134. DOI: 10.7883/yoken.JJID.2014.106 PMID: 25420643

Cianconi, P., Betrò, S., & Janiri, L. (2020). The Impact of Climate Change on Mental Health: A Systematic Descriptive Review. *Frontiers in Psychiatry, Volume 11 - 2020*. DOI: 10.3389/fpsyt.2020.00074

Ciotti, M., & Bernardini, S. (2025). One health and laboratory medicine: Challenges and perspectives for antimicrobial resistance, emerging infections, and food/water security. *Critical Reviews in Clinical Laboratory Sciences*, •••, 1–13. DOI: 10.1080/10408363.2025.2597189 PMID: 41395681

Clayton, S., Manning, C., Krygsman, K., & Speiser, M. (2017). Mental health and our changing climate: Impacts, implications, and guidance. Retrieved from American Psychological Association, 1-69.

Clayton, S., Manning, C. M., Krygsman, K., & Speiser, M. (2017). *Mental health and our changing climate: Impacts, implications, and guidance*. American Psychological Association.

Coker, R., Rushton, J., Mounier-Jack, S., Karimuribo, E., Lutumba, P., Kambarage, D., Pfeiffer, D. U., Stärk, K., & Rweyemamu, M. (2011). Towards a conceptual framework to support one-health research for policy on emerging zoonoses. *The Lancet. Infectious Diseases*, *11*(4), 326–331. DOI: 10.1016/S1473-3099(10)70312-1 PMID: 21376670

Craig, P., Katikireddi, S. V., Leyland, A., & Popham, F. (2017). Natural experiments: An overview of methods, approaches, and contributions to public health intervention research. *Annual Review of Public Health*, *38*(1), 429–451. DOI: 10.1146/annurev-publhealth-031816-044327 PMID: 28125392

Crawford, M., Nordgren, T., & McCole, D. (2021). Every breath you take: Impacts of environmental dust exposure on intestinal barrier function–from the gut-lung axis to COVID-19. *American Journal of Physiology. Gastrointestinal and Liver Physiology*, *320*(4), G586–G600. DOI: 10.1152/ajpgi.00423.2020 PMID: 33501887

Črepinšek, Z., Žnidaršič, Z., & Pogačar, T. (2023). Spatio-temporal analysis of the universal thermal climate index (UTCI) for the summertime in the period 2000–2021 in Slovenia: The implication of heat stress for agricultural workers. *Agronomy (Basel), 13*(2), 331. DOI: 10.3390/agronomy13020331

Cruz, R., Martínez, P., Silveira, H., Cruz-Hipólito, H., Palma-Bautista, C., Vázquez-García, J., & Prado, R. (2019). Management of Glyphosate-Resistant Weeds in Mexican Citrus Groves: Chemical Alternatives and Economic Viability. *Plants, 8*(9), 325. DOI: 10.3390/plants8090325 PMID: 31487903

Cunsolo, A., & Ellis, N. R. (2018). Ecological grief as a mental health response to climate change-related loss. *Nature Climate Change, 8*(4), 275–281. DOI: 10.1038/s41558-018-0092-2

Damtew, Y. T., Tong, M., Varghese, B. M., Anikeeva, O., Hansen, A., Dear, K., Zhang, Y., Morgan, G., Driscoll, T., Capon, T., & Bi, P. (2023). Effects of high temperatures and heatwaves on dengue fever: A systematic review and meta-analysis. *EBioMedicine, 91*, 104582. DOI: 10.1016/j.ebiom.2023.104582 PMID: 37088034

Danasekaran, R. (2024). One health: A holistic approach to tackling global health issues. *Indian Journal of Community Medicine : Official Publication of Indian Association of Preventive & Social Medicine, 49*(2), 260–263. DOI: 10.4103/ijcm.ijcm_521_23 PMID: 38665439

Darabont, D. C., Bejinariu, C., Ionita, I., Bernevig-Sava, M. A., Baciu, C., & Baciu, E. R. (2018). CONSIDERATIONS ON IMPROVING OCCUPATIONAL HEALTH AND SAFETY PERFORMANCE IN COMPANIES USING ISO 45001 STANDARD. [EEMJ]. *Environmental Engineering and Management Journal, 17*(11).

de Crom, T. O., Scholten, B., Traini, E., van der Sanden, K., Kingma, B., Pekel, F., & Pronk, A. (2025). Exposure to heat at work: Development of a quantitative European job exposure matrix (heat JEM). *Scandinavian Journal of Work, Environment & Health, 52*(1), 7. PMID: 40782101

De Keukeleire, M., Vanwambeke, S. O., Somers, B., & Luyasu, V. (2017). Lyme borreliosis in forestry workers: A systematic review. *European Journal of Public Health, 27*(3), 514–519.

de la Fuente, J., Ghosh, S., Lempereur, L., Garrison, A., Sprong, H., Lopez-Camacho, C., Maritz-Olivier, C., Contreras, M., Moraga-Fernández, A., & Bente, D. A. (2024). Interventions for the control of Crimean-Congo hemorrhagic fever and tick vectors. *NPJ Vaccines, 9*(1), 1–10. DOI: 10.1038/s41541-024-00970-5 PMID: 39353956

De Santos, S., & Skinner, M. K. (2026). Environmental Transgenerational Epigenetics: Effects on Fertility and Offspring Health. *Advances in Experimental Medicine and Biology*, *1506*, 153–167. DOI: 10.1007/978-3-032-18376-7_8 PMID: 42036612

de Souza, W. M., & Weaver, S. C. (2024). Effects of climate change and human activities on vector-borne diseases. *Nature Reviews. Microbiology*, *22*(8), 476–491. DOI: 10.1038/s41579-024-01026-0 PMID: 38486116

Dean, A., & McCallum, J. (2025). California's Heat Standard And Heat-Related Deaths Among Outdoor Workers: Article examines California's heat standard and heat-related deaths among outdoor workers. *Health Affairs (Project Hope)*, *44*(12), 1490–1496. DOI: 10.1377/hlthaff.2025.00096 PMID: 41329892

Deguine, J., Aubertot, J., Flor, R., Lescourret, F., Wyckhuys, K., & Ratnadass, A. (2021). Integrated pest management: Good intentions, hard realities. A review. *Agronomy for Sustainable Development*, *41*(3), 38. DOI: 10.1007/s13593-021-00689-w

Desai, S., Patel, V., Nagare, S. D., Patil, R., & More, M. J. (2025). Microbial genomics and infectious disease forecasting. In *Host Adaptation, Virulence, and Evolution* (pp. 193-223). Academic Press. DOI: 10.1016/B978-0-443-31554-1.00012-1

Destoumieux-Garzón, D., Mavingui, P., Boetsch, G., Boissier, J., Darriet, F., Duboz, P., Fritsch, C., Giraudoux, P., Le Roux, F., Morand, S., Paillard, C., Pontier, D., Sueur, C., & Voituron, Y. (2018). The one health concept: 10 years old and a long road ahead. *Frontiers in Veterinary Science*, *5*, 14. DOI: 10.3389/fvets.2018.00014 PMID: 29484301

Desvars-Larrive, A., Vogl, A. E., Puspitarani, G. A., Yang, L., Joachim, A., & Käsbohrer, A. (2024). A One Health framework for exploring zoonotic interactions demonstrated through a case study. *Nature Communications*, *15*(1), 5650. DOI: 10.1038/s41467-024-49967-7 PMID: 39009576

Dhaigude, A. S., & Kamath, G. B. (2025). Mapping responsible artificial intelligence in business and management: Trends, influence, and emerging research directions. *Journal of Open Innovation*, *11*(4), 100640. DOI: 10.1016/j.joitmc.2025.100640

Di Bari, C., Venkateswaran, N., Fastl, C., Gabriël, S., Grace, D., Havelaar, A. H., Huntington, B., Patterson, G. T., Rushton, J., Speybroeck, N., Torgerson, P., Pigott, D. M., & Devleesschauwer, B. (2023). The global burden of neglected zoonotic diseases: Current state of evidence. *One Health*, *17*, 100595. DOI: 10.1016/j.onehlt.2023.100595 PMID: 37545541

Di Napoli, C., McGushin, A., Romanello, M., Ayeb-Karlsson, S., Cai, W., Chambers, J., Dasgupta, S., Escobar, L. E., Kelman, I., Kjellstrom, T., Kniveton, D., Liu, Y., Liu, Z., Lowe, R., Martinez-Urtaza, J., McMichael, C., Moradi-Lakeh, M., Murray, K. A., Rabbaniha, M., & Robinson, E. J. (2022). Tracking the impacts of climate change on human health via indicators: Lessons from the Lancet Countdown. *BMC Public Health*, *22*(1), 663. DOI: 10.1186/s12889-022-13055-6 PMID: 35387618

Diarra, A. Z., Kelly, P., Davoust, B., & Parola, P. (2023). Tick-Borne Diseases of Humans and Animals in West Africa. *Pathogens (Basel, Switzerland)*, *12*(11), 1276. DOI: 10.3390/pathogens12111276 PMID: 38003741

Diprose, K., Liu, C., Valentine, G., Vanderbeck, R. M., & McQuaid, K. (2019). Caring for the future: Climate change and intergenerational responsibility in China and the UK. *Geoforum*, *105*, 158–167. DOI: 10.1016/j.geoforum.2019.05.019

Djiman, T. A., Biguezoton, A. S., & Saegerman, C. (2024). Tick-Borne Diseases in Sub-Saharan Africa: A Systematic Review of Pathogens, Research Focus, and Implications for Public Health. *Pathogens (Basel, Switzerland)*, *13*(8), 697. DOI: 10.3390/pathogens13080697 PMID: 39204297

Domingo, A., Little, M., Beggs, B., Brubacher, L. J., Lau, L. L., & Dodd, W. (2024). Examining the role of community health workers amid extreme weather events in low- and middle-income countries: A scoping review. *Public Health*, *236*, 133–143. DOI: 10.1016/j.puhe.2024.07.023 PMID: 39182471

Domingo, J. L. (2026). Developmental toxicity of metals: Three decades of insights into mechanisms, epigenetics, and transgenerational inheritance. *Environmental Research*, *292*, 123703. DOI: 10.1016/j.envres.2026.123703 PMID: 41506432

Dong, X. S., Largay, J. A., Wang, X., Sokas, R., & Mohr, D. (2019). Heat-related deaths among construction workers in the United States. *American Journal of Industrial Medicine*, *62*(12), 1047–1057. DOI: 10.1002/ajim.23024 PMID: 31328819

Dou, J., Cui, Q., Li, W., Jia, K., & Wang, F. (2026). Heat stroke and the liver: Mechanisms of injury and therapeutic strategies. *European Journal of Medical Research*, *31*(1), 263. Advance online publication. DOI: 10.1186/s40001-025-03775-x PMID: 41530884

Do, V., Parks, R. M., Casey, J. A., Goin, D. E., & Kioumourtzoglou, M.-A. (2024). Use, limitations, and future directions of mixtures approaches to understand the health impacts of weather- and climate change-related exposures, an under-studied aspect of the exposome. *Exposome*, *4*(1), osae007. Advance online publication. DOI: 10.1093/exposome/osae007 PMID: 39444644

Druckman, J. N., & Lupia, A. (2017). Visual Communication in Politics. *Annual Review of Political Science, 20*, 369–388.

Durand-Billaud, E. BERTHOL, N., BIJAOUI, A., BOUSELHAM, M., PERIS, E., & BOUSLAMA, M. (2024). *Maladies zoonotiques et vectorielles en milieu professionnel: Impacts du changement climatique - Article - INRS*. https://www.inrs.fr/media.html?refINRS=TC%20181

Eaton, M. (2017). Navigating anger, fear, grief, and despair. In *Contemplative approaches to sustainability in higher education: Theory and practice* (pp. 40–54). Taylor & Francis.

Ebi, K. L., Teisberg, T. J., Kalkstein, L. S., Robinson, L., & Weiher, R. F. (2004). Heat watch warning systems save lives: Estimated costs and benefits for Philadelphia 1995–1998. *Bulletin of the American Meteorological Society, 85*(8), 1067–1073. https://www1.udel.edu/SynClim/BAMS_Ebi_Kalkstein.pdf. DOI: 10.1175/BAMS-85-8-1067

ECDC. (2024, June 10). *Historical data on local transmission in Europe for West Nile virus*. https://www.ecdc.europa.eu/en/west-nile-fever/surveillance-and-disease-data/historical

Edney, J. M., Kovats, S., Filippi, V., & Nakstad, B. (2022). A systematic review of hot weather impacts on infant feeding practices in low- and middle-income countries. *Frontiers in Pediatrics, 10*, 930348. DOI: 10.3389/fped.2022.930348 PMID: 36147803

Egbendewe-Mondzozo, A., Musumba, M., McCarl, B. A., & Wu, X. (2011). Climate Change and Vector-borne Diseases: An Economic Impact Analysis of Malaria in Africa. *International Journal of Environmental Research and Public Health, 8*(3), 913–930. DOI: 10.3390/ijerph8030913 PMID: 21556186

Ekenze, S. O., Huezo-Garcia, M., Kerr, S. M., Fisher, S. C., Nestoridi, E., Werler, M. M., & the National Birth Defects Prevention Study. (2026). Periconceptional Use of Analgesics and Severity of Structural Birth Defects. *Clinical Pharmacology & Therapeutics, n/a*(n/a). DOI: 10.1002/cpt.70259

El-Sherif, G. (2025). Occupational and Environmental Health Problems Related to Climate Change in Egypt. *Egyptian Journal of Occupational Medicine, 49*(2), 1–16. DOI: 10.21608/ejom.2025.371549.1366

Ergonul, O., & Whitehouse, C. A. (2007). *Crimean-Congo Hemorrhagic Fever: A Global Perspective*. Springer Science & Business Media. DOI: 10.1007/978-1-4020-6106-6

Estrada-Peña, A. (2025). *Hyalomma marginatum* (the Mediterranean Hyalomma). *Trends in Parasitology, 41*(9), 822–823. DOI: 10.1016/j.pt.2025.06.007 PMID: 40617740

Estrada-Peña, A., D'Amico, G., & Fernández-Ruiz, N. (2021). Modelling the potential spread of Hyalomma marginatum ticks in Europe by migratory birds. *International Journal for Parasitology, 51*(1), 1–11. DOI: 10.1016/j.ijpara.2020.08.004 PMID: 32991918

Estrela, M., Leitão, C., Neto, V., Martins, B., Santos, J., Branquinho, A., Figueiras, A., Roque, F., & Herdeiro, M. T. (2025). Educational Interventions for the Adoption of Healthy Lifestyles and Improvement of Health Literacy: A Systematic Review. *Public Health, 245*, 105788. DOI: 10.1016/j.puhe.2025.105788 PMID: 40466525

Evangelinos, K., Fotiadis, S., Skouloudis, A., Khan, N., Konstandakopoulou, F., Nikolaou, I., & Lundy, S. (2018). Occupational health and safety disclosures in sustainability reports: An overview of trends among corporate leaders. *Corporate Social Responsibility and Environmental Management, 25*(5), 961–970. DOI: 10.1002/csr.1512

Fakoorziba, M. R., Golmohammadi, P., Moradzadeh, R., Moemenbellah-Fard, M. D., Azizi, K., Davari, B., Alipour, H., Ahmadnia, S., & Chinikar, S. (2012). Reverse Transcription PCR-Based Detection of Crimean-Congo Hemorrhagic Fever Virus Isolated from Ticks of Domestic Ruminants in Kurdistan Province of Iran. *Vector Borne and Zoonotic Diseases, 12*(9), 794–799. DOI: 10.1089/vbz.2011.0743 PMID: 22651389

Fanelli, A., & Buonavoglia, D. (2021). Risk of Crimean Congo haemorrhagic fever virus (CCHFV) introduction and spread in CCHF-free countries in southern and Western Europe: A semi-quantitative risk assessment. *One Health, 13*, 100290. DOI: 10.1016/j.onehlt.2021.100290 PMID: 34307823

FAO-UNEP-WHO-OIE. (2022). *One Health Joint Plan of Action, 2022–2026*. FAO; UNEP; WHO; World Organisation for Animal Health (WOAH) (founded as OIE); DOI: 10.4060/cc2289en

Fasina, F. O., Fasanmi, O. G., Makonnen, Y. J., Bebay, C., Bett, B., & Roesel, K. (2021). The one health landscape in Sub-Saharan African countries. *One Health, 13*, 100325. DOI: 10.1016/j.onehlt.2021.100325 PMID: 34584927

Fatima, S. H., Giles, L. C., Rothmore, P., Varghese, B. M., & Bi, P. (2025). Heatwaves and occupational injuries and illnesses risk varied at localised spatial scale: A national study in Australia. *Safety Science, 181*, 106684. DOI: 10.1016/j.ssci.2024.106684

Fatima, S. H., Rothmore, P., Giles, L. C., & Bi, P. (2022). Outdoor ambient temperatures and occupational injuries and illnesses: Are there risk differences in various regions within a city? *The Science of the Total Environment, 826*, 153945. DOI: 10.1016/j.scitotenv.2022.153945 PMID: 35189241

Ferrari, G. N., Leal, G. C. L., Ossani, P. C., & Galdamez, E. V. C. (2025). Investigation of the usage of machine learning to explore the impacts of climate change on occupational health: A systematic review and research agenda. *Frontiers in Public Health, 13*, 1578558. DOI: 10.3389/fpubh.2025.1578558 PMID: 40589827

Ferrari, G., Leal, G., Souza, R., & Galdámez, E. (2023). Impact of climate change on occupational health and safety: A review of methodological approaches. *Work : A Journal of Prevention, Assessment, and Rehabilitation, 74*(2), 485–499. DOI: 10.3233/WOR-211303 PMID: 36314181

Flaherty, M., Carley, S., & Konisky, D. M. (2020). Electric utility disconnection policy and vulnerable populations. *The Electricity Journal, 33*(10), 106859. DOI: 10.1016/j.tej.2020.106859

Fletcher, R. (2018). Beyond the end of the world: Breaking attachment to a dying planet. In Kapoor, I. (Ed.), *Psychoanalysis and the Global* (pp. 48–69). University of Nebraska Press. DOI: 10.2307/j.ctv47w9gt.9

Florez-Acevedo, S., Blancas, M. T., & Spector, J. T. (2025). Occupational heat exposure & mental health outcomes: A review and framework incorporating social determinants of health to guide future research. *Current Environmental Health Reports, 12*(1), 15. DOI: 10.1007/s40572-025-00479-6 PMID: 40014259

Flouris, A. D., Dinas, P. C., Ioannou, L. G., Nybo, L., Havenith, G., Kenny, G. P., & Kjellstrom, T. (2018). Workers' health and productivity under climate change: The emerging role of heat stress. *Temperature, 5*(3), 210–221.

Flouris, A. D., Dinas, P. C., Ioannou, L. G., Nybo, L., Havenith, G., Kenny, G. P., & Kjellstrom, T. (2018). Workers' health and productivity under occupational heat strain: A systematic review and meta-analysis. *The Lancet. Planetary Health, 2*(12), e521–e531. DOI: 10.1016/S2542-5196(18)30237-7 PMID: 30526938

Ford, J. D., Cameron, L., Rubis, J., Maillet, M., Nakashima, D., Willox, A. C., & Pearce, T. (2016). Including Indigenous Knowledge and Experience in IPCC Assessment Reports. *Nature Climate Change, 6*(4), 349–353. DOI: 10.1038/nclimate2954

Franzon, H., & Rose, L. M. (2021, May). The SRA Index (Sustainable Risk Awareness Index): A new KPI for Management Support. In *Congress of the International Ergonomics Association* (pp. 457–466). Springer International Publishing. DOI: 10.1007/978-3-030-74608-7_57

Friedrich, J., Schick, T. S., Mess, F., & Blaschke, S. (2025). Wearable device-based interventions in heat-exposed outdoor workers-a scoping review and an explanatory intervention model. *BMC Public Health, 25*(1), 2893. DOI: 10.1186/s12889-025-24262-2 PMID: 40846922

Gage, K., Krausz, R., & Walters, S. (2019). Emerging Challenges for Weed Management in Herbicide-Resistant Crops. *Agriculture, 9*(8), 180. DOI: 10.3390/agriculture9080180

Gagnon, D., Schlader, Z. J., & Jay, O. (2026). The Physiology behind the Epidemiology of Heat-Related Health Impacts. *Physiology (Bethesda, MD), 41*(1), 30–42. DOI: 10.1152/physiol.00012.2025 PMID: 40602791

Gahn, M. C. B., Diouf, G., Cissé, N., Ciss, M., Bordier, M., Ndiaye, M., Bakhoum, M. T., Djiba, M. L., Brown, C., Faburay, B., Fall, A. G., & Lo, M. M. (2024). Large-Scale Serological Survey of Crimean-Congo Hemorrhagic Fever Virus and Rift Valley Fever Virus in Small Ruminants in Senegal. *Pathogens (Basel, Switzerland), 13*(8), 8. Advance online publication. DOI: 10.3390/pathogens13080689 PMID: 39204289

Gale, P., Stephenson, B., Brouwer, A., Martinez, M., de la Torre, A., Bosch, J., Foley-Fisher, M., Bonilauri, P., Lindström, A., Ulrich, R. G., de Vos, C. J., Scremin, M., Liu, Z., Kelly, L., & Muñoz, M. J. (2012). Impact of climate change on risk of incursion of Crimean-Congo haemorrhagic fever virus in livestock in Europe through migratory birds. *Journal of Applied Microbiology, 112*(2), 246–257. DOI: 10.1111/j.1365-2672.2011.05203.x PMID: 22118269

Gándara, L., Jacoby, R., Laurent, F., Spatuzzi, M., Vlachopoulos, N., Borst, N., … & Crocker, J. (2024). Pervasive sublethal effects of agrochemicals as contributing factors to insect decline.

Gangneux, J.-P., Belaz, S., & Robert-Gangneux, F. (2015). Mise au point et actualités sur la leishmaniose viscérale méditerranéenne. *Journal des Anti-Infectieux, 17*(1), 25–28. DOI: 10.1016/j.antinf.2014.12.005

Gao, C. X., Fava, N., Browne, V., Patrick, R., Gunasiri, H., & Menssink, J. M. (2025). Climate Change and Youth: Fast-Tracking Mental Health Solutions. *Journal of Applied Youth Studies*. Advance online publication. DOI: 10.1007/s43151-025-00174-1

Gardiner, S. M. (2011). *A perfect moral storm: the ethical tragedy of climate change*. Oxford University Press., DOI: 10.1093/acprof:oso/9780195379440.001.0001

Garg, S., Rumjit, N. P., & Arora, P. (2024). Addressing climate change impacts through sustainable agricultural solutions: A review. *Environment, Development and Sustainability*. Advance online publication. DOI: 10.1007/s10668-024-05728-6

Gashema, P., Sesonga, P., Iradukunda, P. G., Muvunyi, R., Mugisha, J. C., Ndayisenga, J., Musafiri, T., Habimana, R., Bigirimana, R., Kabanda, A., Gashegu, M., Gahamanyi, N., Izudi, J., Siddig, E. E., Ngabonziza, J. C. S., Ahmed, A., Dzinamarira, T., Mutesa, L., & Muvunyi, C. M. (2025). Enhancing Global Health Security in Sub-Saharan Africa: The case for integrated One Health surveillance against zoonotic diseases and environmental threats. *One Health*, *21*, 101136. DOI: 10.1016/j.onehlt.2025.101136 PMID: 40687600

Gaviria, D. A. G., Martínez, E. M. C., Polo, O. C. C., & Arcila, J. O. S. (2023). Accounting mechanism to measure the environmental impact of industries in the Aburrá Valley and its application of the general disclosure requirements related to financial information on sustainability (IFRS S1) and climate-related disclosures (IFRS S2). *Revista de Gestao Social e Ambiental*, *17*(5), 1–16. DOI: 10.24857/rgsa.v17n5-013

Gellert, G., Montgomery, S., Gellert, T., Gillett, J., Kerekes, A., & Hole, J. (2022). Employee Wellness in a Changing Climate: Environmental Heat Stress Driving Need for Targeted Health Promotion and Risk Reduction. *Journal of Ecophysiology and Occupational Health*, 60–66.

Gibbs, E. P. J. (2014). The evolution of One Health: A decade of progress and challenges for the future. *The Veterinary Record*, *174*(4), 85–91. DOI: 10.1136/vr.g143 PMID: 24464377

Gilbert, M., Golding, N., Zhou, H., Wint, G. R., Robinson, T. P., Tatem, A. J., Lai, S., Zhou, S., Jiang, H., Guo, D., Huang, Z., Messina, J. P., Xiao, X., Linard, C., Van Boeckel, T. P., Martin, V., Bhatt, S., Gething, P. W., Farrar, J. J., & Yu, H. (2014). Predicting the risk of avian influenza A H7N9 infection in live-poultry markets across Asia. *Nature Communications*, *5*(1), 4116. DOI: 10.1038/ncomms5116 PMID: 24937647

Gorelick, N., Hancher, M., Dixon, M., Ilyushchenko, S., Thau, D., & Moore, R. (2017). Google Earth Engine: Planetary-scale geospatial analysis for everyone. *Remote Sensing of Environment*, *202*, 18–27. DOI: 10.1016/j.rse.2017.06.031

Goshua, A., Akdis, C. A., & Nadeau, K. C. (2022). World Health Organization global air quality guideline recommendations: Executive summary. *Allergy*, *77*(7), 1955–1960. DOI: 10.1111/all.15224 PMID: 35060140

Gossner, C. M., Ducheyne, E., & Schaffner, F. (2021). Increased risk of autochthonous dengue transmission in Europe. *Eurosurveillance*, *26*(21), 2100431.

Gottdenker, N. L., Streicker, D. G., Faust, C. L., & Carroll, C. R. (2014). Anthropogenic Land Use Change and Infectious Diseases: A Review of the Evidence. *EcoHealth*, *11*(4), 619–632. DOI: 10.1007/s10393-014-0941-z PMID: 24854248

Grant, M. J., & Booth, A. (2009). A typology of reviews: An analysis of 14 review types and associated methodologies. *Health Information and Libraries Journal*, *26*(2), 91–108. DOI: 10.1111/j.1471-1842.2009.00848.x PMID: 19490148

Green Climate Fund. (2025). *Green Climate Fund*. Available online: https://www.greenclimate.fund/ (accessed on 30 March 2025).

Green Climate Fund. (2025). *Sectoral guide: Health & wellbeing*. Incheon City, Republic of Korea. Available online: https://www.greenclimate.fund/document/sectoral-guide-health-wellbeing (accessed on 30 March 2025).

Gretchev, A., Temples, H. S., England-Mason, G., Dewey, D., Sarasua, S. M., Farrell, C., & Haley-Zitlin, V. (2026). Unpacking developmental programming: A conceptual and historical analysis in the context of DOHaD. *Journal of Developmental Origins of Health and Disease*, *17*, e17. DOI: 10.1017/S2040174426100518 PMID: 41919437

Gribble, K. D., & Berry, N. J. (2011). Emergency preparedness for those who care for infants in developed country contexts. *International Breastfeeding Journal*, *6*(1), 16. DOI: 10.1186/1746-4358-6-16 PMID: 22059481

Gubernot, D. M., Anderson, G. B., & Hunting, K. L. (2015). Characterizing occupational heat-related mortality in the United States, 2000–2010: An analysis using the census of fatal occupational injuries database. *American Journal of Industrial Medicine*, *58*(2), 203–211. DOI: 10.1002/ajim.22381 PMID: 25603942

Guo, Y., Zhang, Y., Yu, P., Wu, Y., Wen, B., Yang, Z., & Li, S. (2023). Strategies to reduce the health impacts of heat exposure. In *Heat Exposure and Human Health in the Context of Climate Change* (pp. 293–322). Elsevier. DOI: 10.1016/B978-0-12-819080-7.00009-4

Haagsma, J. A., Tariq, L., Heederik, D. J., & Havelaar, A. H. (2012). Infectious disease risks associated with occupational exposure: A systematic review of the literature. *Occupational and Environmental Medicine*, *69*(2), 140–146. DOI: 10.1136/oemed-2011-100068 PMID: 22006935

Habibi, P., Heydari, A., Dehghan, H., Moradi, A., & Moradi, G. (2024a). Climate Change and Occupational Heat Strain Among Women Workers: A Systematic Review. *Indian Journal of Occupational and Environmental Medicine*, *28*(1), 4–17. DOI: 10.4103/ijoem.ijoem_320_21 PMID: 38783874

Habibi, P., Moradi, G., Dehghan, H., Moradi, A., & Heydari, A. (2021). The impacts of climate change on occupational heat strain in outdoor workers: A systematic review. *Urban Climate*, *36*, 100770. Advance online publication. DOI: 10.1016/j.uclim.2021.100770

Habibi, P., Razmjouei, J., Badzohreh, A., & Heydari, A. (2025). Climate change and chronic kidney disease (CKD) among outdoor workers: A systematic review. *International Journal of Biometeorology*, *69*(6), 1207–1236. DOI: 10.1007/s00484-025-02896-6 PMID: 40186676

Habibi, P., Razmjouei, J., Moradi, A., Mahdavi, F., Fallah-Aliabadi, S., & Heydari, A. (2024b). Climate change and heat stress resilient outdoor workers: Findings from systematic literature review. *BMC Public Health*, *24*(1), 1711. DOI: 10.1186/s12889-024-19212-3 PMID: 38926816

Haddaji, N. (2022). Environmental contaminants and antibiotic resistance as a One Health threat. In *One Health* (pp. 231–252). Academic Press. DOI: 10.1016/B978-0-12-822794-7.00010-1

Haghighi, A., Cheraghi, M., Pocachard, J., Botta-Genoulaz, V., Jocelyn, S., & Pourzarei, H. (2025). A comprehensive review and bibliometric analysis on collaborative robotics for industry: Safety emerging as a core focus. *Frontiers in Robotics and AI*, *12*, 1605682. DOI: 10.3389/frobt.2025.1605682 PMID: 41019776

Hancock, P. A., Ross, J. M., & Szalma, J. L. (2007). A meta-analysis of performance response under thermal stressors. *Human Factors*, *49*(5), 851–877. DOI: 10.1518/001872007X230226 PMID: 17915603

Hansen, A., Williams, S., Hanson-Easey, S., Varghese, B., Bi, P., Heyworth, J., Nitschke, M., Rowett, S., Sim, M. R., & Pisaniello, D. (2020). Using a Qualitative Phenomenological Approach to Inform the Etiology and Prevention of Occupational Heat-Related Injuries in Australia. *International Journal of Environmental Research and Public Health*, *17*(3), 846. DOI: 10.3390/ijerph17030846 PMID: 32013180

Hardeman, R. R., Karbeah, J., & Kozhimannil, K. B. (2020). Applying a critical race lens to relationship-centered care in pregnancy and childbirth: An antidote to structural racism. *Birth (Berkeley, Calif.)*, *47*(1), 3–7. DOI: 10.1111/birt.12462 PMID: 31630454

Hariry, A., Mardawati, E., Iswanto, A., Karliati, T., Zaini, L., & Adly, M. (2025). Nanocellulose-Based Adhesives for Sustainable Wood-Polymer Composites: Recent Advancement and Future Perspective. *Journal of Renewable Materials*, *13*(4), 773–798. DOI: 10.32604/jrm.2025.058359

Hasan, M. A. E., Islam, M. S., Fahim, N. A. I., Antor, M. T. H., Salam, S., Jahan, R., Masud, R. I., Bakhtiyar, Z., Jany, D. A., Rana, M. L., Punom, S. A., Rahman, A. M. M. T., & Rahman, M. T. (2025). Impact of open markets on zoonotic threats and antimicrobial resistance: A one health concern. *One Health*, *21*, 101228. DOI: 10.1016/j.onehlt.2025.101228 PMID: 41127627

Heikkilä, A.-M., Malmén, Y., Nissilä, M., & Kortelainen, H. (2010). Challenges in Risk Management in Multi-Company Industrial Parks. *Safety Science*, *48*(4), 430–435. DOI: 10.1016/j.ssci.2009.12.008

Heinzerling, A., Laws, R., Frederick, M., Jackson, R., Windham, G., Materna, B., & Harrison, R. (2020). Risk factors for occupational heat-related illness among California workers, 2000–2017. *American Journal of Industrial Medicine*, *63*(12), 1145–1154. DOI: 10.1002/ajim.23191 PMID: 33075156

Hernández, D. (2013). Energy insecurity: A framework for understanding energy, the built environment, and health among vulnerable populations in the context of climate change. *American Journal of Public Health*, *103*(4), e32–e34. DOI: 10.2105/AJPH.2012.301179 PMID: 23409876

Hernández-Duarte, W. A., Guerrero-Calderon, D. P., Duarte-Gómez, W. J., Fonseca-Paipa, A. N., Rodríguez-Muñoz, M., Tellez-Morales, M. A., & Garnica, L. C. C. (2024). Health conditions in land transport workers and climate change. Exploratory systematic review. *Revista Brasileira de Medicina do Trabalho : Publicação Oficial da Associaação Nacional de Medicina do Trabalho-ANAMT*, *22*(2), e20241268. DOI: 10.47626/1679-4435-2024-1268 PMID: 39371277

Hicks, H., Comont, D., Coutts, S., Crook, L., Hull, R., Norris, K., Neve, P., Childs, D. Z., & Freckleton, R. (2018). The factors driving evolved herbicide resistance at a national scale. *Nature Ecology & Evolution*, *2*(3), 529–536. DOI: 10.1038/s41559-018-0470-1 PMID: 29434350

Hoffman, T., Lindeborg, M., Barboutis, C., Erciyas-Yavuz, K., Evander, M., Fransson, T., Figuerola, J., Jaenson, T. G. T., Kiat, Y., Lindgren, P.-E., Lundkvist, Å., Mohamed, N., Moutailler, S., Nyström, F., Olsen, B., & Salaneck, E. (2018). *Alkhurma Hemorrhagic Fever Virus RNA in Hyalomma rufipes Ticks Infesting Migratory Birds, Europe and Asia Minor—Volume 24, Number 5—May 2018—Emerging Infectious Diseases journal—CDC*. DOI: 10.3201/eid2405.171369

Hoffman, J. S., Shandas, V., & Pendleton, N. (2020). The effects of historical housing policies on resident exposure to intra-urban heat: A study of 108 US urban areas. *Climate (Basel)*, *8*(1), 12. DOI: 10.3390/cli8010012

Hoffman, T., Carra, L. G., Öhagen, P., Fransson, T., Barboutis, C., Piacentini, D., Figuerola, J., Kiat, Y., Onrubia, A., Jaenson, T. G. T., Nilsson, K., Lundkvist, Å., & Olsen, B. (2021). Association between guilds of birds in the African-Western Palaearctic region and the tick species *Hyalomma rufipes,* one of the main vectors of Crimean-Congo hemorrhagic fever virus. *One Health*, *13*, 100349. DOI: 10.1016/j.onehlt.2021.100349 PMID: 34825045

Hondula, D. M., Kuras, E. R., Betzel, S., Drake, L., Eneboe, J., Kaml, M., Munoz, M., Sevig, M., Singh, M., Ruddell, B. L., & Harlan, S. L. (2021). Novel metrics for relating personal heat exposure to social risk factors and outdoor ambient temperature. *Environment International*, *146*, 106271. DOI: 10.1016/j.envint.2020.106271 PMID: 33395929

Honles, J., Cerapio, J. P., Monge, C., Marchio, A., Ruiz, E., Fernández, R., Casavilca-Zambrano, S., Contreras-Mancilla, J., Vidaurre, T., Condom, T., Zerathe, S., Dangles, O., Deharo, É., Herrera-Zuñiga, J., Pineau, P., & Bertani, S. (2026). Mapping pesticide mixtures to cancer risk at the country scale with spatial exposomics. *Nature Health*, *1*(5), 520–531. DOI: 10.1038/s44360-026-00087-0 PMID: 42111112

Hornborg, A. (2019). *Nature, Society, and Justice in the Anthropocene: Unraveling the Money-Energy-Technology Complex*. Cambridge University Press., DOI: 10.1017/9781108554985

Hou, J. (2022). Wildfire Smoke Exposure in California: Public Health Impacts and Mitigation Strategies.

Hsu, A., Emerson, J. W., Levy, M. A., de Sherbinin, A., Johnson, L., & Malik, O. (2014). *The 2014 Environmental Performance Index*. Yale Center for Environmental Law & Policy.

Huang, Y., Gong, D., Dang, K., Zhu, L., Guo, J., Yang, W., & Wang, J. (2024). The applications of anterior segment optical coherence tomography in glaucoma: A 20-year bibliometric analysis. *PeerJ*, *12*, e18611. DOI: 10.7717/peerj.18611 PMID: 39619196

Humphrys, E. (2024). Inertia in transformed times: Work health and safety amid climate change. *The Journal of Industrial Relations*, *66*(5), 685–702. DOI: 10.1177/00221856241295497

Hutter, H., Khan, A., Lemmerer, K., Wallner, P., Kundi, M., & Moshammer, H. (2018). Cytotoxic and Genotoxic Effects of Pesticide Exposure in Male Coffee Farmworkers of the Jarabacoa Region, Dominican Republic. *International Journal of Environmental Research and Public Health*, *15*(8), 1641. DOI: 10.3390/ijerph15081641 PMID: 30081446

Hwang, C. H., Iellamo, A., & Ververs, M. (2021). Barriers and challenges of infant feeding in disasters in middle- and high-income countries. *International Breastfeeding Journal*, *16*(1), 62. DOI: 10.1186/s13006-021-00398-w PMID: 34425848

Ianosi, E. S., Tomoroga, D. M., Văsie iu, A. M., Grigorescu, B. L., Vultur, M., & Ianosi, M. B. (2025). Epigenetic Alterations Induced by Smoking and Their Intersection with Artificial Intelligence: A Narrative Review. *International Journal of Environmental Research and Public Health*, *22*(11), 1622. DOI: 10.3390/ijerph22111622 PMID: 41302568

Idris, M. A., Mena, K. D., Markham, C., & Perkison, W. B. (2025). Heat Stress Prevention and Management in the Construction Industry: A Gap Analysis. *Annals of Global Health*, *91*(1), 75. DOI: 10.5334/aogh.4833 PMID: 41179855

ILO. I. (2019). Working on a warmer planet: the impact of heat stress on labour productivity and decent work. *Geneva: International Labour Organization.*

International Labour Office. (2021). [The role of digital labour platforms in transforming the world of work. International Labour Organisation] [ILO]. *World Employment and Social Outlook*, •••, 2021.

International Labour Organization. (2015). Guidelines for a just transition towards environmentally sustainable economies and societies for all.

International Labour Organization. (2018). *Women and men in the informal economy: A statistical picture* (3rd ed.).

Ioannou, L. G., Foster, J., Morris, N. B., Piil, J. F., Havenith, G., Mekjavic, I. B., Kenny, G. P., Nybo, L., & Flouris, A. D. (2022). Occupational heat strain in outdoor workers: A comprehensive review and meta-analysis. *Temperature*, *9*(1), 67–102. DOI: 10.1080/23328940.2022.2030634 PMID: 35655665

IPCC. (2022). *Climate Change 2022: Impacts, Adaptation and Vulnerability*. Contribution of Working Group II to the Sixth Assessment Report of the Intergovernmental Panel on Climate Change. Cambridge University Press.

IPCC. 2023: Climate Change 2023: Synthesis Report. Contribution of Working Groups I, II and III to the Sixth Assessment Report of the Intergovernmental Panel on Climate Change [Core Writing Team, H. Lee and J. Romero (eds.)]. IPCC, Geneva, Switzerland, 184 pp., DOI: 10.59327/IPCC/AR6-9789291691647

IPCC. Climate Change (2020). Synthesis Report. Cambridge University Press, Cambridge and New York

IPCC. ntergovernmental P. O. C. C. (2023). *Climate Change 2022 – Impacts, Adaptation and Vulnerability: Working Group II Contribution to the Sixth Assessment Report of the Intergovernmental Panel on Climate Change* (1st ed.). Cambridge University Press. DOI: 10.1017/9781009325844

Islam, M. A., Begum, S., Solaiman, M., Zohora, F.-T., Datta, S. K., Ahmed, K. A., & Ahmed, M. S. (2025). Complete mitogenomes of two major dengue vectors Aedes aegypti and Aedes albopictus from Bangladesh: Insights from comparative genomics with global mitogenome diversity and phylogenetics. *PLoS One*, *20*(9), e0333693. DOI: 10.1371/journal.pone.0333693 PMID: 41026721

Iwamura, T., Guzman-Holst, A., & Murray, K. A. (2020). Accelerating invasion potential of disease vector Aedes aegypti under climate change. *Nature Communications*, *11*(1), 2130. DOI: 10.1038/s41467-020-16010-4 PMID: 32358588

Iyer, K. A., Koynova-Tenchov, R., Sasso, J. M., Thite, T., Deng, Y., & Zhou, Q. A. (2026). The Future of Epigenetics: Emerging Technologies and Clinical Applications. *ACS Pharmacology & Translational Science*, *9*(3), 506–544. DOI: 10.1021/acsptsci.5c00729 PMID: 41852637

Izabayo, P., Hakizimana, J. C., Uwineza, A., & Alagbonsi, A. I. (2026). Environmental factors associated with microcephaly in Africa: A systematic review. *Neurotoxicology and Teratology*, *113*, 107577. DOI: 10.1016/j.ntt.2025.107577 PMID: 41419017

Jacklitsch, B. L., Williams, W. J., Musolin, K., Coca, A., Kim, J. H., & Turner, N. (2016). Criteria for a recommended standard: occupational exposure to heat and hot environments.

Jain, M., & Nagpal, A. (2020). Analysing the nexus of environmental performance and human well-being: An empirical study of selected emerging economies. *International Journal of Green Economics*, *14*(3), 266–277. DOI: 10.1504/IJGE.2020.111470

Jameton, A. (1984). *Nursing practice: The ethical issues*. Prentice-Hall.

Jhun, I., Mata, D. A., Nordio, F., Lee, M., Schwartz, J., & Zanobetti, A. (2017). Ambient temperature and sudden infant death syndrome in the United States. *Epidemiology (Cambridge, Mass.)*, *28*(5), 728–734. DOI: 10.1097/EDE.0000000000000703 PMID: 28661937

Jia, J. A., & Jia, N. I. (2024). Impacts of Climate Change on Occupational Health and Safety: A Comprehensive Review. *Asian Journal of Advanced Research and Reports*, *18*(10), 136–154. DOI: 10.9734/ajarr/2024/v18i10761

Jiao, A., Sun, Y., Avila, C., Chiu, V., Slezak, J., Sacks, D. A., Abatzoglou, J. T., Molitor, J., Chen, J.-C., Benmarhnia, T., Getahun, D., & Wu, J. (2023). Analysis of heat exposure during pregnancy and severe maternal morbidity. *JAMA Network Open*, *6*(9), e2332780. DOI: 10.1001/jamanetworkopen.2023.32780 PMID: 37676659

Jo, H., Reddy, V., Parkinson, J., Boulton, D., & Tang, W. (2021). Model-Informed Pediatric Dose Selection for Dapagliflozin by Incorporating Developmental Changes. *CPT: Pharmacometrics & Systems Pharmacology*, *10*(2), 108–118. DOI: 10.1002/psp4.12577 PMID: 33439535

Kamguem, I. S., Kirschvink, N., Wade, A., & Linard, C. (2025). Determinants of viral haemorrhagic fever risk in Africa's tropical moist forests: A scoping review of spatial, socio-economic, and environmental factors. *PLoS Neglected Tropical Diseases*, *19*(1), e0012817. DOI: 10.1371/journal.pntd.0012817 PMID: 39820141

Karanikas, N., Foster, C., Beltran Hernandez, A., Harvey, A., Targal, O., & Horswill, N. (2021a). Conventional and Alternative Aviation Fuels: Occupational Exposure and Health Effects. *Journal of Chemical Health and Safety*, *28*(3), 159–170. DOI: 10.1021/acs.chas.0c00120

Karanikas, N., Steele, S., Bruschi, K., Robertson, C., Kass, J., Popovich, A., & MacFadyen, C. (2021b). Occupational health hazards and risks in the wind industry. *Energy Reports*, *7*, 3750–3759. DOI: 10.1016/j.egyr.2021.06.066

Karliner, L. S., Jacobs, E. A., Chen, A. H., & Mutha, S. (2007). Do professional interpreters improve clinical care for patients with limited English proficiency? *Health Services Research*, *42*(2), 727–754. DOI: 10.1111/j.1475-6773.2006.00629.x PMID: 17362215

Kayusi, F., Vallejo, R. G., Juma, L., Omwenga, M. K., & Chavula, P. (2025). A bibliometric analysis of Segment Anything (SA) research: Global trends, key contributors, and thematic insights. *Babylonian Journal of Machine Learning*, *2025*, 27–41. DOI: 10.58496/BJML/2025/003

Kazemi Shariat Panahi, H., Dehhaghi, M., Kinder, J. E., & Ezeji, T. C. (2019). A review on green liquid fuels for the transportation sector: A prospect of microbial solutions to climate change. *Biofuel Research Journal*, *6*(3), 995–1024. DOI: 10.18331/BRJ2019.6.3.2

Kennedy, J., & Lekshmi, J. (2023). *Holistic Pest Management Strategies in Tropical Plant Species*. IntechOpen.

Khayat, M., Halwani, D., Hneiny, L., Alameddine, I., Haidar, M., & Habib, R. (2022). Impacts of Climate Change and Heat Stress on Farmworkers' Health: A Scoping Review. *Frontiers in Public Health*, *10*, 10. DOI: 10.3389/fpubh.2022.782811 PMID: 35211437

Khunkaew, R., Wichianrak, J., & Suttipun, M. (2023). Sustainability reporting, gender diversity, firm value and corporate performance in ASEAN region. *Cogent Business & Management*, *10*(1), 2200608. DOI: 10.1080/23311975.2023.2200608

Kilpatrick, A. M., & Randolph, S. E. (2012). Drivers, dynamics, and control of emerging vector-borne zoonotic diseases. *Lancet*, *380*(9857), 1946–1955. DOI: 10.1016/S0140-6736(12)61151-9 PMID: 23200503

Kjellstrom, T., Briggs, D., Freyberg, C., Lemke, B., Otto, M., & Hyatt, O. (2016). Heat, human performance, and occupational health: A key issue for the assessment of global climate change impacts. *Annual Review of Public Health*, *37*(1), 97–112. DOI: 10.1146/annurev-publhealth-032315-021740 PMID: 26989826

Kjellstrom, T., Holmer, I., & Lemke, B. (2009). Workplace Heat Stress, Health and Productivity – an Increasing Challenge for Low and Middle-Income Countries during Climate Change. *Global Health Action*, *2*(1), 2047. DOI: 10.3402/gha.v2i0.2047 PMID: 20052422

Kodama, C., El Rifay, A. S., Badra, R., Abu Salbi, R., Abubakar, A., & Kayali, G. (2023). Operationalizing One Health: Strategic guidance for prevention and control of emerging and re-emerging vector-borne and zoonotic diseases in the Eastern Mediterranean Region. *Frontiers in Tropical Diseases*, *4*, 1237688. Advance online publication. DOI: 10.3389/fitd.2023.1237688

Korolenko, A. A., Nilsson, E. E., De Santos, S., & Skinner, M. K. (2025). Generational stability of environmentally induced epigenetic transgenerational inheritance of adult-onset disease over ten mammalian generations. *Environmental Epigenetics*, *11*(1), dvaf033. Advance online publication. DOI: 10.1093/eep/dvaf033 PMID: 41438350

Korolenko, A. A., Nilsson, E. E., De Santos, S., & Skinner, M. K. (2026). Stability of epigenetic transgenerational inheritance of adult-onset disease and parturition abnormalities. *Proceedings of the National Academy of Sciences of the United States of America*, *123*(8), e2523071123. DOI: 10.1073/pnas.2523071123 PMID: 41701841

Korrapati, C., Chowdareddy, S., Suresh, C., & Raghavan, R. (2026). Cancer Cell Membrane-Coated Nanoparticles: Therapeutic Advancements and Hybrid Systems. *Applications of Nanomaterials for Oncology*, 131–160. DOI: 10.4018/979-8-3373-3765-4.ch005

Kraemer, M. U., Sinka, M. E., Duda, K. A., Mylne, A. Q., Shearer, F. M., Barker, C. M., Moore, C. G., Carvalho, R. G., Coelho, G. E., Van Bortel, W., Hendrickx, G., Schaffner, F., Elyazar, I. R., Teng, H.-J., Brady, O. J., Messina, J. P., Pigott, D. M., Scott, T. W., Smith, D. L., & Hay, S. I. (2015). The global distribution of the arbovirus vectors Aedes aegypti and Ae. Albopictus. *eLife*, *4*, e08347. DOI: 10.7554/eLife.08347 PMID: 26126267

Krieger, N. (2001). Theories for social epidemiology in the 21st century: An ecosocial perspective. *International Journal of Epidemiology*, *30*(4), 668–677. DOI: 10.1093/ije/30.4.668 PMID: 11511581

Kriswardhana, W., Toaza, B., Esztergar-Kiss, D., & Duleba, S. (2025). Analytic hierarchy process in transportation decision-making: A two-staged review on the themes and trends of two decades. *Expert Systems with Applications*, *261*, 125491. DOI: 10.1016/j.eswa.2024.125491

Kuehnert, P. A., Stefan, C. P., Badger, C. V., & Ricks, K. M. (2021). Crimean-Congo Hemorrhagic Fever Virus (CCHFV): A Silent but Widespread Threat. *Current Tropical Medicine Reports*, *8*(2), 141–147. DOI: 10.1007/s40475-021-00235-4 PMID: 33747715

Kumar, M., Khan, S. Y., Liu, S., Ji, W., Shen, Y., Sohrabi, A., & Mazhar, A. R. (2024). Scientific mapping and data analysis of the research landscape in perovskite solar cell technology. *Solar Energy*, *273*, 112509. DOI: 10.1016/j.solener.2024.112509

Lakhoo, D. P., Brink, N., Radebe, L., Craig, M. H., Pham, M. D., Haghighi, M. M., Wise, A., Solarin, I., Luchters, S., Maimela, G., Chersich, M. F., Chikandiwa, A., Nakstad, B., Wright, C. Y., Brimicombe, C., Technau, K.-G., Harden, L., Boeckmann, M., Strehlau, R., & Brink, N. (2025). A systematic review and meta-analysis of heat exposure impacts on maternal, fetal and neonatal health. *Nature Medicine*, *31*(2), 684–694. DOI: 10.1038/s41591-024-03395-8 PMID: 39500369

Lakshmaiah, A., & Raghavan, R. (2026). Deep Learning-Enhanced Hyperspectral Imaging: Applications in Oncology Cancer Diagnosis and Tumor Margin Identification. *Biomedical Applications in Deep Learning-Enhanced Hyperspectral Imaging*, 335–368. DOI: 10.4018/979-8-3373-7183-2.ch011

Lampropoulos, I. C., Aggelopoulos, E., Giannopoulos, V., Tsameti, A., & Karifyllas, S. G. (2026). Performance Management and Accounting Frameworks for Assessing Corporate Sustainability in the Energy Transition Era. In *Economic, Social, and Environmental Insights on Energy Transition* (pp. 65-96). IGI Global Scientific Publishing.

Langer, C. E., Mitchell, D. C., Armitage, T. L., Moyce, S. C., Tancredi, D. J., Castro, J., Vega-Arroyo, A. J., Bennett, D. H., & Schenker, M. B. (2021). Are Cal/OSHA regulations protecting farmworkers in California from heat-related illness? *Journal of Occupational and Environmental Medicine*, *63*(6), 532–539. DOI: 10.1097/JOM.0000000000002189 PMID: 33741829

Lan, T., Goh, Y. M., Jensen, O., & Asmone, A. S. (2022). The impact of climate change on workplace safety and health hazard in facilities management: An in-depth review. *Safety Science*, *151*, 105745. Advance online publication. DOI: 10.1016/j.ssci.2022.105745

Lan, Y.-L., Huang, W.-T., Kao, C.-L., & Wang, H.-J. (2020). The Relationship between Organizational Climate, Job Stress, Workplace Burnout, and Retention of Pharmacists. *Journal of Occupational Health*, *62*(1), e12079. DOI: 10.1002/1348-9585.12079 PMID: 31407446

Lau, C. L., Smythe, L. D., & Weinstein, P. (2010). Leptospirosis: An emerging disease in travelers? *Travel Medicine and Infectious Disease*, *8*(1), 33–39. DOI: 10.1016/j.tmaid.2009.12.002 PMID: 20188303

Lauer, E. A., Armenti, K., Henning, M., & Sirois, L. (2019). Identifying barriers and supports to breastfeeding in the workplace experienced by mothers in the New Hampshire Special Supplemental Nutrition Program for Women, Infants, and Children utilizing the Total Worker Health framework. *International Journal of Environmental Research and Public Health*, *16*(4), 529. DOI: 10.3390/ijerph16040529 PMID: 30781764

Lawler, O. K., Allan, H. L., Baxter, P. W. J., Castagnino, R., Tor, M. C., Dann, L. E., Hungerford, J., Karmacharya, D., Lloyd, T. J., López-Jara, M. J., Massie, G. N., Novera, J., Rogers, A. M., & Kark, S. (2021). The COVID-19 pandemic is intricately linked to biodiversity loss and ecosystem health. *The Lancet. Planetary Health*, *5*(11), e840–e850. DOI: 10.1016/S2542-5196(21)00258-8 PMID: 34774124

Lazaro, P., & Momayez, M. (2020). Development of a modified predicted heat strain model for hot work environments. *International Journal of Mining Science and Technology*, *30*(4), 477–481. DOI: 10.1016/j.ijmst.2020.05.009

Leal Filho, W., Schmidberger, I., Sharifi, A., Vargas, V. R., Rampasso, I. S., Dibbern, T., Liakh, O., Aina, Y. A., Trevisan, L. V., Mbah, M. F., Anholon, R., & Kozlova, V. (2024). Design thinking for sustainable development: A bibliometric analysis and case study research. *Journal of Cleaner Production*, *455*, 142285. DOI: 10.1016/j.jclepro.2024.142285

Lebel, L., Paquin, V., Kenny, T. A., Fletcher, C., Nadeau, L., Chachamovich, E., & Lemire, M. (2022). Climate change and Indigenous mental health in the Circumpolar North: A systematic review to inform clinical practice. *Transcultural Psychiatry*, *59*(3), 312–336. DOI: 10.1177/13634615211066698 PMID: 34989262

Lee, G. (2024). Impact of Smoke from Wildfire and Agricultural Burning on Farmworker Health and Behavior (Doctoral dissertation, University of California, Davis).

Lee, S. H., Nam, K. W., Jeong, J. Y., Yoo, S. J., Koh, Y.-S., Lee, S., Heo, S. T., Seong, S.-Y., & Lee, K. H. (2013). The Effects of Climate Change and Globalization on Mosquito Vectors: Evidence from Jeju Island, South Korea on the Potential for Asian Tiger Mosquito (Aedes albopictus) Influxes and Survival from Vietnam Rather Than Japan. *PLoS One*, *8*(7), e68512. DOI: 10.1371/journal.pone.0068512 PMID: 23894312

Lefrançois, T., Lina, B., & Autran, B. (2023). One Health approach at the heart of the French Committee for monitoring and anticipating health risks. *Nature Communications*, *14*(1), 7540. DOI: 10.1038/s41467-023-43089-2 PMID: 37990001

Leiserowitz, A., Maibach, E., Rosenthal, S., Kotcher, J., Ballew, M., Goldberg, M., & Gustafson, A. (2020). *Climate Change in the American Mind: April 2020*. Yale University and George Mason University.

Leiter, M. P., & Maslach, C. (2017). Burnout and engagement: Contributions to a new vision. *Burnout Research*, *5*, 55–57. DOI: 10.1016/j.burn.2017.04.003

Lesne, J. (2021). L'émergence de maladies infectieuses d'origine zoonotique: Complexité écologique et responsabilités socio-économiques. *Environnement, Risques & Santé : ERS*, *20*(3), 244–257. DOI: 10.1684/ers.2021.1560

Levi, M., Kjellstrom, T., & Baldasseroni, A. (2018). Impact of climate change on occupational health and productivity: A systematic literature review focusing on workplace heat. *La Medicina del Lavoro*, *109*(3), 163–179. DOI: 10.23749/mdl.v109i3.6851 PMID: 29943748

Levy, B. S., Wegman, D. H., Baron, S. L., & Sokas, R. K. (2017). *Occupational and environmental health: Recognizing and preventing disease and injury* (7th ed.). Oxford University Press. DOI: 10.1093/oso/9780190662677.001.0001

Levy, K., Woster, A. P., Goldstein, R. S., & Carlton, E. J. (2016). Untangling the impacts of climate change on waterborne diseases: A systematic review of relationships between diarrheal diseases and temperature, rainfall, flooding, and drought. *Environmental Science & Technology*, *50*(10), 4905–4922. DOI: 10.1021/acs.est.5b06186 PMID: 27058059

Linnenluecke, M. K., Birt, J., & Griffiths, A. (2015). The role of accounting in supporting adaptation to climate change. *Accounting and Finance*, *55*(3), 607–625. DOI: 10.1111/acfi.12120

Li, P. F., Xu, Y., & Chen, Q. B. (2024). Global trends and research characteristics on CO2 capture and conversion from 2000 to 2023: A bibliometric review. *Sustainable Energy Technologies and Assessments*, *67*, 103836. DOI: 10.1016/j.seta.2024.103836

Lipman, G. S., Gaudio, F. G., Eifling, K. P., Ellis, M. A., Grissom, C. K., Morris, A., & Otten, E. M. (2019). Wilderness Medical Society Clinical Practice Guidelines for the prevention and treatment of heat illness: 2019 update. *Wilderness & Environmental Medicine*, *30*(4), S33–S46. DOI: 10.1016/j.wem.2018.10.004 PMID: 31221601

Litwan, K., Tran, V., Nyhan, K., & Pérez-Escamilla, R. (2021). How do breastfeeding workplace interventions work?: A realist review. *International Journal for Equity in Health*, *20*(1), 148. DOI: 10.1186/s12939-021-01490-7 PMID: 34172068

Liu, Y. G., Jiang, S. T., Zhang, L., Zheng, H., Zhang, T., Zhang, J. W., Zhao, H.-T., Sang, X.-T., Xu, Y.-Y., & Lu, X. (2023). Worldwide productivity and research trend of publications concerning tumor immune microenvironment (TIME): A bibliometric study. *European Journal of Medical Research*, *28*(1), 229. DOI: 10.1186/s40001-023-01195-3 PMID: 37430294

Li, Z., Chen, X., Yang, X., Huang, X., Ye, R., Du, J., Tang, S., Qiu, A., Huang, Y., Deng, Y., Zhong, Y., Liang, B., Hui, C.-Y., Chen, D., & Huang, Z. (2026). Endogenous 15(S)-hydroxyeicosatetraenoic acid mediates amplified estrogenic responses under nanoplastic–homosalate coexposure. *Environment International*, *208*, 110083. DOI: 10.1016/j.envint.2026.110083 PMID: 41592362

Lobus, N. V., Knyazeva, M. A., Popova, A. F., & Kulikovskiy, M. S. (2023). Carbon Footprint Reduction and Climate Change Mitigation: A Review of the Approaches, Technologies, and Implementation Challenges. *C*, *9*(4), 120. Advance online publication. DOI: 10.3390/c9040120

Lopez, H., Lee, S.-K., West, R., Kim, D., & Jia, L. (2025). The longest-lasting 2023 western North American heat wave was fueled by the record-warm Atlantic Ocean. *Nature Communications*, *16*(1), 6544. DOI: 10.1038/s41467-025-61859-y PMID: 40670333

Lubis, A., & Priwiratama, H. (2025). Effectiveness of 356 g/l and 363 g/l glyphosate in controlling *Ageratum conyzoides, Panicum repens*, and *Cyperus kyllingia* in young mature oil palm areas. *IOP Conference Series. Earth and Environmental Science*, *1494*(1), 012006. DOI: 10.1088/1755-1315/1494/1/012006

Ludwig, A., Zheng, H., Vrbova, L., Drebot, M., Iranpour, M., & Lindsay, L. (2019). Increased risk of endemic mosquito-borne diseases in Canada due to climate change. *Canada Communicable Disease Report = Relevé des Maladies Transmissibles au Canada*, *45*(4), 91–97. DOI: 10.14745/ccdr.v45i04a03 PMID: 31285698

Lukács, B., Molnár, P., & Tóth, Á. (2026). Categorisation of SDG targets into ESG pillars based on ESRS taxonomy. *Discover Sustainability*, *7*(1), 120. DOI: 10.1007/s43621-025-02550-6

Luna, Z., & Luker, K. (2013). Reproductive justice. *Annual Review of Law and Social Science*, *9*(1), 327–352. DOI: 10.1146/annurev-lawsocsci-102612-134037

Lundgren, Karin, Kuklane, Kalev, & Gao, Chuansi, and Ingvar HOLM^|^Eacute;R. (. (2013). Effects of Heat Stress on Working Populations When Facing Climate Change. *Industrial Health*, *51*(1), 3–15. DOI: 10.2486/indhealth.2012-0089 PMID: 23411752

Lun, Y., Liu, L., Cheng, L., Li, X., Li, H., & Xu, Z. (2021). Assessment of GCMs simulation performance for precipitation and temperature from CMIP5 to CMIP6 over the Tibetan Plateau. *International Journal of Climatology*, *41*(7), 3994–4018. DOI: 10.1002/joc.7055

Luyten, A., Winkler, M. S., Ammann, P., & Dietler, D. (2023). Health impact studies of climate change adaptation and mitigation measures – A scoping review. *The Journal of Climate Change and Health*, *9*, 100186. Advance online publication. DOI: 10.1016/j.joclim.2022.100186

M, M. K., P, D. A., J, S. P., Moger, S. S., & Raghavan, R. (2026). Spatial Multi-Omics in Cancers: Integrating Molecular Landscapes for Precision Oncology. *Spatially Resolved Single-Cell and Multiple Omics in Human Diseases*, 217–248. DOI: 10.4018/979-8-3373-4849-0.ch007

Mackenzie, J. S., & Jeggo, M. (2019). The One Health approach—Why is it so important? *Tropical Medicine and Infectious Disease*, *4*(2), 88. DOI: 10.3390/tropicalmed4020088 PMID: 31159338

Magnavita, N., Capitanelli, I., Ilesanmi, O., & Chirico, F. (2022). Occupational Lyme Disease: A Systematic Review and Meta-Analysis. *Diagnostics (Basel)*, *12*(2), 296. DOI: 10.3390/diagnostics12020296 PMID: 35204387

Maier, K. J., Whitehead, G. I., & Walter, M. I. (2018). Teaching psychology and climate change. *Teaching of Psychology*, *45*(3), 226–234. DOI: 10.1177/0098628318779261

Mancuso, E., Toma, L., Polci, A., d'Alessio, S. G., Luca, M. D., Orsini, M., Domenico, M. D., Marcacci, M., Mancini, G., Spina, F., Goffredo, M., & Monaco, F. (2019). *Crimean-Congo Hemorrhagic Fever Virus Genome in Tick from Migratory Bird, Italy—Volume 25, Number 7—July 2019—Emerging Infectious Diseases journal—CDC*. DOI: 10.3201/eid2507.181345

Mandlazi, N. B., Vilakati, G. D., Madikizela, L., & Kgomo, H. (2026). Non-targeted screening of emerging pollutants across wastewater and receiving waters: A case study in Eswatini. *The Science of the Total Environment*, *1014*, 181353. DOI: 10.1016/j.scitotenv.2026.181353 PMID: 41512340

Maria, R. L., & Voss, J. (2019). American climate perspectives survey 2019. In In ecoAmerica and Lake Research Partners 2019. 1-4.

Mariappanadar, S., Maurer, I., Kramar, R., & Muller-Camen, M. (2022). Is it a sententious claim? An examination of the quality of occupational health, safety and well-being disclosures in global reporting initiative reports across industries and countries. *International Business Review*, *31*(2), 101922. DOI: 10.1016/j.ibusrev.2021.101922

Martin, N. A., Hulland, K. R. S., Dreibelbis, R., Sultana, F., & Winch, P. J. (2018). Sustained Adoption of Water, Sanitation and Hygiene Interventions: Systematic Review. *Tropical Medicine & International Health : TM & IH*, *23*(2), 122–135. DOI: 10.1111/tmi.13011 PMID: 29160921

Martins-Santos, E., & Oliveira, C. A. (2026). A Review on the Effects of Atrazine on Male Rat Reproductive System Cytoarchitecture, Steroidogenesis and Oxidative Pathway. *Journal of Molecular Histology*, *57*(1), 45. DOI: 10.1007/s10735-025-10671-5 PMID: 41528512

Maslach, C., Jackson, S. E., & Leiter, M. P. (1996). *Maslach Burnout Inventory manual* (3rd ed.). Consulting Psychologists Press.

Ma, T., Moore, J., & Cleary, A. (2022). Climate change impacts on the mental health and wellbeing of young people: A scoping review of risk and protective factors. *Social Science & Medicine, 301*, 114888. DOI: 10.1016/j.socscimed.2022.114888 PMID: 35367905

Mavroulidis, M., Vouros, P., Fotiadis, S., Konstantakopoulou, F., Fountoulakis, G., Nikolaou, I., & Evangelinos, K. (2022). Occupational health and safety of multinational construction companies through evaluation of corporate social responsibility reports. *Journal of Safety Research, 81*, 45–54. DOI: 10.1016/j.jsr.2022.01.005 PMID: 35589305

Mawcha, K., Kinyanjui, G., Simiyu, S., Babalola, O., & Ndolo, D. (2024). *Biopesticides for Sustainable Agriculture: A Review of Their Role in Integrated Pest Management*. IntechOpen.

Ma, X., & Huo, Y. (2023). Are users willing to embrace ChatGPT? Exploring the factors on the acceptance of chatbots from the perspective of the AIDUA framework. *Technology in Society, 75*, 102362. DOI: 10.1016/j.techsoc.2023.102362

McCauley, D., & Heffron, R. (2018). Just transition: Integrating climate, energy and environmental justice. *Energy Policy, 119*, 1–7. DOI: 10.1016/j.enpol.2018.04.014

McClure, E., Ranapurwala, S., Nocera, M., & Richardson, D. (2024). Heat-related fatalities in North Carolina, 1999–2017. *American Journal of Industrial Medicine, 67*(6), 551–555. DOI: 10.1002/ajim.23587 PMID: 38624268

McKay, V. R., Dolcini, M. M., & Hoffer, L. D. (2017). The Dynamics of De-Adoption: A Case Study of Policy Change, de-Adoption, and Replacement of an Evidence-Based HIV Intervention. *Translational Behavioral Medicine*, 7(4), 821–831. DOI: 10.1007/s13142-017-0493-1 PMID: 28397157

Medlock, J. M., & Leach, S. A. (2015). Effect of climate change on vector-borne disease risk in the UK. *The Lancet. Infectious Diseases, 15*(6), 721–730. DOI: 10.1016/S1473-3099(15)70091-5 PMID: 25808458

Mencattelli, G., Ndione, M. H. D., Rosà, R., Marini, G., Diagne, C. T., Diagne, M. M., Fall, G., Faye, O., Diallo, M., Faye, O., Savini, G., & Rizzoli, A. (2022). Epidemiology of West Nile virus in Africa: An underestimated threat. *PLoS Neglected Tropical Diseases, 16*(1), e0010075. DOI: 10.1371/journal.pntd.0010075 PMID: 35007285

Mengistu, D. A., Geremew, A., & Tessema, R. A. (2024). Pesticide safety practice and its public health risk in African regions: Systematic review and meta-analysis. *BMC Public Health, 24*(1), 2295. DOI: 10.1186/s12889-024-19764-4 PMID: 39180029

Meng, X., Xiong, H., Yang, H., & Cao, Y. (2021). Dynamic prediction of indoor wet bulb globe temperature in an industrial workshop. *Applied Thermal Engineering*, *195*, 117219. DOI: 10.1016/j.applthermaleng.2021.117219

Metz, D., Prier, S., & Miller, B. M. (2021). *Prevention of heat-related illness in indoor places of employment: Standardized regulatory impact assessment (SRIA) of the proposed California regulation for heat illness prevention in indoor places of employment (WR-A1386-1)*. RAND Corporation., DOI: 10.7249/WRA1386-1

Mezzacappo, N. F., Inada, N. M., Vollet-Filho, J. D., Denton, M. L., Yakovlev, V. V., & Bagnato, V. S. (2025). Uncovering the time-temperature thresholds of in vitro mitochondrial bioenergetics dysfunction under hyperthermic stress. *International Journal of Hyperthermia*, *42*(1), 2560017. DOI: 10.1080/02656736.2025.2560017 PMID: 40977260

Middleton, J., Cunsolo, A., Jones-Bitton, A., Wright, C. J., & Harper, S. L. (2020). Indigenous mental health in a changing climate: A systematic scoping review of the global literature. *Environmental Research Letters*, *15*(5), 053001. DOI: 10.1088/1748-9326/ab68a9

Mizelle, E., Larson, K., Bolin, L., & Kearney, G. (2022). Fluid Intake and Hydration Status Among North Carolina Farmworkers: A Mixed Methods Study. *Workplace Health & Safety*, *70*(12), 532–541. DOI: 10.1177/21650799221117273 PMID: 36002982

Moda, H., Filho, W., & Minhas, A. (2019). Impacts of Climate Change on Outdoor Workers and Their Safety: Some Research Priorities. *International Journal of Environmental Research and Public Health*, *16*(18), 3458. DOI: 10.3390/ijerph16183458 PMID: 31533360

Mohai, P., Pellow, D., & Roberts, J. T. (2009). Environmental justice. *Annual Review of Environment and Resources*, *34*(1), 405–430. DOI: 10.1146/annurev-environ-082508-094348

Mora, C., McKenzie, T., Gaw, I. M., Dean, J. M., von Hammerstein, H., Knudson, T. A., Setter, R. O., Smith, C. Z., Webster, K. M., Patz, J. A., & Franklin, E. C. (2022). Over half of known human pathogenic diseases can be aggravated by climate change. *Nature Climate Change*, *12*(9), 869–875. DOI: 10.1038/s41558-022-01426-1 PMID: 35968032

Mordecai, E. A., Ryan, S. J., Caldwell, J. M., Shah, M. M., & LaBeaud, A. D. (2020). Climate change could shift disease burden from malaria to arboviruses in Africa. *The Lancet. Planetary Health*, *4*(9), e416–e423. DOI: 10.1016/S2542-5196(20)30178-9 PMID: 32918887

Morea, A., Schino, V., Bartolomeo, N., Ravallese, R., Sacino, G., Ravallese, R., Iatta, R., Otranto, D., Lovreglio, P., & Stufano, A. (2025). Seroprevalence of tick-borne diseases in Europe in occupational settings: A systematic review and metanalysis. *Public Health, 248*, 105923. DOI: 10.1016/j.puhe.2025.105923 PMID: 40907304

Morelli, G., Gurrieri, A. R., Drago, C., Gatto, A., & Magazzino, C. (2026). A Biblio-Ecological Study on Sustainable Land Planning Research: Evidence from Machine Learning Methods. *Socio-Economic Planning Sciences, 105*, 102431. DOI: 10.1016/j.seps.2026.102431

Morris, K. I., Chan, A., Morris, K. J. K., & Walker, T. R. (2022). Climate change and workers' health: A scoping review. *International Journal of Environmental Research and Public Health, 19*(15), 9445. PMID: 35954801

Morris, N. B., Piil, J. F., Morabito, M., Messeri, A., Levi, M., Ioannou, L. G., Ciuha, U., Pogačar, T., Kajfež Bogataj, L., Kingma, B., Casanueva, A., Kotlarski, S., Spirig, C., Foster, J., Havenith, G., Sotto Mayor, T., Flouris, A. D., & Nybo, L. (2021). The HEAT-SHIELD project-Perspectives from an inter-sectoral approach to occupational heat stress. *Journal of Science and Medicine in Sport, 24*(8), 747–755. DOI: 10.1016/j.jsams.2021.03.001 PMID: 33757698

Moser, S. C. (2010). Communicating climate change: History, challenges, process and future directions. *Wiley Interdisciplinary Reviews: Climate Change, 1*(1), 31–53. DOI: 10.1002/wcc.11

Mousavi, S. M., Dehghan, H., & Yazdanirad, S. (2026). A critical systematic narrative review of the WBGT index for heat stress assessment in diverse climatic regions: Insights from Iran. *International Journal of Biometeorology, 70*(3), 63. DOI: 10.1007/s00484-025-03093-1 PMID: 41703112

Moyce, S., Armitage, T., Mitchell, D., & Schenker, M. (2019). Acute kidney injury and workload in a sample of California agricultural workers. *American Journal of Industrial Medicine, 63*(3), 258–268. DOI: 10.1002/ajim.23076 PMID: 31773783

Moyce, S., Mitchell, D., Armitage, T., Tancredi, D., Joseph, J., & Schenker, M. (2017). Heat strain, volume depletion and kidney function in California agricultural workers. *Occupational and Environmental Medicine, 74*(6), 402–409. DOI: 10.1136/oemed-2016-103848 PMID: 28093502

Msellemu, D., Tanner, M., Yadav, R., & Moore, S. J. (2024). Occupational exposure to malaria, leishmaniasis and arbovirus vectors in endemic regions: A systematic review. *Current Research in Parasitology & Vector-borne Diseases, 6*, 100185. DOI: 10.1016/j.crpvbd.2024.100185 PMID: 39027087

Mubemba, B., Mburu, M. M., Changula, K., Muleya, W., Moonga, L. C., Chambaro, H. M., Kajihara, M., Qiu, Y., Orba, Y., Hayashida, K., Sutcliffe, C. G., Norris, D. E., Thuma, P. E., Ndubani, P., Chitanga, S., Sawa, H., Takada, A., & Simulundu, E. (2022). Current knowledge of vector-borne zoonotic pathogens in Zambia: A clarion call to scaling-up "One Health" research in the wake of emerging and re-emerging infectious diseases. *PLoS Neglected Tropical Diseases*, *16*(2), e0010193. DOI: 10.1371/journal.pntd.0010193 PMID: 35120135

Mu, L., Li, Y., Liu, H., & Wang, Q. (2025). Implementation or Inaction: How Governmental Multi-Interventions Catalyze Farmers' Adoption for Climate Adaption Technology? *Journal of Cleaner Production*, *521*, 146193. DOI: 10.1016/j.jclepro.2025.146193

Mullins-Jaime, C., Stefancik, J., Gannon, A., Bourbai, C., & Ismyana, M. (2026). What states are most at risk? An epidemiologic analysis of state-level occupational heat-related illness and fatality trends over three decades. *Journal of Safety Research*. Advance online publication. DOI: 10.1016/j.jsr.2026.01.013

Muramoto, J., Parr, D., Perez, J., & Wong, D. (2022). Integrated Soil Health Management for Plant Health and One Health: Lessons From Histories of Soil-borne Disease Management in California Strawberries and Arthropod Pest Management. *Frontiers in Sustainable Food Systems*, *6*, 6. DOI: 10.3389/fsufs.2022.839648

Musa, S. F. P. D., & Ariff Lim, S. (2025). Revitalising agriculture through climate change mitigation: A systematic literature review on smart technologies and sustainable practices. *International Journal of Climate Change Strategies and Management*, *17*(1), 481–499. DOI: 10.1108/IJCCSM-05-2024-0071

Muzammil, K., Rayyani, S., Abbas Sahib, A., Gholizadeh, O., Naji Sameer, H., Jwad Kazem, T., Badran Mohammed, H., Ghafouri Kalajahi, H., Zainul, R., & Yasamineh, S. (2024). Recent Advances in Crimean-Congo Hemorrhagic Fever Virus Detection, Treatment, and Vaccination: Overview of Current Status and Challenges. *Biological Procedures Online*, *26*(1), 20. DOI: 10.1186/s12575-024-00244-3 PMID: 38926669

Nagashri, N., Archana, L., & Raghavan, R. (2024). Microbial Volatile Compounds (MVOCs) in Food Industries and Food Safety Applications. *Innovations, Securities, and Case Studies Across Healthcare, Business, and Technology*, 75–96. DOI: 10.4018/979-8-3693-1906-2.ch005

Naguib, M. M., Li, R., Ling, J., Grace, D., Nguyen-Viet, H., & Lindahl, J. F. (2021). Live and wet markets: Food access versus the risk of disease emergence. *Trends in Microbiology*, *29*(7), 573–581. DOI: 10.1016/j.tim.2021.02.007 PMID: 33712334

Nanjundeshwara, N., Lakshmaiah, A., & Raghavan, R. (2025). Aflatoxigenic Fungi. *Aflatoxigenic Fungi*, 225–239. DOI: 10.1201/9781003487043-29

Naser, S. A., Zaqout, M. A., Ghosh, R., Atallah, E., & Alajrami, E. (2015). Predicting student performance using artificial neural network: In the faculty of engineering and information technology. *International Journal of Hybrid Information Technology*, *8*(2), 221–228. DOI: 10.14257/ijhit.2015.8.2.20

Nassikas, N., Spangler, K. R., Fann, N., Nolte, C. G., Dolwick, P., Spero, T. L., Sheffield, P., & Wellenius, G. A. (2020). Ozone-related asthma emergency department visits in the US in a warming climate. *Environmental Research*, *183*, 109206. DOI: 10.1016/j.envres.2020.109206 PMID: 32035409

Nath, A., Sahu, S., & Lee, J. K. W. (2026). An Umbrella Review of Systematic Reviews and Meta-Analyses on Occupational Heat Exposure, Health Risks, and Productivity Losses Globally. *Current Environmental Health Reports*, *13*(1), 2. DOI: 10.1007/s40572-025-00520-8 PMID: 41499074

Naylor, C., Canvin, S., Dunk, L., Gopaldasani, V., Liebenberg, A., Lux, G., Macfarlane, F., Panyam, S., Sharp, V., Thomas, A., Uru, E., & Whitelaw, J. L. (2025). What does good health look like? Development of leading occupational health performance indicators. *Annals of Work Exposures and Health*, *69*(9), 904–916. DOI: 10.1093/annweh/wxaf050 PMID: 40973106

Nicholas, P. K., Breakey, S., & Blank, P. (2020). Roles of nurse practitioners: Health consequences of climate change in vulnerable older adults. *The Journal for Nurse Practitioners*, *16*(6), 433–437. DOI: 10.1016/j.nurpra.2020.03.011

Njeshi, C. N., Robertson, A. P., & Martin, R. J. (2026). Onchocerciasis overview: Pathogenesis, clinical spectrum, and therapeutic challenges of a neglected tropical disease. *Frontiers in Parasitology*, *5*, 1690973. Advance online publication. DOI: 10.3389/fpara.2026.1690973 PMID: 41908972

Nnko, H. J., Gwakisa, P. S., Ngonyoka, A., Sindato, C., & Estes, A. B. (2021). Potential impacts of climate change on geographical distribution of three primary vectors of African Trypanosomiasis in Tanzania's Maasai Steppe: G. m. morsitans, G. pallidipes and G. swynnertoni. *PLoS Neglected Tropical Diseases*, *15*(2), e0009081. DOI: 10.1371/journal.pntd.0009081 PMID: 33571190

NOAA. (2019). When was the first Earth Day? Retrieved from https://oceanservice.noaa.gov/facts/earth-day.html

Nowrouzi-Kia, B., Choi, C., Premji, R., Sathananthan, A., Balakrishnar, K., Haritos, A., Long, B.-Z. S., Mazur, M., & Fatemi, A.-B. 2026. "The Impacts of Climate Change on Occupational Health and Work among Outdoor Workers: A Scoping Review" edited by J. Robinson. *PLOS Global Public Health* 6(2):e0005888. DOI: 10.1371/journal.pgph.0005888

Nunfam, V. F., Afrifa-Yamoah, E., Adusei-Asante, K., Van Etten, E. J., Frimpong, K., Mensah, I. A., & Oosthuizen, J. (2021). Construct validity and invariance assessment of the social impacts of occupational heat stress scale (SIOHSS) among Ghanaian mining workers. *The Science of the Total Environment, 771*, 144911. DOI: 10.1016/j.scitotenv.2020.144911 PMID: 33736178

Nunfam, V., Adusei-Asante, K., Etten, E., Oosthuizen, J., & Frimpong, K. (2018). Social impacts of occupational heat stress and adaptation strategies of workers: A narrative synthesis of the literature. *The Science of the Total Environment, 643*, 1542–1552. DOI: 10.1016/j.scitotenv.2018.06.255 PMID: 30189570

Nyakuwanika, M., & Panicker, M. (2025). The Role of Environmental Accounting in Mitigating Climate Change: ESG Disclosures and Effective Reporting-A Systematic Literature Review. *Journal of Risk and Financial Management, 18*(9), 480. DOI: 10.3390/jrfm18090480

Odigie, A., Stufano, A., Schino, V., Zarea, A., Ndiana, L., Mrenoshki, D., Ugochukwu, I. C. I., Lovreglio, P., Greco, G., Pratelli, A., Camero, M., & Tempesta, M. (2024). West Nile Virus Infection in Occupational Settings—A Systematic Review. *Pathogens (Basel, Switzerland), 13*(2), 157. DOI: 10.3390/pathogens13020157 PMID: 38392895

Ogden, N. H., Radojević, M., Wu, X., Duvvuri, V. R., Leighton, P. A., & Wu, J. (2014). Estimated effects of projected climate change on the basic reproductive number of the Lyme disease vector Ixodes scapularis. *Environmental Health Perspectives, 122*(6), 631–638. DOI: 10.1289/ehp.1307799 PMID: 24627295

Ojala, M., & Bengtsson, H. (2019). Young people's coping strategies concerning climate change: Relations to perceived communication with parents and friends. *Environment and Behavior, 51*(8), 907–935. DOI: 10.1177/0013916518763894

Oliveira, S., Rocha, J., Sousa, C. A., & Capinha, C. (2021). Wide and increasing suitability for Aedes albopictus in Europe is congruent across distribution models. *Scientific Reports, 11*(1), 9916. DOI: 10.1038/s41598-021-89096-5 PMID: 33972597

Omassoli, J., Hill, N., Woods, D., Delves, S., Fallowfield, J., Brett, S., Wilson, D., Corbett, R. W., Allsopp, A. J., & Stacey, M. (2019). Variation in renal responses to exercise in the heat with progressive acclimatisation. *Journal of Science and Medicine in Sport*, *22*(9), 1004–1009. DOI: 10.1016/j.jsams.2019.04.010 PMID: 31085130

Omwenga, M., Anika, A., Mwangi, S., Omwenga, M., Anika, A., & Mwangi, S. (2024). Prevalence of conduct disorder indicators in young children from public primary schools in Masaba South, Kenya. *International Journal of Science and Research Archive*, *13*(2), 2.

Omwenga, M., & Kayusi, F. (2024). Perception of parents of children with mental health in Kisii County, Kenya. *Asian Journal of Advanced Research and Reports*, *18*(11), 264–278. DOI: 10.9734/ajarr/2024/v18i11794

Omwenga, M., & W, M. S. (2024). The influence of sociodemographic factors on anti-social behavior among children: A case of Bobaracho area, Nyaribari Chache, Kenya. *Archives of Current Research International*, *24*(11), 375–385. DOI: 10.9734/acri/2024/v24i11979

Orikpete, O. F., Kikanme, K. N., Falade, T. D. O., Dennis, N. M., Ejike Ewim, D. R., & Fadare, O. O. (2025). Neonicotinoid pesticides in African agriculture: What do we know and what should be the focus for future research? *Chemosphere*, *372*, 144057. DOI: 10.1016/j.chemosphere.2024.144057 PMID: 39746486

Osaka, S., Bellamy, R., & Castree, N. (2021). Framing "nature-based" solutions to climate change. *Wiley Interdisciplinary Reviews: Climate Change*, *12*(5), e729. Advance online publication. DOI: 10.1002/wcc.729

P., A., D., K. M., M., J., P., S., & Raghavan, R. (2025). The Cancer Exposome. *Cancer Exposomics and Environmental Influences on Carcinogenesis*, 161–192. DOI: 10.4018/979-8-3373-2165-3.ch004

Paivio, A. (1991). Dual Coding Theory: Retrospect and Current Status. *Canadian Journal of Psychology*, *45*(3), 255–27. DOI: 10.1037/h0084295

Palasz, J., Farooqi, W., Musharraf, M. B., Rippon, B., Jedlicka, C., & Razzak, J. (2025). Diagnostic Biomarkers for Heat Stroke and Heat Exhaustion: A Scoping Review. *Disaster Medicine and Public Health Preparedness*, *19*, e153. DOI: 10.1017/dmp.2025.10069 PMID: 40524268

Pal, G., Patel, T., & Banik, T. (2021). Effect of climate change associated hazards on agricultural workers and approaches for assessing heat stress and its mitigation strategies–review of some research significances. *International Journal of Current Microbiology and Applied Sciences*, *10*(02), 2947–2975. DOI: 10.20546/ijcmas.2021.1002.325

Palinkas, L. A., & Wong, M. (2020). Global climate change and mental health. *Current Opinion in Psychology*, *32*, 12–16. DOI: 10.1016/j.copsyc.2019.06.023 PMID: 31349129

Pallavi, V. N., Chowdareddy, S., Byalahunasishetter, B. I., & Raghavan, R. (2026). CRISPR-Cas9 Technology for Precision Cancer Medicine: Target Discovery, Disease Modeling, and Therapeutics. *CRISPR-Cas9 Technology for Precision Medicine*, 35–72. DOI: 10.4018/979-8-3373-4857-5.ch002

Palomar, A. M., Portillo, A., Santibáñez, P., Mazuelas, D., Arizaga, J., Crespo, A., Gutiérrez, Ó., Cuadrado, J. F., & Oteo, J. A. (2013). Crimean-Congo Hemorrhagic Fever Virus in Ticks from Migratory Birds, Morocco1. *Emerging Infectious Diseases*, *19*(2), 260–263. DOI: 10.3201/eid1902.121193 PMID: 23347801

Pandey, D. K., Lucey, B. M., & Kumar, S. (2023). Border disputes, conflicts, war, and financial markets research: A systematic review. *Research in International Business and Finance*, *65*, 101972. DOI: 10.1016/j.ribaf.2023.101972

Pan, J., Zhang, Y., Zhao, B., Shao, W., & Chen, X. (2025). Current perspectives and trend of ferroptosis in head and neck cancer: A bibliometric analysis. *Frontiers in Oral Health*, *6*, 1601962. DOI: 10.3389/froh.2025.1601962 PMID: 40860793

Pan, Q., Sumner, D. A., Mitchell, D. C., & Schenker, M. (2021). Compensation incentives and heat exposure affect farm worker effort. *PLoS One*, *16*(11), e0259459. DOI: 10.1371/journal.pone.0259459 PMID: 34727122

Part, C., Filippi, V., Cresswell, J. A., Ganaba, R., Hajat, S., Nakstad, B., Roos, N., Kadio, K., Chersich, M., Lusambili, A., Kouanda, S., & Kovats, S. (2022). How do high ambient temperatures affect infant feeding practices? A prospective cohort study of postpartum women in Bobo-Dioulasso, Burkina Faso. *BMJ Open*, *12*(10), e061297. DOI: 10.1136/bmjopen-2022-061297 PMID: 36198451

Pascual, M., Bouma, M. J., & Dobson, A. P. (2002). Cholera and climate: Revisiting the quantitative evidence. *Microbes and Infection*, *4*(2), 237–245. DOI: 10.1016/S1286-4579(01)01533-7 PMID: 11880057

Patel, C. J., Ioannidis, J. P. A., & Manrai, A. K. (2026). An atlas of exposome–phenome associations in health and disease risk. *Nature Medicine*, *32*(4), 1501–1510. DOI: 10.1038/s41591-026-04266-0 PMID: 41851348

Paterson, J., Berry, P., Ebi, K., & Varangu, L. (2014). Health and safety aspects of flood cleanup. *Canadian Medical Association Journal*, *186*(3), 199.

Patz, J. A., Frumkin, H., Holloway, T., Vimont, D. J., & Haines, A. (2014). Climate change: Challenges and opportunities for global health. *Journal of the American Medical Association*, *312*(15), 1565–1580. DOI: 10.1001/jama.2014.13186 PMID: 25244362

Pawirosumarto, S., Sarjana, P. K., & Muchtar, M. (2017). Factors Affecting Employee Performance of PT.Kiyokuni Indonesia. *International Journal of Law and Management*, *59*(4), 602–614. DOI: 10.1108/IJLMA-03-2016-0031

Pawłowska, Z. (2015). Using lagging and leading indicators for the evaluation of occupational safety and health performance in industry. *International Journal of Occupational Safety and Ergonomics*, *21*(3), 284–290. DOI: 10.1080/10803548.2015.1081769 PMID: 26647949

Paz, S. (2020). Climate change impacts on vector-borne diseases in Europe: Risks, predictions and actions. *The Lancet Regional Health. Europe*, *1*, 100017. DOI: 10.1016/j.lanepe.2020.100017 PMID: 35104838

Paz, S., & Semenza, J. C. (2013). El Niño and climate change—Contributing factors in the dispersal of West Nile virus in the United States. *Global Health Action*, *6*(1), 21830.

Pedersen, B. N., Jenkins, A., & Kjelland, V. (2020). Tick-borne pathogens in Ixodes ricinus ticks collected from migratory birds in southern Norway. *PLoS One*, *15*(4), e0230579. DOI: 10.1371/journal.pone.0230579 PMID: 32271774

Perkins-Kirkpatrick, S. E., & Lewis, S. C. (2020). Increasing trends in regional heatwaves. *Nature Communications*, *11*(1), 3357. DOI: 10.1038/s41467-020-16970-7 PMID: 32620857

Pervin, L., & Gan, T. Y. (2021). Sensitivity of physical parameterization schemes in WRF model for dynamic downscaling of climatic variables over the MRB. *Journal of Water and Climate Change*, *12*(4), 1043–1058. DOI: 10.2166/wcc.2020.036

Piacentino, J. D., & Schwartz, B. S. (2002). Occupational risk for Lyme disease in the United States. *Journal of Occupational and Environmental Medicine*, *44*(6), 491–500. PMID: 12085472

Pihkala, P. (2018). Eco-anxiety, tragedy, and hope: Psychological and spiritual dimensions of climate change. *Zygon, 53*(2), 545–569. DOI: 10.1111/zygo.12407

Pihkala, P. (2020). Eco-anxiety and environmental education. *Sustainability (Basel), 12*(23), 10149. DOI: 10.3390/su122310149

Poirier, B., Haag, D., Soares, G. H., & Jamieson, L. M. (2023). Whose values, what bias, which subjectivity?: The need for reflexivity and positionality in epidemiological health equity scholarship. *Australian and New Zealand Journal of Public Health, 47*(4), 100079. Advance online publication. DOI: 10.1016/j.anzjph.2023.100079 PMID: 37633183

Porretta, D., Mastrantonio, V., Amendolia, S., Gaiarsa, S., Epis, S., Genchi, C., Bandi, C., Otranto, D., & Urbanelli, S. (2013). Effects of global changes on the climatic niche of the tick Ixodes ricinus inferred by species distribution modelling. *Parasites & Vectors, 6*(1), 271. DOI: 10.1186/1756-3305-6-271 PMID: 24330500

Poulos, A., & Wilson, K. (2025). Effect of heat on outdoor recess in Arizona elementary schools, July–September 2023. *Public Health Reports (Washington, D.C.), 140*(5–6), 437–440. Advance online publication. DOI: 10.1177/00333549251342205 PMID: 40616399

Pousa, C., & Mathieu, A. (2014). The Influence of Coaching on Employee Performance: Results From Two International Quantitative Studies. *Performance Improvement Quarterly, 27*(3), 75–92. DOI: 10.1002/piq.21175

Prasetya, A. (2018). ANALYSIS OF FACTORS THAT INFLUENCE EMPLOYEE PERFORMANCE (Study on Permanent Employees in Operational Section of PT WIMCycle Indonesia - Surabaya). *Profit, 12*(1), 1–12. DOI: 10.21776/ub.profit.2018.012.01.1

Prata, J. C., da Costa, J. P., Lopes, I., Andrady, A. L., Duarte, A. C., & Rocha-Santos, T. (2021). A One Health perspective of the impacts of microplastics on animal, human and environmental health. *The Science of the Total Environment, 777*, 146094. DOI: 10.1016/j.scitotenv.2021.146094 PMID: 33677304

Priya, G., Kotwal, N., Madhu, S. V., Patra, S., Pandit, K., Anne, B., Shenoy, M. T., Swain, J., Das, S., Kopalle, L. N., Nagendra, L., Bhattacharya, S., Shaikh, S. S., Mondal, S., Bhat, G. K., Joshi, A., Kalra, S., Biswas, K., Kumar, H. K. V. S., & Mittal, S. (2026). Endocrine Society of India's National White Paper on Endocrine Disrupting Chemicals – A Call for Action. *Indian Journal of Endocrinology and Metabolism, 30*(1), 4–19. DOI: 10.4103/ijem.ijem_976_25 PMID: 41918600

Proctor, E., Silmere, H., Raghavan, R., Hovmand, P., Aarons, G., Bunger, A., Griffey, R., & Hensley, M. (2011). Outcomes for implementation research: Conceptual distinctions, measurement challenges, and research agenda. *Administration and Policy in Mental Health, 38*(2), 65–76. DOI: 10.1007/s10488-010-0319-7 PMID: 20957426

Quesada, J., Hart, L. K., & Bourgois, P. (2011). Structural Vulnerability and Health: Latino Migrant Laborers in the United States. *Medical Anthropology, 30*(4), 339–362. DOI: 10.1080/01459740.2011.576725 PMID: 21777121

Quinn, A., Tamerius, J. D., Perzanowski, M., Jacobson, J. S., Goldstein, I., Acosta, L., & Shaman, J. (2014). Predicting indoor heat exposure risk during extreme heat events. *The Science of the Total Environment, 490*, 686–693. DOI: 10.1016/j.scitotenv.2014.05.039 PMID: 24893319

Quintarelli, V., Ben Hassine, M., Radicetti, E., Stazi, S. R., Bratti, A., Allevato, E., Mancinelli, R., Jamal, A., Ahsan, M., Mirzaei, M., & Borgatti, D. (2024). Advances in nanotechnology for sustainable agriculture: A review of climate change mitigation. *Sustainability (Basel), 16*(21), 9280. DOI: 10.3390/su16219280

Raghavan, R., & Ahamad, S. (2025). Developmental Origins of Health and Disease. *Cancer Exposomics and Environmental Influences on Carcinogenesis*, 1–34. DOI: 10.4018/979-8-3373-2165-3.ch001

Rakotoarinia, M. R. (2025). *Effets combinés des changements climatiques et des changements d'utilisation des sols sur les populations de moustiques en Ontario.* https://hdl.handle.net/1866/41127

Ramadan, M., & Mourad, B. (2022). Impact of climate change on occupational health and its possible preventive strategies. *Egyptian Journal of Occupational Medicine, 46*(3), 47–62. DOI: 10.21608/ejom.2022.133585.1272

Randazza, J. M., Hess, J. J., Bostrom, A., Hartwell, C., Nori-Sarma, A., Spangler, K. R., Sun, Y., Weinberger, K. R., Wellenius, G. A., & Errett, N. A. (2023). Planning to reduce the health impacts of extreme heat: A content analysis of heat action plans in local United States jurisdictions. *American Journal of Public Health, 113*(5), 559–567. DOI: 10.2105/AJPH.2022.307217 PMID: 36926967

Randolph, S. E., Gern, L., & Nuttall, P. A. (1996). Co-feeding ticks: Epidemiological significance for tick-borne pathogen transmission. *Parasitology Today (Personal Ed.), 12*(12), 472–479. DOI: 10.1016/S0169-4758(96)10072-7 PMID: 15275266

Randolph, S. E., & Rogers, D. J. (2000). Fragile transmission cycles of tick-borne encephalitis virus may be disrupted by predicted climate change. *Proceedings. Biological Sciences*, *267*(1454), 1741–1744. DOI: 10.1098/rspb.2000.1204 PMID: 12233771

Ray, S. J. (2021). Climate anxiety is an overwhelmingly white phenomenon. *Scientific American*. Retrieved from https://www.scientificamerican.com/article/the-unbearable-whiteness-of-climate-anxiety/

Ray, S. J. (2018). Coming of age at the end of the world: The affective arc of undergraduate environmental studies curricula. In *Affective ecocriticism: Emotion, embodiment, environment*. University of Nebraska. DOI: 10.2307/j.ctv75d0g8.19

Read, D. J., & Wainger, L. (2023). Assessing Intervention Effectiveness at Promoting Voluntary Conservation Practice Adoption in Agrienvironments. *Conservation Biology : The Journal of the Society for Conservation Biology*, *37*(1), e14009. DOI: 10.1111/cobi.14009 PMID: 36178035

Ready, P. D. (2010). Leishmaniasis emergence and climate change. [OIE]. *Revue Scientifique et Technique (International Office of Epizootics)*, *29*(3), 399–412. PMID: 18819668

Reid, C. E., Brauer, M., Johnston, F. H., Jerrett, M., Balmes, J. R., & Elliott, C. T. (2016). Critical review of health impacts of wildfire smoke exposure. *Environmental Health Perspectives*, *124*(9), 1334–1343. DOI: 10.1289/ehp.1409277 PMID: 27082891

Reid, C. E., O'Neill, M. S., Gronlund, C. J., Brines, S. J., Brown, D. G., Diez-Roux, A. V., & Schwartz, J. (2009). Mapping community determinants of heat vulnerability. *Environmental Health Perspectives*, *117*(11), 1730–1736. DOI: 10.1289/ehp.0900683 PMID: 20049125

Reporting, S. (2019). Corporate Reporting Practices Concerning Non-financial Aspects: A Possible Prolix?. *Integrated Reporting*, 241.

Richard, S., & Oppliger, A. (2015). Zoonotic occupational diseases in forestry workers – Lyme borreliosis, tularemia and leptospirosis in Europe. *Annals of Agricultural and Environmental Medicine : AAEM*, *22*(1), 43–50. DOI: 10.5604/12321966.1141368 PMID: 25780827

Richards, C., Brumley, C., Graves, J., Parker, M., Khot, L., & Postma, J. (2025). Mapping Research Priorities for Climate Change Adaptation in Agriculture: A One Health Perspective. *Workplace Health & Safety*, *73*(9), 454–465. DOI: 10.1177/21650799251334147 PMID: 40359014

Richey, M., Golightly, Y., Marshall, S., Novicoff, W., Keil, A., Nocera, M., & Richardson, D. (2022). Trends in fatal occupational injuries in Latino/a workers relative to other groups, North Carolina 2000–2017. *American Journal of Industrial Medicine, 65*(4), 242–247. DOI: 10.1002/ajim.23331 PMID: 35128690

Rizzoli, A., Hauffe, H. C., Carpi, G., Vourc'h, G., Neteler, M., & Rosà, R. (2011). Lyme borreliosis in Europe. *Eurosurveillance, 16*(27), 19906. DOI: 10.2807/ese.16.27.19906-en PMID: 21794218

Robusti, E., Androcioli, H., Hata, F., Júnior, D., & Júnior, A. (2023). Farmers' and rural extension workers' perceptions on the adoption of the integrated pest management. *Delos Desarrollo Local Sostenible, 16*(46), 2294–2307. DOI: 10.55905/rdelosv16.n46-018

Rocklöv, J., & Dubrow, R. (2020). Climate change: An enduring challenge for vector-borne disease prevention and control. *Nature Immunology, 21*(5), 479–483. DOI: 10.1038/s41590-020-0648-y PMID: 32313242

Rocque, R. J., Beaudoin, C., Ndjaboue, R., Cameron, L., Poirier-Bergeron, L., Poulin-Rheault, R. A., & Witteman, H. O. (2021). Health effects of climate change: An overview of systematic reviews. *BMJ Open, 11*(6), e046333. DOI: 10.1136/bmjopen-2020-046333 PMID: 34108165

Rolnick, D., Donti, P. L., Kaack, L. H., Kochanski, K., Lacoste, A., Sankaran, K., . . . Bengio, Y. (2019). Tackling Climate Change with Machine Learning. arXiv preprint arXiv:1906.05433.

Romanello, M., Di Napoli, C., Drummond, P., Green, C., Kennard, H., Lampard, P., Scamman, D., Arnell, N., Ayeb-Karlsson, S., Ford, L. B., Belesova, K., Bowen, K., Cai, W., Callaghan, M., Campbell-Lendrum, D., Chambers, J., van Daalen, K. R., Dalin, C., Dasandi, N., & Costello, A. (2022). The 2022 report of the Lancet Countdown on health and climate change: Health at the mercy of fossil fuels. *Lancet, 400*(10363), 1619–1654. DOI: 10.1016/S0140-6736(22)01540-9 PMID: 36306815

Roome, A., Gouli, S., Yodsuwan, R., Victory, J., Collins, C., Jenkins, P., Scribani, M., Krupa, N., Freilich, D., & Gadomski, A. (2022). Tick magnets: The occupational risk of tick-borne disease exposure in forestry workers in New York. *Health Science Reports, 5*(2), e509. DOI: 10.1002/hsr2.509 PMID: 35229053

Rose, D., & Birchall, S. J. (2025). Investigating the extent and quality of health-focused climate adaptation planning: Insights from Western Canadian cities. *Environmental Science & Policy, 172*, 104192. DOI: 10.1016/j.envsci.2025.104192

Rossi, M. F., Leone, R., & Moscato, U. (2025). Climate Change and Occupational Risks in Outdoor Workers: A Systematic Review of the Health Effects of Extreme Temperatures. *Atmosphere (Basel), 16*(7), 839. Advance online publication. DOI: 10.3390/atmos16070839

Rother, H. A., John, J., Wright, C. Y., Irlam, J., Oosthuizen, R., & Garland, R. M. (2019). Perceptions of occupational heat, sun exposure, and health risk prevention: A qualitative study of forestry workers in South Africa. *Atmosphere (Basel), 11*(1), 37. DOI: 10.3390/atmos11010037

Rüegg, S. R., McMahon, B. J., Häsler, B., Esposito, R., Nielsen, L. R., Ifejika Speranza, C., Ehlinger, T., Peyre, M., Aragrande, M., Zinsstag, J., Davies, P., Mihalca, A. D., Buttigieg, S. C., Rushton, J., Carmo, L. P., De Meneghi, D., Canali, M., Filippitzi, M. E., Goutard, F. L., & Lindberg, A. (2017). A Blueprint to Evaluate One Health. *Frontiers in Public Health, 5*. Advance online publication. DOI: 10.3389/fpubh.2017.00020 PMID: 28261580

Rufo, J., Qiu, C., Han, D., Baxter, N., Daley, G., Dhillon, J., Wong, F., Collins, J. J., & Wilson, M. Z. (2026). FATE-MAP predicts teratogenicity and human gastrulation failure modes by integrating deep learning and mechanistic modeling. *Nature Communications, 17*(1), 3327. Advance online publication. DOI: 10.1038/s41467-026-69596-6 PMID: 41714622

Rupasinghe, R., Chomel, B. B., & Martínez-López, B. (2022). Climate change and zoonoses: A review of the current status, knowledge gaps, and future trends. *Acta Tropica, 226*, 106225. DOI: 10.1016/j.actatropica.2021.106225 PMID: 34758355

Rustia, D., Chiu, L., Lu, C., Wu, Y., Chen, S., Chung, J., Hsu, J.-C., & Lin, T. (2022). Towards intelligent and integrated pest management through an AIoT-based monitoring system. *Pest Management Science, 78*(10), 4288–4302. DOI: 10.1002/ps.7048 PMID: 35716088

Ryan, S. J., Carlson, C. J., Mordecai, E. A., & Johnson, L. R. (2019). Global expansion and redistribution of Aedes-borne virus transmission risk with climate change. *PLoS Neglected Tropical Diseases, 13*(3), e0007213. DOI: 10.1371/journal.pntd.0007213 PMID: 30921321

Sabando-Vera, D., Yonfa-Medranda, M., Montalván-Burbano, N., Albors-Garrigos, J., & Parrales-Guerrero, K. (2022). Worldwide research on open innovation in SMEs. *Journal of Open Innovation, 8*(1), 20. DOI: 10.3390/joitmc8010020

Saheb, T., Sidaoui, M., & Schmarzo, B. (2024). Convergence of artificial intelligence with social media: A bibliometric & qualitative analysis. *Telematics and Informatics Reports, 14*, 100146. DOI: 10.1016/j.teler.2024.100146

Salas-Pilco, S. Z., & Yang, Y. (2022). Artificial intelligence applications in Latin American higher education: A systematic review. *International Journal of Educational Technology in Higher Education, 19*(1), 1–20. DOI: 10.1186/s41239-022-00326-w PMID: 35013716

Samuels, L., Nakstad, B., Roos, N., Bonell, A., Chersich, M., Havenith, G., Luchters, S., Day, L.-T., Hirst, J. E., Singh, T., Elliott-Sale, K., Hetem, R., Part, C., Sawry, S., Le Roux, J., & Kovats, S. (2022). Physiological mechanisms of the impact of heat during pregnancy and the clinical implications: Review of the evidence from an expert group meeting. *International Journal of Biometeorology, 66*(8), 1505–1513. DOI: 10.1007/s00484-022-02301-6 PMID: 35554684

Sánchez, A., García-Galán, A., García, E., Gómez-Martín, Á., de La Fe, C., Corrales, J. C., & Contreras, A. (2020). Occupational exposure to influenza virus of the wild birds. *Revista Española de Salud Pública, 94*, 202003022. PMID: 32381999

Sandall, J., Fernandez Turienzo, C., Devane, D., Soltani, H., Gillespie, P., Gates, S., Jones, L. V., Shennan, A. H., & Rayment-Jones, H. (2024). Midwife continuity of care models versus other models of care for childbearing women. *Cochrane Database of Systematic Reviews, 2024*(4), CD004667. Advance online publication. DOI: 10.1002/14651858.CD004667.pub6 PMID: 38597126

Šavrič, J., Ladkin, A., Cooper, C., & Turnšek, M. (2025). Climate change and tourism work: A perspective paper. *Tourism Review*. Advance online publication. DOI: 10.1108/TR-10-2024-0956

Schaltegger, S., Zvezdov, D., Etxeberria, I. A., Csutora, M., & Günther, E. (Eds.). (2015). *Corporate carbon and climate accounting*. Springer International Publishing. DOI: 10.1007/978-3-319-27718-9

Schapiro, L. H., McShane, M. A., Marwah, H. K., Callaghan, M. E., & Neudecker, M. L. (2024). Impact of extreme heat and heatwaves on children's health: A scoping review. *The Journal of Climate Change and Health, 19*, 100335. DOI: 10.1016/j.joclim.2024.100335 PMID: 41647315

Schenk, L. (2020). A scoping survey of attitudes towards occupational exposure limits and REACH-derived no-effect levels for workers among chemical risk managers at Swedish workplaces. *International Journal of Occupational Medicine and Environmental Health, 33*(5), 611–620. DOI: 10.13075/ijomeh.1896.01493 PMID: 32699425

Schoemaker, N. K., Wentholt, W. G. M., Goemans, A., Vermeer, H. J., Juffer, F., & Alink, L. R. A. (2020). A Meta-Analytic Review of Parenting Interventions in Foster Care and Adoption. *Development and Psychopathology*, *32*(3), 1149–1172. DOI: 10.1017/S0954579419000798 PMID: 31366418

Schulte, P. A., Bhattacharya, A., Butler, C. R., Chun, H. K., Jacklitsch, B., Jacobs, T., & Wagner, G. R. (2016). Advancing the framework for considering the effects of climate change on worker safety and health. *Journal of Occupational and Environmental Hygiene*, *13*(11), 847–865. DOI: 10.1080/15459624.2016.1179388 PMID: 27115294

Schulte, P. A., & Chun, H. (2009). Climate change and occupational safety and health: Establishing a preliminary framework. *Journal of Occupational and Environmental Hygiene*, *6*(9), 542–554. DOI: 10.1080/15459620903066008 PMID: 19551548

Schulte, P., Jacklitsch, B., Bhattacharya, A., Chun, H., Edwards, N., Elliott, K., Flynn, M., Guerin, R., Hodson, L., Lincoln, J., MacMahon, K. L., Pendergrass, S., Siven, J., & Vietas, J. (2023). Updated assessment of occupational safety and health hazards of climate change. *Journal of Occupational and Environmental Hygiene*, *20*(5-6), 183–206. DOI: 10.1080/15459624.2023.2205468 PMID: 37104117

Schwartz, S. E. O., Benoit, L., Clayton, S., Parnes, M. F., Swenson, L., & Lowe, S. R. (2022). Climate change anxiety and mental health: Environmental activism as buffer. *Current Psychology*, •••, 1–14. DOI: 10.1007/s12144-022-02735-6 PMID: 35250241

Scott, N., Graczyk, H., & Papandrea, D. (2024). *Ensuring safety and health at work in a changing climate: Global report—International Labour Organization* (No. 995361942702676; p. 132). https://s3.eu-central-1.amazonaws.com/eu-st01.ext.exlibrisgroup.com/41ILO_INST/storage/alma/5C/8E/93/00/A8/1F/1F/D3/44/0E/55/11/5E/A1/2F/FF/wcms_922850.pdf?response-content-type=application%2Fpdf&X-Amz-Algorithm=AWS4-HMAC-SHA256&X-Amz-Date=20260504T071608Z&X-Amz-SignedHeaders=host&X-Amz-Credential=AKIAJUYRSPBSIJGGDX6A%2F20260504%2Feu-central-1%2Fs3%2Faws4_request&X-Amz-Expires=119&X-Amz-Signature=5edfaf9b8da827053cdb37cea40de514c7ceacb85ce9d2cd66312f495bbd6922

Scotti, K. L., Jiao, C., Gainey, M. A., Bongiovanni, E. A., & Slayton, E. R. (2025). Charting open science landscapes: A systematized review of US academic libraries' engagement in open research practices. *Journal of Academic Librarianship*, *51*(3), 103054. DOI: 10.1016/j.acalib.2025.103054

Sekamatte, M., Krishnasamy, V., Bulage, L., Kihembo, C., Nantima, N., Monje, F., Ndumu, D., Sentumbwe, J., Mbolanyi, B., Aruho, R., Kaboyo, W., Mutonga, D., Basler, C., Paige, S., & Behravesh, C. B. (2018). Multisectoral prioritization of zoonotic diseases in Uganda, 2017: A One Health perspective. *PLoS One*, *13*(5), e0196799. DOI: 10.1371/journal.pone.0196799 PMID: 29715287

Semenza, J. C., Rocklöv, J., & Ebi, K. L. (2022). Climate change and cascading risks for infectious disease transmission. *The Lancet. Planetary Health*, *6*(4), e350–e358. PMID: 35397223

Semenza, J. C., & Suk, J. E. (2018). Vector-borne diseases and climate change: A European perspective. *FEMS Microbiology Letters*, *365*(2), fnx244. Advance online publication. DOI: 10.1093/femsle/fnx244 PMID: 29149298

Sepúlveda-Oviedo, E. H., Travé-Massuyès, L., Subias, A., Pavlov, M., & Alonso, C. (2023). Fault diagnosis of photovoltaic systems using artificial intelligence: A bibliometric approach. *Heliyon*, *9*(11), e21491. DOI: 10.1016/j.heliyon.2023.e21491 PMID: 37954345

Sharma, M., Suri, N. M., & Kant, S. (2022). Analyzing occupational heat stress using sensor-based monitoring: A wearable approach with environmental ergonomics perspective. *International Journal of Environmental Science and Technology*, *19*(11), 11421–11434. DOI: 10.1007/s13762-021-03862-6 PMID: 35106066

Sharpe, I., & Davison, C. M. (2022). A Scoping Review of Climate Change, Climate-Related Disasters, and Mental Disorders among Children in Low- and Middle-Income Countries. *International Journal of Environmental Research and Public Health*, *19*(5), 2896. Advance online publication. DOI: 10.3390/ijerph19052896 PMID: 35270593

Sheeran, P., Wright, C. E., Listrom, O., Klein, W. M. P., & Rothman, A. J. (2023). Which Intervention Strategies Promote the Adoption and Maintenance of Physical Activity? Evidence From Behavioral Trials With Cancer Survivors. *Annals of Behavioral Medicine : A Publication of the Society of Behavioral Medicine*, *57*(9), 708–721. DOI: 10.1093/abm/kaad002 PMID: 37078961

Silvério, A., Martins, I., Nogueira, D., Mello, M., Loyola, E., & Graciano, M. (2020). Assessment of Primary Health Care for rural workers exposed to pesticides. *Revista de Saúde Pública*, *54*, 9. DOI: 10.11606/s1518-8787.2020054001455 PMID: 31967278

Simonato, M., Brener, M. I., Ferrari, G., Graves, C., Yan, B., & George, I. (2026). Impact of microplastics and nanoplastics on cardiovascular health. *Trends in Cardiovascular Medicine*. Advance online publication. DOI: 10.1016/j.tcm.2026.02.008 PMID: 41720227

Singh, A., Rejeb, A., Nangru, H., & Pathak, S. (2024). Global research trends on cyberbullying: A bibliometric study. *Computers in Human Behavior Reports*, *16*, 100499. DOI: 10.1016/j.chbr.2024.100499

Socha, W., Kwasnik, M., Larska, M., Rola, J., & Rozek, W. (2022). Vector-Borne Viral Diseases as a Current Threat for Human and Animal Health—One Health Perspective. *Journal of Clinical Medicine*, *11*(11), 3026. Advance online publication. DOI: 10.3390/jcm11113026 PMID: 35683413

Soe, B. K., Huertas-Lopez, A., Sato, H., & Adisakwattana, P. (2026). Ticks and tick-borne pathogens with zoonotic importance in animals in Southeast Asia: A decade of evidence from systematic review and meta-analysis. *Acta Tropica*, *274*, 107974. DOI: 10.1016/j.actatropica.2026.107974 PMID: 41529770

Somboon, S., Jirapongsuwan, A., Kalampakorn, S., & Tipyamongkolkul, M. (2022). Acute Pesticide Poisoning among Mixed-crop Agricultural Workers in Thailand. *Trends in Sciences*, *19*(3), 2154. DOI: 10.48048/tis.2022.2154

Sorensen, C., Murray, V., Lemery, J., & Balbus, J. (2018). Climate change and women's health: Impacts and policy directions. *PLoS Medicine*, *15*(7), e1002603. DOI: 10.1371/journal.pmed.1002603 PMID: 29990343

Soto-Méndez, M. J., Maldonado, A., Burgunder, L., Scieszka, L., Gil, Á., & Solomons, N. W. (2016). Effects of maternal hydration status on the osmolality of maternal milk. *Nutrición Hospitalaria*, *33*(Suppl. 3). Advance online publication. DOI: 10.20960/nh.318 PMID: 27491583

Souza, M. C. O., Barbosa, F.Jr, & Domingo, J. L. (2026). Mercury contamination in a changing climate: One health approaches to human, wildlife, and ecosystem protection. *Chemosphere*, *394*, 144821. DOI: 10.1016/j.chemosphere.2025.144821 PMID: 41478202

Spector, J. T., Krenz, J., & Blank, K. N. (2015). Risk factors for heat-related illness in Washington crop workers. *Journal of Agromedicine*, *20*(3), 349–359. DOI: 10.1080/1059924X.2015.1047107 PMID: 26237726

Spector, J. T., Sheffield, P. E., & Kinney, P. L. (2019). Climate change and occupational health: Current evidence and future directions. *Journal of Occupational and Environmental Medicine*, *61*(4), e155–e157. PMID: 30694882

Spengler, J. R., Bergeron, É., & Rollin, P. E. (2016). Seroepidemiological Studies of Crimean-Congo Hemorrhagic Fever Virus in Domestic and Wild Animals. *PLoS Neglected Tropical Diseases*, *10*(1), e0004210. DOI: 10.1371/journal.pntd.0004210 PMID: 26741652

Stassen, R. C., Benedek, R., Gabbe, B., De Witte, S., Mertens, L., & Glazer, K. B. (2024). Extreme heat and pregnancy: A content analysis of heat health risk communication in English language public health guidance. *Disaster Medicine and Public Health Preparedness*, *18*, e44. DOI: 10.1017/dmp.2024.44 PMID: 38497500

Stevenson, K. T., Peterson, M. N., & Bondell, H. D. (2018). Developing a model of climate change behavior among adolescents. *Climatic Change*, *151*(3/4), 589–603. DOI: 10.1007/s10584-018-2313-0

Stone, J., Mayo, J. A., Shaw, G. M., & Gonzalez, D. J. X. (2026). Structural birth defects and exposure to oil and gas wells during pregnancy. *Pediatric Research*, •••, 1–7. DOI: 10.1038/s41390-025-04719-1 PMID: 41540200

Stowell, J. D., Sun, Y., Spangler, K. R., Milando, C. W., Bernstein, A., Weinberger, K. R., Sun, S., & Wellenius, G. A. (2023). Warm-season temperatures and emergency department visits among children with health insurance. *Environmental Research, Health : ERH*, *1*(1), 015002. DOI: 10.1088/2752-5309/ac78fa PMID: 36337257

Stufano, A., Schino, V., Plantone, D., & Lucchese, G. (2025). Occupational zoonoses, neurological diseases, and public health: A one health approach. *Infectious Medicine*, *4*(2), 100184. DOI: 10.1016/j.imj.2025.100184 PMID: 40530081

Stufano, A., Schino, V., Sacino, G., Ravallese, R., Ravallese, R., Benedictis, L., & Lovreglio, P. (2025). West Nile Virus Seroprevalence Among Outdoor Workers in Southern Italy: Unveiling Occupational Risks and Public Health Implications. *Viruses*, *17*(3), 310. DOI: 10.3390/v17030310 PMID: 40143241

Suman, A. (2021). Role of renewable energy technologies in climate change adaptation and mitigation: A brief review from Nepal. *Renewable & Sustainable Energy Reviews*, *151*, 111524. Advance online publication. DOI: 10.1016/j.rser.2021.111524

Sun, C., Rao, Q., Chen, B., Liu, X., Adnan Ikram, R. M., Li, J., Wang, M., & Zhang, D. (2024). Mechanisms and Applications of Nature-Based Solutions for Stormwater Control in the Context of Climate Change: A Review. *Atmosphere (Basel)*, *15*(4), 403. Advance online publication. DOI: 10.3390/atmos15040403

Suu-Ire, R. D., Gyebi, G., Baum, S., Abugri, H. A., Breen, R. W. B., Karesh, W. B., & Machalaba, C. (2025). The effects of climate change on zoonotic disease risks in Ghana: Reviewing the state of knowledge from a One Health perspective. *PAMJ - One Health : Pan African Medical Journal*, *18*(9). Advance online publication. DOI: 10.11604/pamj-oh.2025.18.9.46717

Syahrir, M. S., & Falah, F. (2021). Occupational Burnout and Workload: Mental Health Challenges in the Workplace and Solutions. *Journal of Health Literacy and Qualitative Research, 1*(2), 69–83. DOI: 10.61194/jhlqr.v1i2.528

Tanana, H., Combs, J., & Hoss, A. (2021). Water is life: Law, systemic racism, and water security in Indian Country. *Health Security, 19*(1_suppl, S1), S78–S82. DOI: 10.1089/hs.2021.0034 PMID: 33944613

Tang, B., Hu, W., & Duan, A. (2021). Assessment of extreme precipitation indices over Indochina and South China in CMIP6 models. *Journal of Climate, 34*(18), 7507–7524. DOI: 10.1175/JCLI-D-20-0948.1

Tang, W. S. W., & Ho, C. S. H. (2025). A systematic review on the impact of climate change on occupational mental health: A focus on vulnerable industries. *Social Psychiatry and Psychiatric Epidemiology, 60*(10), 2275–2287. DOI: 10.1007/s00127-025-02936-x PMID: 40473944

Tappura, S., Sievänen, M., Heikkilä, J., Jussila, A., & Nenonen, N. (2015). A management accounting perspective on safety. *Safety Science, 71*, 151–159. DOI: 10.1016/j.ssci.2014.01.011

Tawatsupa, B., Yiengprugsawan, V., Kjellstrom, T., Berecki-Gisolf, J., Seubsman, S. A., & Sleigh, A. (2013). Association between heat stress and occupational injury among Thai workers: Findings of the Thai Cohort Study. *Industrial Health, 51*(1), 34–46. DOI: 10.2486/indhealth.2012-0138 PMID: 23411755

Taylor, L. H., Latham, S. M., & woolhouse, M. E. J. (2001). Risk factors for human disease emergence. *Philosophical Transactions of the Royal Society of London. Series B, Biological Sciences, 356*(1411), 983–989. DOI: 10.1098/rstb.2001.0888 PMID: 11516376

Tesfaye, A. H., Prior, J., & McIntyre, E. (2025). Impact of climate change on health workers: A scoping review. *Journal of Public Health (Berlin).* Advance online publication. DOI: 10.1007/s10389-025-02418-z

Thomson, M. C., & Stanberry, L. R. (2022). Climate Change and Vectorborne Diseases. *The New England Journal of Medicine, 387*(21), 1969–1978. DOI: 10.1056/NEJMra2200092 PMID: 36416768

Tiffin, H. S., Gordon, J. R., & Poh, K. C. (2025). One Health, many approaches: Integrated vector management strategies support One Health goals. *Frontiers in Insect Science, 5*, 1549348. DOI: 10.3389/finsc.2025.1549348 PMID: 40530168

Tigchelaar, M., Battisti, D., & Spector, J. (2020). Work adaptations are insufficient to address the growing heat risk for U.S. agricultural workers. *Environmental Research Letters*, *15*(9), 094035. DOI: 10.1088/1748-9326/ab86f4 PMID: 33133229

Tokarska-Rodak, M., Plewik, D., Kozioł-Montewka, M., Szepeluk, A., & Paszkiewicz, J. (2014).. . *Medycyna Pracy*, *65*(1), 109–117. DOI: 10.13075/mp.5893.2014.017 PMID: 24834698

Trombley, J., Chalupka, S., & Anderko, L. (2017). Climate Change and Mental Health. *The American Journal of Nursing*, *117*(4), 44–52. DOI: 10.1097/01.NAJ.0000515232.51795.fa PMID: 28333743

Tsai, M. L., Ong, C. W., & Chen, C. L. (2023). Exploring the use of large language models (LLMs) in chemical engineering education: Building core course problem models with ChatGPT. *Education for Chemical Engineers*, *44*, 71–95. DOI: 10.1016/j.ece.2023.05.001

Tsakonas, K., Badyal, S., Takaro, T., & Buse, C. (2025). Occupational Health Impacts of Climate-Related Emergency Events on Health System Workers in British Columbia. *International Journal of Disaster Risk Reduction*, *118*, 105227. DOI: 10.1016/j.ijdrr.2025.105227

Tsakonas, K., Badyal, S., Takaro, T., & Buse, C. G. (2024). Rapid review of the impacts of climate change on the health system workforce and implications for action. *The Journal of Climate Change and Health*, *19*, 100337. DOI: 10.1016/j.joclim.2024.100337 PMID: 41647318

Tsalis, T. A., Stylianou, M. S., & Nikolaou, I. E. (2018). Evaluating the quality of corporate social responsibility reports: The case of occupational health and safety disclosures. *Safety Science*, *109*, 313–323. DOI: 10.1016/j.ssci.2018.06.015

Tuholske, C., Lynch, V. D., Spriggs, R., Ahn, Y., Raymond, C., Nigra, A. E., & Parks, R. M. (2024). Hazardous heat exposure among incarcerated people in the United States. *Nature Sustainability*, *7*(4), 394–398. Advance online publication. DOI: 10.1038/s41893-024-01293-y PMID: 41727418

Turner, M. C., Basagaña, X., Albin, M., Broberg, K., Burdorf, A., van Daalen, K. R., Guseva Canu, I., Kolstad, H. A., Kogevinas, M., Lowe, R., Pearce, N., Pega, F., Saget, C., Schubauer-Berigan, M. K., Svensson, S., Vineis, P., & Straif, K. (2025). Occupational health in the era of climate change and the green transition: A call for research. *The Lancet Regional Health. Europe*, *54*, 54. DOI: 10.1016/j.lanepe.2025.101353 PMID: 40613104

Tustin, A. W., Lamson, G. E., Jacklitsch, B. L., Thomas, R. J., Arbury, S. B., Cannon, D. L., Gonzales, R. G., & Hodgson, M. J. (2018). Evaluation of occupational exposure limits for heat stress in outdoor workers—United States, 2011–2016. *MMWR. Morbidity and Mortality Weekly Report*, *67*(26), 67. DOI: 10.15585/mmwr.mm6726a1 PMID: 29975679

Urasa, S. J., Dekker, M. C. J., Howlett, W. P., Mwezi, R. J., Dorsey, E. R., & Bloem, B. R. (2024). Parkinson's Disease in Sub-Saharan Africa: Pesticides as a Double-Edged Sword. *Journal of Parkinson's Disease*, *14*(3), 437–449. DOI: 10.3233/JPD-230409 PMID: 38517806

Utomo, B. (2025). Connection of Human, Animal, and Environmental Health: A One Health Perspective. *One Health Integration: Global Perspectives on Animal Health and Sustainable Agriculture*, 63-80.

Utzet, M., Ayala-Garcia, A., Benavides, F. G., & Basagaña, X. (2023). Extreme temperatures and sickness absence in the Mediterranean province of Barcelona: An occupational health issue. *Frontiers in Public Health*, *11*, 1129027. DOI: 10.3389/fpubh.2023.1129027 PMID: 36891341

UVM Office of Institutional Research. (2021). Enrollment. Retrieved from https://www.uvm.edu/oir/enrollment

Van der Horst, L., Bouaddi, O., Williams, S., Jago, M., Lau, K., Furber, B., Zaini, A. Z., El-Ghitany, E. M., Loganathan, T., Flouris, A. D., Testa, D. J., Zimmerman, C., & Hargreaves, S. (2025). International migrant workers, heat exposure, and climate change: A systematic review of health risks and protective interventions. *BMC Global and Public Health*, *3*(1), 107. DOI: 10.1186/s44263-025-00224-z PMID: 41320785

Van Hecke, O., Cameron, A., & Cruickshank, M. (2018). Seroprevalence of Borrelia burgdorferi in occupational groups: A systematic review and meta-analysis of European studies. *BMJ Open*, *8*(3), e020310.

Vandenberg, L., Pierce, E., & Arsenault, R. (2025). Pesticides are an urgent challenge to global environmental health and planetary boundaries. *Frontiers in Toxicology*, *7*, 7. DOI: 10.3389/ftox.2025.1656297 PMID: 41114400

Varghese, B., Hansen, A., Bi, P., & Pisaniello, D. (2018). Are workers at risk of occupational injuries due to heat exposure? A comprehensive literature review. *Safety Science*, *110*, 380–392. DOI: 10.1016/j.ssci.2018.04.027

Vendl, C., Taylor, M. D., Bräunig, J., Gibson, M. J., Hesselson, D., Neely, G. G., & Nakagawa, S. (2022). PFAS exposure of humans, animals and the environment: Protocol of an evidence review map and bibliometric analysis. *Environment International, 158*, 106973. DOI: 10.1016/j.envint.2021.106973

Vonesch, N., Binazzi, A., Bonafede, M., Melis, P., Ruggieri, A., Iavicoli, S., & Tomao, P. (2019). Emerging zoonotic viral infections of occupational health importance. *Pathogens and Disease, 77*(2), ftz018. Advance online publication. DOI: 10.1093/femspd/ftz018 PMID: 30916772

Vonesch, N., D'Ovidio, M. C., Melis, P., Remoli, M. E., Ciufolini, M. G., & Tomao, P. (2016). Climate change, vector-borne diseases and working population. *Annali dell'Istituto Superiore di Sanità, 52*(3), 397–405. DOI: 10.4415/ANN_16_03_11 PMID: 27698298

Wallace, R. L., Greenburg, J., & Clark, S. G. (2020). Confronting anxiety and despair in environmental studies and sciences: An analysis and guide for students and faculty. *Journal of Environmental Studies and Sciences, 10*(2), 148–155. DOI: 10.1007/s13412-020-00609-6

Wallerstein, N., & Duran, B. (2010). Community-based participatory research contributions to intervention research: The intersection of science and practice to improve health equity. *American Journal of Public Health, 100*(S1), S40–S46. DOI: 10.2105/AJPH.2009.184036 PMID: 20147663

Wang, H., Wang, M., Fan, C., Sun, Y., Feng, J., He, M., Li, N., Shang, F., & Liu, B. (2026). Potential association between endocrine disrupting chemicals (EDCs) and gynecomastia: A systematic review based on partial experimental evidenceendocrine-disrupting chemicals. *Frontiers in Endocrinology, 17*, 1734644. Advance online publication. DOI: 10.3389/fendo.2026.1734644 PMID: 41710411

Wang, X., & Qin, W. (2026). Effect of gut microbiome and blood metabolites on colorectal cancer: A bidirectional Mendelian randomization and mediation analysis based on STROBE-MR guidelines. *Medicine, 105*(1), e46936. DOI: 10.1097/MD.0000000000046936 PMID: 41496083

Wang, Y., & Zhang, L. (2025, June). Bibliometric Analysis and Visualization of Research Trends and Hotspots in the Body Donation Field. In *Transplantation Proceedings*. Elsevier. DOI: 10.1016/j.transproceed.2025.05.015

Watts, N., Amann, M., Arnell, N., Ayeb-Karlsson, S., Beagley, J., Belesova, K., & Costello, A. (2021). The 2020 report of the Lancet Countdown on health and climate change: Responding to converging crises. *Lancet, 397*(10269), 129–170. DOI: 10.1016/S0140-6736(20)32290-X PMID: 33278353

Wegner, G. I., Murray, K. A., Springmann, M., Muller, A., Sokolow, S. H., Saylors, K., & Morens, D. M. (2022). Averting wildlife-borne infectious disease epidemics requires a focus on socio-ecological drivers and a redesign of the global food system. *EClinicalMedicine*, *47*, 101386. Advance online publication. DOI: 10.1016/j.eclinm.2022.101386 PMID: 35465645

Weheba, A., Vertigan, A., Abdelsayad, A., & Tarlo, S. M. (2024). Respiratory diseases associated with wildfire exposure in outdoor workers. *The Journal of Allergy and Clinical Immunology. In Practice*, *12*(8), 1989–1996. DOI: 10.1016/j.jaip.2024.03.033 PMID: 38548173

Wen, X., Liu, J., Zhao, W., Liu, X., Liu, L., Zeng, Z., & Wei, D. (2026). Photodynamic Therapy for Colorectal Cancer: A Bibliometric and Visualization Analysis (1986–2025). *Photodiagnosis and Photodynamic Therapy*, *58*, 105368. DOI: 10.1016/j.pdpdt.2026.105368 PMID: 41587715

Werth, J., Thornby, D., Keenan, M., Hereward, J., & Chauhan, B. (2021). Effectiveness of glufosinate, dicamba, and clethodim on glyphosate-resistant and -susceptible populations of five key weeds in Australian cotton systems. *Weed Technology*, *35*(6), 967–973. DOI: 10.1017/wet.2021.59

White, B. P., Breakey, S., Brown, M. J., Smith, J. R., Tarbet, A., Nicholas, P. K., & Ros, A. M. V. (2023). Mental Health Impacts of Climate Change Among Vulnerable Populations Globally: An Integrative Review. *Annals of Global Health*, *89*(1), 66. DOI: 10.5334/aogh.4105 PMID: 37810609

WHO & ILO. (2021). *Climate change and occupational health: Joint policy brief.* World Health Organization and International Labour Organization.

WHO. (2017). *One Health: Operational framework.* World Health Organization.

WHO. (2017a). *Global vector control response 2017–2030.* https://www.who.int/publications/i/item/9789241512978

WHO. (2017b). *West Nile virus.* https://www.who.int/news-room/fact-sheets/detail/west-nile-virus

WHO. (2021). *2021 WHO health and climate change global survey report.* World Health Organization.

WHO. (2023a). *Leishmaniasis.* https://www.who.int/news-room/fact-sheets/detail/leishmaniasis

WHO. (2023b). *Trypanosomiasis, human African (sleeping sickness)*. https://www.who.int/news-room/fact-sheets/detail/trypanosomiasis-human-african-(sleeping-sickness)

WHO. (2025a). *Crimean-Congo haemorrhagic fever*. https://www.who.int/news-room/fact-sheets/detail/crimean-congo-haemorrhagic-fever

WHO. (2025b). *Dengue*. https://www.who.int/news-room/fact-sheets/detail/dengue-and-severe-dengue

WHO. (2025c). *Dengue*. https://www.who.int/news-room/fact-sheets/detail/dengue-and-severe-dengue

WHO. (2025d). *Ebola disease*. https://www.who.int/news-room/fact-sheets/detail/ebola-disease

WHO. (2025e). *Onchocerciasis (river blindness)*. https://www.who.int/news-room/fact-sheets/detail/onchocerciasis

WHO. (2025f). *Public health advice on Crimean-Congo haemorrhagic fever*. https://www.who.int/europe/news-room/questions-and-answers/item/public-health-advice-on-crimean-congo-haemorrhagic-fever

Whyte, K. (2017). Indigenous Climate Change Studies: Indigenizing Futures, Decolonizing the Anthropocene. *English Language Notes*, *55*(1-2), 153–162. DOI: 10.1215/00138282-55.1-2.153

Wibowo, R., Do, V., Quartucci, C., Koller, D., Daanen, H. A., Nowak, D., Bose-O'Reilly, S., & Rakete, S. (2024). Effects of heat and personal protective equipment on thermal strain in healthcare workers: Part B-application of wearable sensors to observe heat strain among healthcare workers under controlled conditions. *International Archives of Occupational and Environmental Health*, *97*(1), 35–43. DOI: 10.1007/s00420-023-02022-2 PMID: 37947815

Wiesner, L., Araya, S., & Barle, E. (2022). Identifying nonhazardous substances in pharmaceutical manufacturing and setting default health-based exposure limits (HBELs). *Journal of Applied Toxicology : JAT*, *42*(9), 1443–1457. DOI: 10.1002/jat.4323 PMID: 35315528

Wilson, N. J., Montoya, T., Arseneault, R., & Curley, A. (2021). Governing water insecurity: Navigating Indigenous water rights and regulatory politics in settler colonial states. *Water International*, *46*(6), 783–801. DOI: 10.1080/02508060.2021.1928972

Wister, A., Fyffe, I., & O'Dea, E. (2021). Technological interventions for loneliness and social isolation among older adults: A scoping review protocol. *Systematic Reviews, 10*(1), 217. DOI: 10.1186/s13643-021-01775-6 PMID: 34362447

Wolfe, N. D., Daszak, P., Kilpatrick, A. M., & Burke, D. S. (2005). *Bushmeat Hunting, Deforestation, and Prediction of Zoonotic Disease—Volume 11, Number 12—December 2005—Emerging Infectious Diseases journal—CDC*. DOI: 10.3201/eid1112.040789

World Health Organization. (2022). *Mental health and climate change: policy brief*. World Health Organization.

World Meteorological Organization & World Health Organization. (2015). *Heat-waves and health: Guidance on warning-system development*. World Meteorological Organization. https://heathealth.info/wp-content/uploads/WMO_WHO_Heat_Health_Guidance_2015.pdf

Wuersch, L., Neher, A., Marino, F. E., Bamberry, L., & Pope, R. (2023). Impacts of Climate Change on Work Health and Safety in Australia: A Scoping Literature Review. *International Journal of Environmental Research and Public Health, 20*(21), 7004. Advance online publication. DOI: 10.3390/ijerph20217004 PMID: 37947561

Xiao, W., Cao, Y., Ma, F., Zhang, H., & Xia, D. (2026). Volatile organic compound exposure and depression risk: Insights from individual, mixture, and sex-stratified analyses. *BMC Public Health, 26*(1), 1748. Advance online publication. DOI: 10.1186/s12889-026-27444-8 PMID: 41998567

Xiong, T., Chen, P., Mu, Y., Li, X., Di, B., Li, J., Qu, Y., Tang, J., Liang, J., & Mu, D. (2020). Association between ambient temperature and hypertensive disorders in pregnancy in China. *Nature Communications, 11*(1), 2925. Advance online publication. DOI: 10.1038/s41467-020-16775-8 PMID: 32522990

Xu, J., Wang, Q., Anikeeva, O., Zhu, P., Bi, P., & Huang, C. (2025). Effects of extreme heat on physiology, morbidity, and mortality under climate change: Mechanisms and clinical implications. *BMJ (Clinical Research Ed.), 391*, e084675. DOI: 10.1136/bmj-2025-084675 PMID: 41161807

Xu, R., Yu, P., Abramson, M. J., Johnston, F. H., Samet, J. M., Bell, M. L., & Li, S. (2020). Wildfires, global climate change, and human health. *The New England Journal of Medicine, 383*(22), 2173–2181. DOI: 10.1056/NEJMsr2028985 PMID: 33034960

Yang, F., Yao, R., Ren, Y., & Guo, L. (2025). Harmony in diversity: Digital literacy research in a multidisciplinary landscape. *Computers & Education, 230*, 105265. DOI: 10.1016/j.compedu.2025.105265

Yao, D., Qiu, J., & Liang, Q. (2025). Bibliometric and Visual Analysis of Inadvertent Perioperative Hypothermia: A Last 20 Years of Research (2004 to 2024). *Journal of PeriAnesthesia Nursing*.

Yea, S. (2025). Worker Protections at Sea: Climate change and life aboard distant water fishing vessels. *Anti-Trafficking Review*, (25), 153–157. DOI: 10.14197/atr.201225259

Yong, W. T. L., Thien, V. Y., Rupert, R., & Rodrigues, K. F. (2022). Seaweed: A potential climate change solution. *Renewable & Sustainable Energy Reviews, 159*, 112222. Advance online publication. DOI: 10.1016/j.rser.2022.112222

Yuan, X., Wang, H., Li, X., Lyu, J., Zhou, S., Ji, Y., Wang, H., Liu, J., Li, Q., & Wang, H.-J. (2026). The association of heat stress with fetal umbilical artery blood flow: A birth cohort study in Beijing, China. *Environmental Research, 288*, 123167. DOI: 10.1016/j.envres.2025.123167

Zawacki-Richter, O., Marín, V. I., Bond, M., & Gouverneur, F. (2019). Systematic review of research on artificial intelligence applications in higher education – Where are the educators? *International Journal of Educational Technology in Higher Education, 16*(1), 1–27. DOI: 10.1186/s41239-019-0171-0

Zhang, Y., Zhang, Y., Que, H., Lu, C., & Zhou, S. (2025). Occupational nanoparticles: Major sources, physicochemical properties, multi-organ toxic effects, and associated mechanisms. *Toxicology Mechanisms and Methods, 35*(8), 847–864. DOI: 10.1080/15376516.2025.2505629 PMID: 40394907

Zhou, R., Sprong, H., Liu, Q., Krafft, T., & Estrada-Peña, A. (2025). Mapping the potential suitable habitats for Hyalomma rufipes (Acari: Ixodidae) in Africa and Western Eurasia. *PLoS Neglected Tropical Diseases, 19*(3), e0012923. DOI: 10.1371/journal.pntd.0012923 PMID: 40100878

Zinsstag, J., Crump, L., Schelling, E., Hattendorf, J., Maidane, Y. O., Ali, K. O., Muhummed, A., Umer, A. A., Aliyi, F., Nooh, F., Abdikadir, M. I., Ali, S. M., Hartinger, S., Mäusezahl, D., de White, M. B. G., Cordon-Rosales, C., Castillo, D. A., McCracken, J., Abakar, F., & Cissé, G. (2018). Climate change and one health. *FEMS Microbiology Letters, 365*(11), fny085. DOI: 10.1093/femsle/fny085 PMID: 29790983

Zinsstag, J., Kaiser-Grolimund, A., Heitz-Tokpa, K., Sreedharan, R., Lubroth, J., Caya, F., Stone, M., Brown, H., Bonfoh, B., Dobell, E., Morgan, D., Homaira, N., Kock, R., Hattendorf, J., Crump, L., Mauti, S., del Rio Vilas, V., Saikat, S., Zumla, A., & de la Rocque, S. (2023). Advancing One human–animal–environment Health for global health security: What does the evidence say? *Lancet*, *401*(10376), 591–604. DOI: 10.1016/S0140-6736(22)01595-1 PMID: 36682371

Zinsstag, J., Schelling, E., Waltner-Toews, D., & Tanner, M. (2011). From "one medicine" to "one health" and systemic approaches to health and well-being. *Preventive Veterinary Medicine*, *101*(3-4), 148–156. DOI: 10.1016/j.prevetmed.2010.07.003 PMID: 20832879

About the Contributors

Ioannis Adamopoulos is a coordinator, advisor, and lecturer (adjunct associate professor, teaching staff) at the Hellenic Open University (HOU) in the School of Social Sciences, Master's Program in Public Health and Policies, and coordinator of the Crisis Management in Health Care thematic course. He is a postdoctoral researcher at the University of West Attica, School of Public Health, Department of Public Health Policy, Occupational and Environmental Health Sector. He holds a PhD in Occupational Safety and Health & Public Health from the European University of Cyprus, with undergraduate studies in Public Health—Public Health Inspector and postgraduate studies in Public Health, as well as a postgraduate specialization in Economics and Health Services Management. He is a professor and principal member of the Register of Principal Teaching Staff at the Institute of Continuing Education (IN.EP.) of the National Center for Administration and Local Government (E.K.D.D.A.) and the National School of Public Administration and Decentralization (ESDDA). He has many years of professional experience, with a significant impact and numerous references and citations in his research activity. He is a member of international organizations and committees and scientific societies, as well as an author and reviewer. He is a contributor-editor to many scientific journals and books and has managed dozens of scientific publications, research articles, research protocols, and research papers and chaired several international conferences and given presentations at many international, European, and Greek conferences. He is a head division public health inspector at the Directorate of Public Health Control and Sanitarian Environmental Hygiene in the Central Sector of the Attica Region, Hellenic Republic, and has managed cases while also gaining work experience and teaching. Primarily in the fields of environmental hygiene inspections, health inspections, food safety, occupational safety and health, environmental safety, climate crisis, hygiene, prevention and promotion of public health, health administration and policy, and crisis management and as a member of relevant expert committees.

Aida Vafae Eslahi is researcher at Medical Microbiology Research Center at Qazvin University of Medical Sciences, Iran. She is a parasitologist specializing in waterborne and foodborne parasitic infections, infectious disease epidemiology, and One Health research. She is a Ph.D. graduate in Medical Parasitology from Tarbiat Modares University in Tehran, with a M.Sc. in Medical Parasitology from Tehran University of Medical Sciences and a B.Sc. in Biology from the University of Guilan. Dr. Vafae Eslahi has co-authored more than 52 peer-reviewed publications in internationally-recognized journals. She is passionate about integrating molecular parasitology with public health strategies to combat emerging zoonotic, waterborne and foodborne infectious diseases through a One Health approach. Her projects have contributed to the first molecular identification of several parasite species in Iran, and she actively collaborates with distinguished researchers and institutions worldwide. In addition to her research, Dr. Vafae Eslahi is an active reviewer for several world-recognized scientific journals, contributing her expertise to the advancement of parasitology and public health research. She collaborates extensively with institutions across Europe, Asia, and the Americas, including universities in Portugal, Germany, Cyprus, Greece, and the United States.

***

Eboni Anderson, PhD, DHEd, MSW, MEd, MA, is an accomplished educator, social worker, and public health professional with over two decades of experience in higher education, health professions training, and community health. She currently serves as Associate Professor of Public Health and Director of Community Oriented Primary Care at A.T. Still University- School of Osteopathic Medicine in Arizona (ATSU- SOMA). She earned a bachelor's degree in English and three master's degrees in counseling, higher and postsecondary education, and social work from Wayne State University and Arizona State University, respectively. She holds two doctoral degrees in health education and health professions education from A.T. Still University and Bellarmine University. Dr. Anderson is also an advocate, researcher, published author, and dedicated mentor. Her research focuses on educator and staff well- being, social determinants of health, and structural inequities in medical education.

Maria Kalogera earned her PhD in International Management and Entrepreneurship (2023). Her research explores startup growth, financial performance, and SME dynamics in turbulent environments, earning international acclaim and multiple "Best Paper" awards. With 17 years of leadership in her consulting firm, she offers expert advisory, accounting, and data analysis services, specializing in European funding for startups. Dr. Kalogera has conducted significant research at the University of Patras and NTUA, contributing to national programs like "Research-Create-Innovate" focused on big data. An experienced educator, she teaches undergraduate and postgraduate courses in economics, taxation, and innovation at the Universities of Patras and Peloponnese.

Nektarios Karanikas BSc, MSc, DProf, SFHEA, AFHEA (Indigenous Knowledges), FAIHS, MHFESA. Dr Karanikas holds a doctorate in Safety and Quality Management from Middlesex University (United Kingdom), an MSc in Human Factors & Safety Assessment from Cranfield University (United Kingdom), and a Bachelor (Hons) in Aeronautical Engineering from the Hellenic Air Force Academy (Greece). Dr Karanikas is a Fellow of the Australian Institute of Health and Safety and the Royal Aeronautical Society, an Associate Editor for Safety Science, and a member of editorial boards of journals from various disciplines. He has delivered numerous invited talks and lectures and has been a guest in several media contributions about safety and human factors. Dr Karanikas is a leading expert in occupational health and safety, human factors, and safety management systems, with over 30 years of experience spanning academia, military aviation and other industries. Dr Karanikas' research combines public health, engineering, and psychology perspectives with a focus on safety management and human factors. His consultancy experience demonstrates a strong track record in reviewing and developing safety management frameworks and regulations. Dr Karanikas has contributed to policy documents and provided expert consultancy services. His work bridges the gap between academic research and practical application, fostering safer and healthier work environments across various sectors.

Pardeep Kumar Assistant Professor, Department of Mathematics, Lovely Professional University, Phagwara, Punjab, India Dr. Pardeep Kumar holds a Ph.D. in Mathematics from Lovely Professional University (2021). His research expertise lies in Stochastic Processes, with a specialized focus on Markov Models, Engineering System Reliability, and Reliability Theory. He actively contributes to the modeling and reliability analysis of IoT-enabled and sensor-integrated systems for next-generation engineering applications. Dr. Kumar has a profound command of Stochastic Modelling, Reliability Testing, and advanced analytical methods for complex system behavior under uncertainty. His work emphasises practical implementations in smart technologies and cyber-physical systems. He has an exceptional academic track record, having qualified the prestigious CSIR-UGC NET twice, securing All India Ranks (AIR) of 15 and 32, respectively. To date, Dr. Kumar has published 26 high-quality research papers in reputed national and international journals, reflecting his continued contribution to the advancement of mathematical reliability and its engineering applications.

Sanjeev Kumar Is a Professor in Lovely Professional University Punjab India. He received Undergraduate in Hotel Management from Osmania University, Postgraduate in Tourism Management from Madurai Kamaraj University. He completed Doctor of Philosophy in Hospitality Management in 2019 from Amity University Jaipur India. His research areas lies in food Service Industry, his research focuses on Alcoholic beverages, Event management and Sustainable Management Practices, Metaverse, AI Machine Learning and Artificial Intelligence. He published more than 80 research papers, articles and chapters in Scopus Indexed, UGC Approved and peer reviewed Journals and books. Dr. Sanjeev Kumar participated and acted as resource person in various National and International conferences, seminars, session chair, research workshops and industry talks and his work has been widely cited. Dr. Sanjeev also edited more than 20 books with various International Publishers such as IGI Global, Cambridge Scholars, Emerald, CABI, Bentham Science, Taylor & Francis. Dr Sanjeev is an active researcher and has published numerous research articles in prestigious National and International journals and books.

Ioannis Ch. Lampropoulos holds doctoral and postdoctoral research experience in managerial accounting, performance measurement, and health economics. His research focuses on performance measurement and management of cost and profit centers in organizations operating in dynamic environments, as well as on cost estimation and big-data analysis in public health systems. He teaches at undergraduate and postgraduate university programs and has an active research profile with publications in international scientific journals. Professionally, he is engaged in management and human resources development.

Maad M. Mijwil is an Iraqi Academician; he was born in Baghdad, Iraq, in 1987. He received his B.Sc. degree in software engineering from Baghdad College of Economic Sciences University, Iraq, in 2009. He received a M.Sc. degree in 2015 from the computer science department, University of Baghdad, in the field of wireless sensor networks, Iraq. Currently, he is a director of the Division of Quality Assurance and Performance at the College of Administration and Economics, Al-Iraqia University. He is working as an Assistant Professor and an academic member of staff in the Computer Techniques Engineering Department at Baghdad College of Economic Sciences University, Iraq. He has over ten years of experience in teaching, guiding projects for undergraduates, and laboratory sessions. He has published 220 works, including research articles, review papers, book chapters, and preprints in national and international journals and conferences. He is also an editor in more than 10 international/national journals and a reviewer in more than 100 international/national journals. He has served on technical program committees for many prestigious conferences. Also, he graduated from Publons Academy as a peer reviewer. According to his Google Scholar profile, he has amassed over 5,900 citations, reflecting the high visibility of his research in the global scientific community. His work frequently explores the intersection of AI with healthcare, cybersecurity, and education, including studies on deep learning for disease detection.

Archile Paguem currently a senior lecturer of Molecular Parasitology. works at the Veterinary Medicine Department, Faculty of Agriculture and Veterinary Medicine at the University of Buea, Cameroon . Archile does research on Livestock and wildlife population genetics and Genomics. Another focus is on Arthropods vectors of zoonotic pathogens such as ticks and tick borne viruses, bacterial and protozoan and tsetse fly transmitting sleeping sicknesses or black flies transmitting onchocerciasis. I also have an interest on taxonomy and phylogeny of Gastrointestinal nematodes and trematodes such as Fasciola, Schistosoma. Another research focus are on discovery of novel parasitic drugs target and development of new anthelmintic Drugs from African medicinal plants secondary metabolites

Ramya Raghavan is currently serving as faculty at Sri Sathya Sai University for Human Excellence, India. Dr Raghavan's research interests include Genetics, Biostatistics, Biophotons, Epigenetic factors, Intellectual disability, and Assistive Technologies. Her expertise extends to the critical intersection of the exposome and Social Determinants of Health (SDOH). Dr. Raghavan's work underscores how a holistic understanding of environmental exposures, from the molecular level to population-wide influences, is crucial for addressing health disparities and omics, exposomics implications for public health.

D. Rajalakshmi is an Assistant Professor in the Department of Data Science and Business Systems, in SRM Institute of Science and Technology. She is having 19 years of teaching Experience in engineering colleges. She has published 30 papers in various National and International journals. She is a professional body member of IEEE, ISTE, CSI and IAENG. She has attended more than 50 FDP/workshop/ training programmes. Her research interests include MANETs, Network Security and cloud computing.

Rakshitha Ramesh, Student in the Department of Life Sciences, Sri Sathya Sai University for Human Excellence, India. 585313.

Sharanya S. is currently working as Associate Professor in the department of Data Science and Business Systems at SRM Institute of Science and Technology, Kattankulathur, Chennai. She has been awarded Doctor of Philosophy by SRM

Institute of Science and Technology. Recently, she was awarded post graduate fellowship from LIPS Lab, European International University. She completed her post graduate and under graduate studies in Computer Science and Engineering at Anna University. She has nearly 10 years of academic experience and has published nearly 40 research articles in reputed journals. She has delivered many guest lectures and acted as resource person for technical events. She holds many granted patents in equipment condition monitoring. She has published several technical books and book chapters. Her research interests includes Machine learning, Deep learning and predictive analytics.

Amrutha Shivaraju completed her Master's Degree in Dept of Life Sciences, Sri Sathya Sai University for Human Excellence, India.

Cierra Swington, MPH, is a Doctor of Public Health (DrPH) candidate at California Baptist University. Her research focuses on maternal mental health, stress, resilience, and health disparities in Southern populations. She has professional experience in oncology clinical operations, quality improvement, and provider support coordination. Her scholarly interests include maternal and perinatal health policy, health equity, and community-based research.

Aishah Johnson, MPH, is a public health practitioner and public policy professional whose work bridges regulatory analysis and legislative policy. Her experience spans local, state, and federal government, where she has contributed to evidence-informed policymaking, environmental health protection, and public health systems improvement. She has supported legislative and oversight functions in the U.S. House of Representatives, including research, bill analysis, hearing preparation, and stakeholder engagement on issues related to labor, health, nutrition, occupational safety and health, and healthcare affordability. Johnson previously served on the White House Council on Environmental Quality's Environmental Justice team, where she supported implementation of Executive Order 14096 and evaluation of Justice40 investments under the Bipartisan Infrastructure Law and Inflation Reduction Act. She also served as a Risk Manager at the U.S. Environmental Protection Agency, where she conducted risk assessments, developed mitigation strategies, and provided science-based recommendations to inform regulatory decision-making affecting consumers and workers. At the state and local levels, she has applied environmental science and spatial analysis to support permitting, enforcement, and community-centered responses to environmental concerns. Across her career, Johnson has translated technical and scientific evidence into policy recommendations, legislative materials, and strategic guidance for public-sector and private-sector stakeholders. She holds a B.S. in Microbiology with a Chemistry minor from the University of Louisiana at Lafayette, an MPH from Louisiana State University Health Sciences Center, and is a Doctor of Public Health (DrPH) candidate at California Baptist University.

Daryl Traylor is a public health educator, researcher, and medical student whose multidisciplinary expertise spans nursing, pharmacology, public health, and healthcare administration. He has over 15 years of experience teaching undergraduate and graduate- level courses in public health, health sciences, and nursing at institutions such as Arizona State University, A.T. Still University, Eastern Washington University, and Chamberlain University. Dr. Traylor's research focuses on HIV prevention, PrEP prescribing practices among primary care providers, diversity and inclusion in health professions education, and the health impacts of climate change. He is committed to mentoring underrepresented students and advancing health equity through academic, clinical, and community- based initiatives. His work bridges the gap between clinical care and public health practice, with a strong emphasis on social justice and systemic reform.

José G. Vargas-Hernández, M.B.A.; Ph.D. Member of the National System of Researchers of Mexico and a research professor at Tecnológico Mario Molina Unidad Zapopan formerly at University Center for Economic and Managerial Sciences, University of Guadalajara. Professor Vargas-Hernández has a Ph. D. in Public Administration and a Ph.D. in Organizational Economics. He has undertaken studies in Organisational Behaviour and has a Master of Business Administration, published four books and more than 200 papers in international journals and reviews (some translated to English, French, German, Portuguese, Farsi, Chinese, etc.) and more than 300 essays in national journals and reviews. He has obtained several international Awards and recognition.

Muhammad Zia Ul Haq is an assistant professor in the Department of Agronomy at the University of Agriculture, Faisalabad, Pakistan. With his extensive knowledge and expertise in weed management and herbicide resistance, he has made significant contributions to the field of agronomy and weed science. He pursued higher education at the University of Agriculture, Faisalabad, Pakistan. His quest for knowledge led him to the Weed Research Laboratory at the College of Life Science, Nanjing Agricultural University, Nanjing, P.R. China. Here, Dr. Zia Ul Haq focused his research on weeds and herbicide resistance, a topic of great relevance in the agricultural landscape. He made remarkable discoveries through rigorous experimentation, data analysis, and scientific inquiry and expanded the understanding of herbicide resistance mechanisms in weeds. After completing his PhD, Dr. Zia Ul Haq returned to his home country, Pakistan, with a vision to share his expertise and impact the agricultural landscape. Joining the University of Agriculture, Faisalabad, he took on the role of an assistant professor in the Department of Agronomy. His engaging teaching style and research background have inspired and mentored numerous students, fostering their interest in agronomy and weed management. Dr. Zia Ul Haq's dedication to his field extends beyond the classroom and laboratory walls. He has actively participated in conferences, workshops, and seminars, presenting his research findings and exchanging ideas with fellow experts. As an assistant professor, researcher, and mentor, Dr. Muhammad Zia Ul Haq continues to contribute significantly to agronomy. Through his research, teaching, and collaborative efforts, he is working towards the sustainable management of weeds and herbicide resistance, ensuring the prosperity and productivity of agricultural systems.

Index

A

B

C

D

E

G

H

J

M

O

P

S

V

W

Z

www.ingramcontent.com/pod-product-compliance
Lightning Source LLC
LaVergne TN
LVHW080908300926
849162LV00003B/601
* 9 7 9 8 3 3 7 3 8 4 7 0 2 *